Chronology of the Evolution-Creationism Controversy

No Prospect of an End . . .

Chronology of the Evolution-Creationism CONTROVERSY

Randy Moore, Mark Decker, and Sehoya Cotner

GREENWOOD PRESS
An Imprint of ABC-CLIO, LLC

A B C CLIO

Santa Barbara, California • Denver, Colorado • Oxford, England

Library of Congress Cataloging-in-Publication Data

Moore, Randy.
Chronology of the evolution-creationism controversy / Randy Moore, Mark Decker, and Sehoya Cotner.
p. cm.
Includes bibliographical references and index.
ISBN 978-0-313-36287-3 (alk. paper) — ISBN 978-0-313-36288-0 (ebook)
1. Human evolution—Philosophy. 2. Creationism—Philosophy. 3. Human evolution—History—Chronology. 4. Creationism—History—Chronology. I. Decker, Mark. II. Cotner, Sehoya. III. Title.
GN281.4.M65 2010
231.7'650202—dc22 2009039784

14 13 12 11 10 1 2 3 4 5

This book is also available on the World Wide Web as an eBook.
Visit www.abc-clio.com for details.

ABC-CLIO, LLC
130 Cremona Drive, P.O. Box 1911
Santa Barbara, California 93116-1911

This book is printed on acid-free paper ∞
Manufactured in the United States of America

Cover art credits: TOP: Famed defense attorney Clarence Darrow (wearing suspenders at the left of the photo) leans against a table during the Scopes "Monkey" Trial, the most famous event in the history of the evolution-creationism controversy. In this trial, which occurred during July, 1925, high school coach and substitute science-teacher John Scopes was prosecuted for the misdemeanor of teaching human evolution. Scopes, wearing a white shirt, has his arms folded and is leaning forward near the center of the photo (Bettmann/CORBIS); BOTTOM: A visitor to London's Natural History Museum walks past a huge picture of Charles Darwin outside the Darwin exhibit. (Andy Rain/epa/CORBIS).

To Mom and Dad, with love and thanks. —R.M.
To Mom and Dad, for always letting me choose my own path. —M.D.
To William C. Harris, a bona fide historian, and Betty G. Harris,
my first science teacher —S.C.

Contents

Preface

Life is the most important thing in the world, and the most important thing about life is evolution.

—George Gaylord Simpson, 1964

The final and conclusive evidence against evolution is the fact that the Bible denies it.

—Henry Morris, 1967

Most Americans expect that science will be able to address societal problems and improve our lives. However, people also often have beliefs—almost always religious—that conflict with what science tells us about the history of life, including our own origins. For example, a poll conducted by the National Science Foundation in 2002 reported that 48% of respondents believed that humans lived contemporaneously with dinosaurs, a belief promoted by young-earth creationists that is incompatible with modern science. A similar poll in 2005 found that 64% of Americans supported teaching both creationism and evolution in public schools, a view that conflicts with a variety of court rulings about what science is and what it is not. And in 2009, the Pew Research Center's Forum on Religion and Public Life reported that only 48% of Americans agree that evolution is the best explanation for the origin of human life on Earth. These and similar conflicts are among the many issues underlying one of the most interesting and lasting controversies in the United States—namely, the evolution-creationism controversy. Indeed, this controversy impinges on virtually every aspect of American society, including religion, politics, public education, and the law.

Unlike many treatments of evolution and creationism, this chronology is not only about surviving ideas. Instead, we have tried to show how some ideas have flourished throughout history (often with a change in names), whereas others have thrived only to later disappear. In this chronology, you will encounter all of these ideas, as well as other lesser-known claims that have been influential, but poorly understood.

We have also tried to give you a glimpse of the people who have produced the evolution-creationism controversy. Although ideas underlie any controversy, the people who have advocated these ideas have been strongly influenced by historical circumstances. Thus, throughout this book, we have provided contextual information about the people, circumstances, and events associated with the controversy. We urge readers to resist judging this history with today's criteria, for doing so will blur the significance of the people, ideas, and events that have led to today's controversy.

Finally, this book is not necessarily meant to overtly endorse or refute any of the diverse claims made throughout the history of the evolution-creationism controversy. We have simply tried to explain what happened, what was said, and what it meant.

Acknowledgments

Our research and writing were assisted by many people and organizations who advised us, opened their homes to us, and granted our requests for interviews, documents, directions, advice, and photographs. We are especially grateful to Charles O'Dale, Sue Hendrickson, and Larry Shaffer of the Black Hills Institute, Trudy Case of Sinclair Oil, Carolyn Belardo and Eileen Mathias of the Academy of Natural Science in Philadelphia, Mindy McNaugher of the Carnegie Museum of Natural History, Sue Wick, Richard Cornelius of Bryan College, Vickie Bryant of Arlington Baptist College, The University of Texas Center for American History, Angus Miller of GeoWalks, The Library of Congress, the Billy Sunday Museum, Jerry Tompkins, Susan and Jon Epperson, Norman Butcher, Jim Cotner, Cissy Ballen, Stephanie Cox, Don Moll, Niles Eldredge, Melany Ethridge, Janice Moore, Creation Evidence Museum, Layton Brueske of First Baptist Church in Minneapolis, Lee Pierce and Lawrence Ford of the Institute for Creation Research, the Creation Museum, Answers in Genesis, Roy Kattschmidt of the Lawrence Berkeley National Laboratory, Darrell and Donna Vodopich, the British Museum (Natural History), Westminster Abbey, Don Aguillard, Rod LeVake, and Eugenie Scott and the other helpful people at the National Center for Science Education. We also thank the many librarians, colleagues, and others who helped us find various documents and books. In every instance, we have tried to accurately quote and/or portray the authors' ideas. We apologize if we have failed to do so anywhere in this book.

Finally, we especially thank David Paige, Ted Young, Kara Witt, and Glenn Branch for their encouragement, advice, and help with our work.

Introduction

Actually, most scientific problems are far better understood by studying their history than their logic.

—Ernst Mayr, 1982

Late in 2006, motorists driving along Interstate Highway 35W in downtown Minneapolis could not help but notice a giant billboard proclaiming an evocative message: "Everyone has an opinion on evolution. Read ours. Post yours." (Figure 1). That and similar billboards in Minnesota about evolution and creationism were paid for by Julie Haberle "to give people the ability to reject the message of evolution." Visitors to Haberle's Web site were not only told about "the serious misrepresentations and lack of scientific proof for the Theory of Evolution and Darwinism," but also "that Jesus is The Way, The Truth, and The Life. Our hope is that we can Advance His Kingdom by countering the false foundations for the faith of evolution." Haberle sponsored the billboards to "keep some kind of a media frenzy (about evolution) going," and she succeeded.

Welcome to the evolution-creationism controversy.

Why would someone go to the trouble and expense of producing billboards attacking evolution? A common starting point for answering this question is the publication in 1859 of Charles Darwin's *On the Origin of Species*, a book that proposed evolution by natural selection as a mechanism to explain the history of life on Earth. Darwin's idea has proven to be remarkably powerful, and Darwin himself has become the global brand of a thriving industry; his image and name appear on stamps, books, and publicity materials for conferences and festivals. More than 150 years after its publication, *Origin* continues to frame many of the questions that still define our understanding of biology. Some people even claim that Darwin's idea is the best idea *ever*.

So why, then, is there a controversy about Darwin's idea? It is not because biologists question whether evolution occurs—university libraries are filled with books and journals describing how evolution continues to be validated by research in molecular biology, geology, paleontology, comparative anatomy, and other scientific disciplines. However, much to scientists' chagrin, this evidence—as impressive as it is—has done little to overcome the resistance to evolution by nonscientists. This is because the evolution-creationism controversy has been driven by the perceived relationship of evolution to issues *outside* of science, including religion, morality, and public education; the Minneapolis billboard's linkage of evolution with religion is merely one of thousands of such examples. Others have linked evolution with more unusual topics, including UFOs, dancing, pornography, the metric system, and capital punishment. The perceived implications of evolution for religious, moral, political, and countless other controversial topics is why—to borrow the words of famed geologist James Hutton—the controversy has "no prospect of an end" (Figure 2).

Evolution, in its purest form, has nothing to do with societal, cultural, or theological issues. However, this has not stopped people from attaching evolution to social, political, and religious agendas unrelated to *On the Origin of Species*: famed biologist Thomas Henry Huxley—who was known as "Darwin's Bulldog"—used evolution to condemn power-hungry clerics, robber baron Andrew Carnegie used it to promote unfettered capitalism, philosopher Herbert Spencer used it to justify ignoring the plight of the poor, Nazis used it to justify genocide, and countless politicians and theologians have used it to explain societal ills while campaigning for moral and religious revival. These claims—and the subsequent furors that they have produced—have long been commonplace in books, magazines, television shows, sermons, courtroom trials, and even proclamations on roadside billboards.

It is tempting to dismiss these events as being little more than local skirmishes and historical curiosities. However, the saga of the evolution-creationism controversy shows otherwise: presidents and presidential candidates who have flaunted their support for creationism, school boards that have forced teachers to promote particular religious beliefs in science classes, state legislatures that have passed laws protecting the teaching of creationism in biology classes, multimillion-dollar museums that have promoted creationism while rejecting modern science, private citizens who have organized antievolution protests, politicians who have blamed evolution for racism and school violence, and famous evangelists who have promised that God will harm people who reject creationism are just a few of the recent events that are best understood in the ever-expanding context of the evolution-creationism controversy. In this chronology, we have discussed these and many related events, for they—as you will see—are not trivial aberrations.

Before Darwin's time, many naturalists explained life's history by invoking a deity. Even Alfred Russel Wallace—the codiscoverer with Darwin of evolution by natural selection—invoked the supernatural (e.g., the intervention of "an Overruling Intelligence" and an "Infinite and Eternal Being") to explain human evolution.

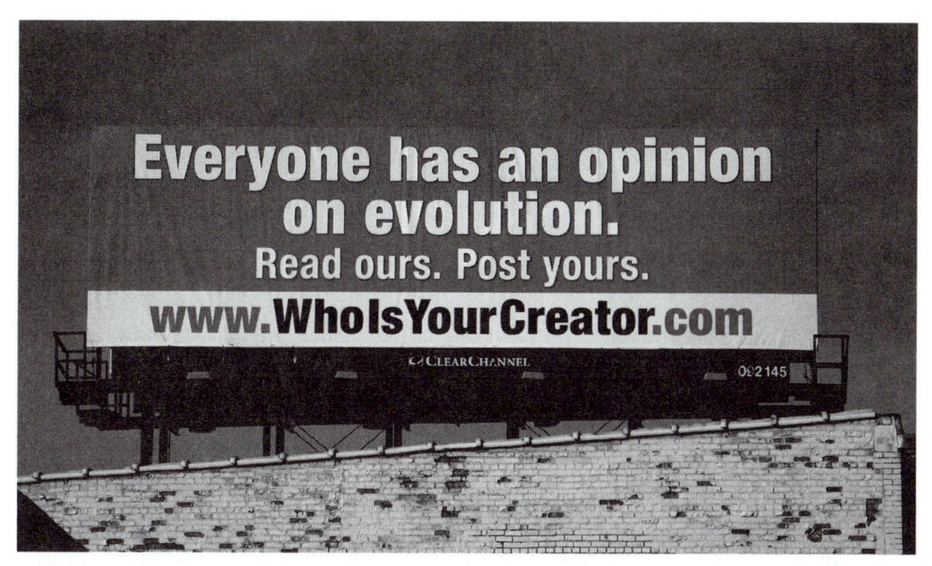

Figure 1 The evolution–creationism controversy appears throughout American society in a variety of ways, including roadside billboards. This billboard in downtown Minneapolis promoted the belief that the validity of evolution is determined by people's opinions. *(Randy Moore)*

Although modern biologists continue to argue about details such as rates of evolution and the identity of common ancestors (arguments that make evolutionary biology a healthy, vibrant science), they do not question whether evolution occurs, for the evidence for its occurrence is overwhelming. The American public, however, is not nearly so convinced. For several decades, dozens of polls have consistently documented the public's belief in and support of creationism. Throughout the United States, creationism is popular, and skepticism about evolution is deeply rooted and diverse.

In this book, you will learn much about creationism—not only about its popularity, but also its diverse claims. The history of creationism is tied to the history of religion. But contrary to popular belief, creationism is not monolithic; the dichotomy that someone is either a creationist or not grossly misrepresents the many diverse, and sometimes incompatible, versions of creationism. Although creationists have often been quick to point out the disagreements among biologists, they have seldom acknowledged the many irreconcilable differences among themselves. For example, young-earth creationism is incompatible with several other types of creationism, including gap creationism and day-age creationism, which both accept that the earth is ancient. In this chronology, you will learn about the history and changing popularity of these many types of creationism. We hope that this will help you appreciate today's version of the evolution-creationism controversy, as well as its likely future.

You will also learn about evolution, the foundation of modern biology. Our treatment of this topic is not meant to be comprehensive; readers wanting more information can consult any modern biology textbook to learn about the mechanics of evolution and its importance in modern science. However, some people and

Figure 2 Scottish geologist James Hutton, who founded the principle of "deep time" in geology, concluded his famous book *Theory of Earth* by noting, "We find no vestige of a beginning,–no prospect of an end." This memorial to Hutton stands at the site of Hutton's home on St. John's Hill in Edinburgh. *(Randy Moore)*

events have been seminal in the evolution-creationism controversy. A few of these people and places—for example, the Galápagos Islands, Charles Darwin, and Darwin's *On the Origin of Species*—are well known. However, even these icons were not isolated and independent. There is much more to the story.

And of course, you will learn about the evolution-creationism controversy itself. While science and society often intersect in ways that lead to conflict, the public's response to evolutionary theory is remarkable not only for its often-visceral nature, but also for its relative stasis. Through time, biologists, geologists, and others have made countless technological and empirical advances that support the Darwinian concept of descent with modification. Yet creationists have continually—and often effectively—countered these advances with arguments similar to rebuttals from centuries past. Throughout this book, we have marked these recurring themes in the controversy with these icons:

Legislation and legal challenges 

Argument from design and Intelligent Design

Scientific understanding of human evolution

Origin of life

Age of Earth

Evidence for evolution

Creation science and flood geology

Advances in technology

Social Darwinism and eugenics

We have also included a glossary and appendices summarizing estimates of the earth's age, the geologic timescale, the major hominins, and the major legal challenges associated with the evolution-creationism controversy. We hope that these resources will help you appreciate the long history of many aspects of the current controversy.

How to Use This Book

Although the evolution-creationism controversy unfolds as a fascinating history of science and society, this book is not a narrative *per se*. Instead, we designed *Chronology of the Evolution-Creationism Controversy* for a variety of uses:

To understand the history of the evolution-creationism controversy. We've included all of the iconic events of the controversy, such as the publication of *On the Origin of Species*, the 1925 prosecution of John Scopes for allegedly teaching human evolution, and the rise of the Intelligent Design movement. We have also included numerous other lesser-known events ranging from legal challenges to the societal influences and cultural implications of the evolution-creationism controversy. Taken together, these events comprise an interesting story in which some themes endure (e.g., the argument from design), discredited ideas are renamed and reintroduced (e.g., creation science), frauds appear and are exposed (e.g., Piltdown Man), and individuals' personalities influence the controversy in important—and even bizarre—ways.

To understand the evolution-creationism controversy during a particular period. Readers may be naturally interested in learning about the major events that have occurred during the controversy. However, this book will help you to consider the historical antecedents of these events as well as their implications and outcomes. We therefore suggest that you examine at least two decades before and after any particular year you're interested in to better understand the context of particular events. For example, most people interested in the evolution-creationism controversy understand the impact of William Paley's *Natural Theology* (1802), but do not know that Paley developed his ideas almost two decades earlier in his *Principles of Moral and Political Philosophy* (1785). Similarly, most biologists understand the impact of Thomas Malthus' *An Essay on the Principle of Population* (1798) on evolutionary thought (biologist Ernst Mayr described it as "the foundation of modern evolutionary theory"), but do not appreciate how it was influenced by Benjamin Franklin's *Observations Concerning the Increase of Mankind* (1751), which accurately predicted the population growth-rates of Great Britain and North America.

To supplement the many books about evolution, creationism, and the evolution-creationism controversy. There are many excellent books about the evolution-creationism controversy. However, most of these books focus on the history of either creationism or evolution, and do not provide a format that encourages easy consideration of events related thematically or temporally. We hope that this chronology will provide historical insights and context that supplement those works in a form that emphasizes related events. Historical events are embedded in societal contexts, and we hope that this chronology will help readers understand these linkages.

To locate events, themes, ideas, and quotations. The chronological format of this book, the icons that trace recurring themes, and the extensive index provide easy access to information about specific events and ideas. We hope these features make this book a useful tool to learn more about the people, places, events, and ideas of the controversy, as well as a means to discover less well-known information related to the original search target.

Finally, we have tried to tell a story that goes beyond simplistic "then and now" comparisons. We have not limited our discussion to only the people and ideas that have endured history's gaze. On the contrary, you will also learn about interesting places, important contexts, the many failed—but nonetheless important—ideas, and the astonishing array of interesting people and curious connections associated with the controversy. Some of the people have been brilliant heroes, some shameless frauds, and others eccentric oddballs, but they have all shaped today's ideas about this long and fascinating controversy.

We hope that you enjoy the story.

Randy Moore, Mark Decker, and Sehoya Cotner
October, 2009

Abbreviations

AAAS	American Association for the Advancement of Science
ACLJ	American Center for Law and Justice
ACLU	American Civil Liberties Union
AIBS	American Institute of Biological Sciences
AiG	Answers in Genesis
AMNH	American Museum of Natural History
ASA	American Scientific Affiliation
BAAS	British Association for the Advancement of Science
BSCS	Biological Sciences Curriculum Study
CMNH	Carnegie Museum of Natural History
CRS	Creation Research Society
CSRC	Creation Science Research Society
ERO	Eugenics Record Office
HMS	His Majesty's Ship
ICR	Institute for Creation Research
ID	Intelligent Design
KKK	Ku Klux Klan
NABT	National Association of Biology Teachers
NAS	National Academy of Sciences

NCSE	National Center for Science Education
NHM	Natural History Museum (London)
NRC	National Research Council
NSF	National Science Foundation
WCFA	World's Christian Fundamentals Association

The Chronology

ca. 2700 BCE Egyptians produce the earliest known document describing creation beliefs. Most of the first-documented creation stories proceeded in stages and involved hatching eggs (that produced the universe and all life), lotus buds (from which Horus the sun god emerged), and gods such as Nun (god of the primordial waters and father of the gods), Shu (god of air), Nut (goddess of the sky), and Ra (god of the risen Sun).

ca. 1100 BCE *Enûma elish*—the Mesopotamian creation story—is recorded on tablets. The story included battles between sea monsters and the gods, after which the Babylonian god Marduk emerged victorious. Marduk split the corpse of Tiamat (the primordial goddess of the ocean) into heaven and earth and used blood from Tiamat's accomplice (Kingus) to create humans.

ca. 800 BCE In the first known Greek account of the universe (written by Hesiod), earth (Gaea) emerges from chaos and gives rise to Uranus, the sky. Subsequent battles involved gods who buried offspring, castrated enemies, ate children, and eventually produced Zeus, who settled on Mt. Olympus and ruled a calm world.

ca. 550 BCE Greek astronomer and philosopher Anaximander of Miletus (610–546 BCE)—one of the first people to comment on evolution—claims that life began in muddy marshes and slowly evolved to colonize drier areas. Near the same time, Thales (ca. 624–546 BCE) suggested that all living things originated in water, and Anaximenes (585–525 BCE) claimed that all things originated from air.

530 BCE Greek poet and philosopher Xenophanes of Colophon (ca. 560–478 BCE), the founder of the Eleatic school of philosophy, discovers fossilized seashells and fish atop mountains and suggests that land atop the mountains was once underwater. This is the first known recognition of fossils as witnesses of geologic change.

ca. 460 BCE Greek philosopher Democritus (ca. 460–370 BCE) is born. Democritus and his teacher Leucippus proposed that all matter consists of various imperishable entities called *atoma* ("indivisible units"). This materialistic view, which became known as *atomism*, strongly influenced Epicurus, the greatest ancient critic of the argument from design. Atomists theorized that everything is made of indivisibly small particles whose random, purposeless interactions did not require divine intervention. Democritus's claim that everything was due to "chance and necessity" was popularized by Nobel laureate Jacques Monod in a 1971 book having that title.

ca. 440 BCE In the 2,000-verse *On Nature*, Greek aristocrat and philosopher Empedocles (ca. 493–433 BCE) claims that the universe and everything in it has been gradually changing since the beginning of time. Empedocles proposed that Love and Hate were the two great forces in nature, and that during the reign of Love, the four elements (earth, wind, fire, and air) produced parts of animals that combined randomly to produce various creatures. Some of these creatures were monsters that died, but others survived and became the first members of modern species. Empedocles, who believed in the concept of spontaneous generation, also suspected that extinctions occur: "Many races of living creatures must have been unable to continue their breed: for in the case of every species that now exists, either craft, or courage, or speed, has from the beginning of its existence protected and preserved it." These ideas, which were echoed by Lucretius (ca. 99–55 BCE), hinted at Charles Darwin's later claims about variation and natural selection.

428 BCE Greek cosmologist and natural philosopher Anaxagoras (ca. 500–428 BCE) dies. He had been charged with heresy for suggesting that the sun is a large, hot stone rather than a deity, that the moon borrows light from the sun, and that the universe is infinitely large and composed of atoms and empty space.

~380 BCE In *The Republic*, Plato (428–348 BCE) advocates selective breeding: "The best men must have intercourse with the best women as frequently as possible, and the opposite is true of the very inferior." (Plato also urged using a fake lottery so feelings were not hurt by the selection criteria.) Selective breeding was fundamental to the eugenics movement of the 20th century.

ca. 371 BCE In *Memorabilia*, Greek writer Xenophon (ca. 431–355 BCE) describes Socrates's (469–399 BCE) belief that humans were uniquely favored by gods as "products of design and not of chance." Socrates invoked the divine craftsmanship of the human eye to make his argument. This was one of the first clear statement of the "argument from design" and the first of many invocations of the human eye to bolster the design claim.

350 BCE In *Physics*, Plato's student Aristotle (384–322 BCE) describes a purposeful universe: "It is plain then that nature is a cause, a cause that operates for a purpose." Unlike Plato, who argued that our imperfect perceptions never allow us to completely understand reality, Aristotle argued that knowledge is based on empirical information gathered from the universe. By basing conclusions on careful observation, Aristotle revolutionized philosophy and became the "father of science." Aristotle interpreted life as an orderly and perfect series of forms, with one type grading into another, and believed that form and function were explained by *teleology* (i.e., that things are designed for, or are directed toward, some final result). This idea, combined with Plato's essentialism, restricted the development of evolutionary thought for centuries.

341 BCE Greek philosopher Epicurus (341–270 BCE) is born. Epicurus, who insisted that nothing should be believed except through direct observation and logical deduction, became the greatest ancient critic of the argument from design. Epicurus—a materialist—did not try to abolish belief in gods, but he did argue that they were unnecessary to understand the natural world. Epicurus's ideas about materialism influenced countless philosophers and theologians associated with the evolution-creationism controversy, including John Ray, David Hume, St. Augustine, Thomas Aquinas, and modern ID-advocate Benjamin Wiker, who noted in

Moral Darwinism (2002) that "all roads [about materialism] lead to Epicurus and the train of thought he set in motion."

~330 BCE Aristotle arranges life from the lowest form to the highest (humans), with each link in this "Great Chain of Being" (also called *scala naturae*) representing an unchanging and closer approximation of perfection (e.g., apes are an incomplete, imperfect realization of the human form; Figure 3). Aristotle believed that this ordering of life was perfect. The "fixity" of species in the Great Chain of Being precluded evolutionary change within a form or the creation of new forms. Aristotle's view of the immutability of life

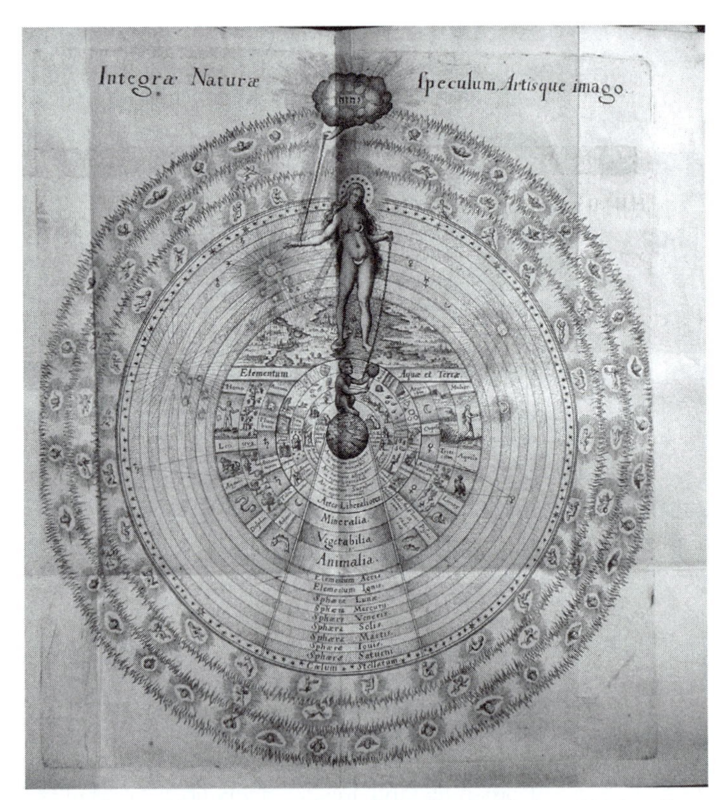

Figure 3 In this representation of the Great Chain of Being from 1617, species were arranged in a hierarchy according to their degree of perfection, with the most perfect organisms at the pinnacle. Each species had a clearly defined position that did not, and could not, change. (*From Robert Fludd,* Utriusque Cosmi Maioris, *1617. Image courtesy History of Science Collections, University of Oklahoma Libraries*)

was accepted for centuries. Christians later extended the Great Chain to include angels and God. The lowest of the animals were snakes, which were relegated to this position because of the serpent's actions in the Garden of Eden. (Similarly, the lowest plant was the demonic yew tree.)

ca. 260 BCE Greek astronomer Aristarchus of Samos (310–230 BCE) becomes the first person to explicitly argue for a heliocentric model of the solar system. His claim was revived almost 1,800 years later by Copernicus.

57 BCE Two years before committing suicide, Roman poet and philosopher Titus Lucretius Carus (ca. 99–55 BCE)—a follower of Epicurus—notes that "the question troubles the mind with doubts, whether there was ever a birth-time of the world and whether likewise there is to be any end." Lucretius believed that whole organisms sprang from earth. In his famous poem *On the Nature of Things*, Lucretius rejected deities and argued that everything in the universe consists of tiny atoms moving in an infinite void, adding that "[t]hose eternal elements became everything that is, without interference from Gods."

45 BCE In *The Nature of the Gods* (*De Natura Deorum*), Roman philosopher and orator Marcus Tullius Cicero (106–43 BCE) claims that "the universe and everything that is in it [was] made for the sake of gods and men." Cicero was also an early advocate for the argument from design: "When you see a sundial or a water-clock, you see that it tells the time by design and not by chance. How then can you imagine that the universe as a whole is devoid of purpose and intelligence?" Cicero's claims foreshadowed those made by creationists for centuries to come. Like Cicero, the most prominent subsequent advocates of natural theology used analogies involving timekeeping devices (e.g., John Ray with a clock, William Paley with a watch).

128 Roman legionnaires complete Hadrian's Wall (Figure 4), a 73-mile stone wall marking the northern boundary of the Roman Empire. Almost 1,700 years later, stones in this wall helped convince Scottish geologist James Hutton that the earth is ancient (Figure 2).

150 Greek astronomer Claudius Ptolemaeus, known in English as Ptolemy (100–178), argues for a geocentric universe in *The*

Figure 4 The rocks of Hadrian's Wall intrigued James Hutton. Hutton noted that although the wall was almost 1,900 years old, the wall's rocks had barely weathered, but nearby mountains had weathered significantly. This helped convince Hutton that the earth is ancient. (*Randy Moore*)

Almagest. Ptolemy's model dominated astronomical thinking for centuries.

169 In *To Autolycus*, Syrian saint Theophilus of Antioch (ca. 115–180) becomes the first scholar to use the Bible as a foundation for geologic chronology when he claims "all the years from the creation of the world amount to a total of 5,698 years, and the odd months and days." Many theologians and others (e.g., astronomer Johannes Kepler [1571–1630]) subsequently used the Bible to make similar claims (see also Appendix A).

ca. 240 Egyptian philosopher Origen (185–254) becomes one of the first people to comment on the logistics of Noah's Ark. Origen claimed that the cubit mentioned in Genesis was the larger Egyptian cubit, thereby making the pyramid-shaped Ark 0.5 miles long. Origen's claims were later disputed by St. Augustine's *City of God* and French monk Hugh of St. Victor's (1096–1141) *De Arca Noe Mystica* (*The Ark of Noah According to the Spiritual Method of Reading*). The logistics of Noah's Ark later became the subject of numerous analyses.

354 Aurelius Augustinus (Augustine; 354–430) is born in Tagaste, at the southern extent of the Roman Empire (now Algeria). Augustine famously stated that "[n]othing is to be accepted save on the authority of Scripture, since greater is that authority than all the powers of the human mind." Many subsequent creationists, including Henry Morris, were strongly influenced by Augustine's ideas. However, Augustine also warned that there is danger in literal interpretations of the Bible. Augustine knew that earth is spherical but argued that its other side could not be inhabited because it would violate the Bible's claim that every eye would see Jesus when he returned (Revelation 1:7). Parts of Augustine's interpretation of Genesis—for example, that the "framework of the six days of creation might seem to imply intervals of time"—reappeared more than 1,400 years later as day-age creationism.

~360 Basil of Caesarea (329–379) links biological similarities to common ancestry: "There is a family link between the creatures that fly and those that swim . . . their common derivation from the waters has made them one family."

380 The Christianization of the Roman Empire is nearly complete when Emperor Theodosius (347–395) makes "Catholic Christianity" the only approved religion in the Roman Empire. Largely as a result of this edict, several Western ideas influential to what would become the discipline of biology can be traced to Genesis, including the idea that life is the direct, intentional, and sudden creation of God and that life has not changed much since the Creation.

415 In *Literal Commentary on Genesis*, Augustine cautions believers not to be dogmatic about specific interpretations of Genesis 1–3 because these books had "been written obscurely for the purpose of stimulating our thought." Augustine expressed dismay at the ignorance of some Christians and hoped that "[n]o Christian will dare say that the narrative [of Genesis] must not be taken in a figurative sense." He also urged readers to "choose the interpretation that he can grasp" because "different interpretations are sometimes possible without prejudice to the faith we have received. In such a case, we should not rush in headlong and so firmly take our stand on one side that, if further progress in the search of truth justly undermines this position, we too fall with it." Despite Augustine's advice, millions of people still use a literal interpretation of Scripture to reject scientific evidence for evolution and an ancient earth.

415 Influential pagan mathematician and philosopher Hypatia of Alexandria (ca. 360–415) is killed by a Christian mob. Accounts of the attack describe a gruesome ordeal, involving flaying Hypatia's body, scattering body parts throughout the city, and setting her corpse on fire. Although it is unclear whether Hypatia was slain for her heretical teachings, she is often cited as an early casualty of science-versus-religion antagonism.

1027 Persian naturalist Ibn Sina's (980–1037) *The Book of Healing* describes fossils and petrification.

1210 As part of a series of condemnations spanning six decades, the University of Paris declares Aristotle's work a pagan threat to Christianity and forbids its teaching: "Neither the books of Aristotle on natural philosophy nor their commentaries are to be read at Paris in public or secret, and this we forbid under penalty of excommunication."

1225 Thomas Aquinas (Thomas of Aquin; 1225–1274) is born. In his most influential works, *Summa contra Gentiles* and *Summa Theologica*, Aquinas—a Dominican scholar and arguably the greatest of medieval theologians—suggested that creation occurred prior to the six days described in Genesis. Aquinas also argued that there is no conflict between Church doctrine and natural science because there exists both a primary cause (the act of the Creator) as well as secondary causes (natural processes). Knowledge of the primary cause is available only via revelation, while secondary causes are open to scientific examination. Aquinas claimed that nature is purposeful and that it operates "as though moved or directed by another thereto, as an arrow directed to the target by the archer, who knows the end unknown to the arrow. Wherefore, as the movement of the arrow toward a definite end shows clearly that it is directed by someone with knowledge, so the unvarying course of natural things which are without knowledge, shows clearly that the world is governed by some reason." Aquinas condemned Epicurus's materialism and claimed that there had to be a pre-existing intelligence that gave nature direction "like an archer giving a definite motion to an arrow." Centuries later, advocates of ID invoked the writings of Aquinas to support their perspective. Aquinas argued that what logic and reason tell us about the world should be taken seriously and should not be rejected by appeals to Scripture. With this view, Aquinas believed that it was acceptable for Christians to reject a literal interpretation of Genesis because "the manner and the order according to which creation took place concerns the faith only incidentally."

1308 Italian poet Dante Alighieri (ca. 1265–1321) begins writing *The Divine Comedy*, an epic poem of Italian literature. In Canto X of *Inferno*, Dante consigned materialist Epicurus and his followers to an eternity of torture in the sixth circle of Hell ("where the heretics lie").

1317 For presuming to control natural processes, 14th-century alchemists are censured by Pope John XXII's (1249–1334) prohibition against transmutation, an early example of the Church exerting control over science.

1506 Leonardo da Vinci (1452–1519) begins writing a manuscript that becomes known as the *Leicester Codex* (named for Lord Leicester, who bought the codex after it was discovered in the late 1690s; it is now owned by Bill Gates). The manuscript deals primarily with the properties and action of water, but it also discusses earth's evolutionary history. Leonardo believed that the human body and earth are interconnected and parallel entities—what he referred to as "microcosm" and "macrocosm," respectively—and claimed that processes in the human body similarly occur within the entire planet. One of these shared processes, the circulation of fluids, was proposed to distribute landmasses, such that parts of the earth were constantly rising and falling to achieve an overall planetary equilibrium as water circulated globally. Leonardo interpreted marine fossils on mountaintops as evidence for this vertical movement of landmasses and may have identified the principle of superposition—"stratified stones of the mountains are all layers of clay, deposited one above the other by various floods"—before it was developed by Nicolas Steno in 1669. Leonardo invoked the idea of a changing planet that experienced mountain-building and erosion, ideas that presaged the uniformitarianism of James Hutton and Charles Lyell. During the Reformation, a literal interpretation of the Bible, including the Flood, prevailed.

1535 Dominican Fray Tomás de Berlanga (1487–1551), the Bishop of Panama, and his crew are instructed by King Charles V (1500–1558) to sail to Peru to settle a dispute between Spanish conquistador Francisco Pizarro (1471–1541) and his officers after their conquest of the Incas. Lacking sufficient winds, Berlanga sailed to the Galápagos Islands, where he noted the tameness of the islands' animals and described the soil as "worthless, because it has not the power of raising a little grass, but only some thistles." Berlanga's comments discouraged colonization of the islands, leaving them in an unspoiled condition when the *Beagle*—with Charles Darwin aboard—arrived in 1835. In his *Journal of Researches* (1839), Darwin was less critical of the islands: "The natural history of these islands is eminently curious, and well deserves attention."

1541 German ecclesiastical reformer Martin Luther (1483–1546), who launched the Reformation in 1517 when he posted his 95 theses on the door of the Wittenberg Cathedral, claims that creation occurred in 3961 BCE. By the mid-1700s, virtually all of the more than 200 biblical chronologies set creation near 4000 BCE. Luther—who believed that fossils originated in the biblical Flood—also used the Bible as a basis for his insistence that earth is flat.

1542 Pope Paul III (1468–1549) establishes the Supreme Sacred Congregation of the Roman and Universal Inquisition, which later became known as the Holy Office, charged with defending faith against heresy, be it philosophical, political, or scientific. In 1616, the Holy Office rejected heliocentrism as "formally heretical . . . foolish and absurd." The Holy Office's most famous case occurred in 1633, when Galileo was tried for his heretical claims about the cosmos. Today, the Holy Office is named the Congregation for the Doctrine of the Faith.

1543 Flemish anatomist Andreas Vesalius (1514–1664), in his seven-volume *De Humani Corporis Favrica*, claims that men and women have the same number of ribs. This conclusion created a public controversy, for it was considered a scientific affront to biblical verse.

1543 Polish prelate Nicholas Copernicus's (1473–1543) *De Revolutionibus Orbium Coelestium* (*On the Revolutions of the Heavenly Spheres*)—with complex geometric diagrams and published just before his death—discards Ptolemy's claims by advocating a heliocentric view of the universe. Copernicus wrote the first account of his heliocentric theory in 1514, but the 1543 work was a more complete description of his ideas. In the book's preface, Copernicus dismissed people who used Scripture to "dare to assail this my work" as being "of no importance to me." Copernicus dedicated his book to Pope Paul III, who in 1542 had established the Supreme Sacred Congregation of the Roman and Universal Inquisition to defend the Church and condemn heresy. This office later arrested astronomer Galileo Galilei, Copernicus's most famous defender. When Martin Luther heard of Copernicus's idea, he responded, "[t]his fool [Copernicus] wishes to reverse the entire science of astronomy; but sacred Scripture tells us that Joshua commanded the sun to stand still, and not the earth" (Joshua 10:12–13).

1551 Swiss naturalist Conrad Gesner (1516–1565) completes the first volume of *Historiae Animalium.* This series was banned by Pope Paul IV (1476–1559), not for the work itself, but because Gesner was a Protestant.

1554 French mathematician and monk Johannes Buteo's (1492–1564) popular *The Shape and Capacity of Noah's Ark* is the first serious discussion of the logistics of the Ark that is not part of a commentary about Genesis. Buteo argued that the Ark was rectangular.

1555 French zoologist Pierre Belon's (1517–1564) descriptions of bird and human skeletons is the first clearly expressed description of homology. Some of Belon's most important work was published a year after his murder in Paris.

1559 The Roman Inquisition begins compiling the first worldwide *Index of Forbidden Books*. The *Index* would later include books by Galileo, Descartes, and others, and would not be abolished until 1966.

1570 The Galápagos Islands first appear on a map as Insulae de los Galopegos ("Islands of the Tortoises").

1572 Danish astronomer Tycho Brahe (1546–1601, born Tyge Ottesen Brahe) observes a nova and concludes that the heavens can change.

1584 Italian philosopher, pantheist, and former Catholic priest Giordano Bruno (1548–1600; Figure 5) claims neither the earth nor the sun is the center of the universe and instead that there are "countless suns and countless earths all rotating around their suns."

1588 In *A Concent of Scripture*, feisty scholar Hugh Broughton (1549–1612) uses biblical texts to claim that creation occurred in 3960 BCE. Broughton's biographers included John Lightfoot, who also used the Bible to produce a chronology of earth's history (see also Appendix A).

Figure 5 Italian philosopher Giordano Bruno was a proponent of heliocentrism and an infinite universe. After a seven-year trial before the Inquisition, Bruno was declared a heretic and burned at the stake. After killing him, the Holy Office consigned Bruno's writings—*omnia scripta* ("every one of them")—to its *Index of Forbidden Books*.

1588 French craftsman and religious reformer Bernard Palissy (1509–1589) is labeled a heretic for claiming that fossils are the remains of ancient organisms. The following year, Palissy died in a Bastille dungeon of "starvation and maltreatment."

1592 Giordano Bruno (Figure 5) is arrested for a variety of offenses, including heresy, blasphemy, and immoral conduct. Bruno claimed that hell did not exist, that praying to saints was a waste of time, and that nothing taught by the church could be proved. The following year he was sent to Rome to stand trial before the Vatican Holy Office (known informally as the Inquisition) for espousing a heliocentric and infinite universe.

1600 After seven years of imprisonment and many interrogations, the recalcitrant Giordano Bruno is convicted of "obstinate and pertinacious heresy." On February 17, 1600, after refusing to refute his claims, Bruno's tongue was cut out, he was stripped naked and burned at the stake, and his ashes were dumped in the Tiber River. The site of his killing—the central Roman market Campo de' Fiori—is now dominated by a statue of Bruno (Figure 5), which was dedicated in 1889 by a group that included Herbert Spencer and Ernst Haeckel. The statue memorializes a manacled Bruno with the inscription: "To Bruno, from the generation he foresaw, here, where the pyre burned." The unveiling of Bruno's statue was attended by 30,000 people, and *The New York Times* reported that "[t]he Pope is much depressed." Every year, atheists and other freethinkers gather in February at Bruno's statue.

1605 British statesman and philosopher Francis Bacon (1561-1626) advocates a new system of knowledge based on empirical and deductive principles that became known as the Baconian method. Opposite the title page of *On the Origin of Species* (1859), Charles Darwin noted Bacon's claim that people should study God's word and nature.

1609 Italian scientist Galileo Galilei (1564–1642) builds his first telescope, which he uses to study astronomy. The following year, in *The Starry Messenger*, Galileo affirmed Copernicus's heliocentric solar system and wrote to powerful Florentine statesman Belisario Vinta that "I give infinite thanks to God, who has been pleased to make me the first observer of marvelous things." *The Starry Messenger* was a success; the first 500 copies sold within months. Prudently, Galileo did not mention Giordano Bruno (Figure 5).

1615 In a letter to the Grand Duchess Christina (1565–1637), Galileo famously quotes Cardinal Caesar Baronius, "the Bible tells us how to go to Heaven, but not how the heavens go. . . . I do not feel obligated to believe that the same God who has endowed us with senses, reason, and intellect has intended us to forgo their use."

1616 The Catholic Church declares Copernican heliocentrism "false and erroneous" and places *On the Revolutions of the Heavenly Spheres* on its *Index of Forbidden Books*.

1619 Italian philosopher Lucilio Vanini (1585–1619) is arrested and condemned for being an atheist and suggesting that humans descended from apes. Vanini's tongue was cut out and he was strangled at the stake, after which his body was burned.

1620 In *Novum Organum*, Francis Bacon formulates the scientific method.

1626 Jardin des Plantes is founded in Paris as a medicinal herbarium for King Louis XIII (1610–1643). The museum, which opened to the public in 1640 and became part of the Muséum National d'Histoire Naturelle, later sponsored the work of Jean-Baptiste Lamarck, Georges Cuvier, Georges-Louis Leclerc (Comte de Buffon), and other accomplished naturalists associated with the evolution-creationism controversy.

1632 Galileo Galilei advocates a heliocentric universe in his popular book *Dialogue of Galileo Galilei, Lycean Special Mathematician of the University of Pisa And Philosopher and Chief Mathematician of the Most Serene Grand Duke of Tuscany. Where, in the meetings of four days, there is discussion concerning the two Chief Systems of the World, Ptolemaic and Copernican, Propounding inconclusively the philosophical and physical reasons as much for one side as for the other*. The following year, Galileo was forced to appear before the Church's Holy Office; he was convicted of "grave suspicion of heresy" but avoided Giordano Bruno's (Figure 5) fate by recanting his conclusions. Galileo, whose work set the stage for the Scientific Revolution, valued empiricism over doctrine; as he noted, "[i]n questions of science, the authority of a thousand is not worth the humble reasoning of a single individual." Galileo spent the last six years of his life under house arrest, and when he died in 1642 his body was buried in an unmarked grave. However, in 1737 his body was moved to its current location in the Basilica of Santa Croce in Florence, Italy. (Before he was reinterred, officials removed vertebra, one tooth, and three fingers from Galileo's right hand; his middle finger is displayed at Italy's Museo di Storia del Scienza.) In 1664, the Church consigned Galileo's book to its *Index of Forbidden Books* and did not lift the ban until 1822.

1637 René Descartes's (1596–1650) anonymous, first-person essay *Discourse on the Method of Rightly Conducting the Reason and Seeking Truth in the Sciences*—often considered the first work of modern philosophy—promotes using reason and skepticism rather than theological teachings as the best methods for understanding the world. Descartes believed that his mechanistic theory of nature was a defense, rather than a repudiation, of Christianity. Descartes's famous statement—*Je pense, donc je suis* (*I think, therefore I am*)—appeared in part IV of *Discourse*. Descartes repeated his famous statement—this time as *Cogito, ergo sum*—in *Principles of Philosophy* (1644) while arguing that the universe is governed by simple laws and that natural processes could have shaped the earth.

1642 John Lightfoot (1602–1675), minister and later the Vice Chancellor of Cambridge University, writes the wonderfully titled *A Few, and New Observations, upon the Booke of Genesis, the most of them certain; the rest, probable; all, harmeless, strange, and rarely heard of before*. In Verse 26 on page 4 of this small, 20-page booklet, Lightfoot—an architect of *The Westminster Confession of Faith* (which established the doctrinal standards of the Church of England)—noted that "Man was created by the *Trinity* about the third houre of the day, or nine of the clocke in the morning." (Two years later, Lightfoot claimed that creation began on Sunday, September 12, 3928 BCE.) Although Lightfoot was often credited with claiming that creation occurred at 9:00 a.m. on October 23, 4004 BCE (a date similar to that claimed eight years later by Irish prelate James Ussher; see also Appendix A), he never wrote that creation occurred on October 23; that date was added by subsequent writers, most notably Andrew Dickson White (1832–1918) in his influential two-volume *A History of the Warfare of Science with Theology in Christendom* (1896). Lightfoot also did not mention the creation of the earth; his "nine of the clocke in the morning" referred only to the creation of humans.

1650 Irish prelate James Ussher (1581–1656) completes his influential *Annals of the Old Testament, deduced from the first origins of the world*. *Annals* placed divine creation on the evening of Saturday, October 22, 4004 BCE (see also Appendix A). Ussher started with the known dates of the reign of Nebuchadnezzar II (ca. 630–562 BCE)

and subtracted the lifespans of the Bible's patriarchs. These calculations, combined with several assumptions (e.g., that creation occurred on an equinox or solstice, that the presence of a ripe fruit in the Garden of Eden meant that creation occurred in the autumn), produced Ussher's claim. Ussher's book accepted the vast ages of biblical patriarchs, including that Adam lived to be 930 years old, Seth to be 912 years old, and Methuselah—Adam's great-great-great-great-great grandson—to be 969 years old. Ussher also provided dates for biblical events such as the Flood (2348 BCE), the Exodus from Egypt (1491 BCE), and the destruction of Israel by Babylon (586 BCE). In the early 1900s, *The Scofield Reference Bible* popularized these dates, and some contemporary Bibles and Christian apologists still cite them as historical facts. Ussher's claim that creation occurred in 4004 BCE became a foundation of young-earth creationism.

1650 René Descartes dies in Stockholm. Sixteen years later, his body was exhumed and his body parts taken by collectors. Descartes's skull had a variety of owners, including a Swedish casino operator, Louis XIV, and, in 1821, famed anatomist Georges Cuvier.

1655 French theologian Isaac La Peyrère's (1596–1676) anonymously published *Prae-Adamitae* (published in English the following year as *Men Before Adam*) questions the authenticity and accuracy of the Bible, claims that the biblical Flood was a local event in Palestine, and formulates the pre-Adamite theory that men and women were on earth before Adam. Peyrère, the most notorious heretic of his age, likened his idea to that of Copernicus. Peyrère's book was publicly burned in Paris, and after being arrested and imprisoned, Peyrère was denounced by Pope Alexander VII (1599–1667) as *un hérétique détestable*. Peyrère was released from prison after he recanted his idea and became a Catholic. Peyrère's claim that some groups of people are not descended from Adam was later used by a variety of individuals—including famed biologist Louis Agassiz—to justify racism.

1659 In the preface of *A Chain of Scripture Chronology*, cleric Thomas Allen (1608–1673) asks, "[t]he world, which his hand made, is aged, but of what age who can justly tell?" Allen believed that the answer to this question was in biblical texts (see also Appendix A).

1662 As part of the four-part *Clarendon Code* (intended to strengthen the authority of the Church of England), the British Parliament passes the "Act of Conformity" ("Act of Uniformity") to ensure that all clergy (including university faculty members) adhere to *The Book of Common Prayer*. Rather than submit to an oath, hundreds of faculty resigned, including influential Cambridge Fellow John Ray.

1663 Botanist and Anglican clergyman John Ray (1628–1705)—England's foremost naturalist—discovers an ancient buried forest and struggles to understand the enormity of geologic time: "Many years before all records of antiquity these places were part of the firm land and covered with wood; afterwards being overwhelmed by the violence of the sea they continued so long under water till the rivers brought down earth and mud enough to cover the trees, full up these shallows and restore them to firm land again . . . that of old time the bottom of the sea lay so deep and that hundred foot thickness of earth arose from the sediment of those great rivers which there emptied themselves into the sea . . . is a strange thing considering the novity of the world, the age whereof, according to the usual account, is not yet 5,500 years" (see also Appendix A).

1663 The Holy Office places four of René Descartes's books on its *Index of Forbidden Books*.

1665 English biologist Robert Hooke (1635–1703) describes the structure of cells (and coins the term *cell*) in his large and exquisitely illustrated popular book *Micrographia*. Hooke offered no suggestions for cells' functions except the buoyancy they provided to the cork he examined. Hooke also used a microscope to examine fossils, including petrified wood and fossil shells; he concluded that shell-like fossils are the remains of once-living organisms and that the cellular structure of petrified wood indicated that it had originally been wood. In 1705, Hooke's *Lectures and Discourse on Earthquakes and Subterranean Eruptions*—which later influenced geologist James Hutton (Figure 2)—argued that shells atop mountains could not be explained by a flood. Hooke's conclusion that past events could be recognized and dated with fossils forecast the science of biostratigraphy.

1665 In *Mundus Subterraneus*, German scholar Athanasius Kircher (1601–1680) attributes giant fossil bones to extinct races of giant humans.

1666 After examining a giant shark beached near Livorno, Italy, Danish physician Nicolas Steno (1638–1686) concludes that shark teeth are the same thing as glossopetrae, the tongue-shaped fossils suspected of having curative powers. Steno recognized that the shark teeth were an example of the many forms of marine life embedded in rocks far from the sea. The following year, in an appendix to his anatomical description of the shark, Steno published his claim about glossopetrae and a description of how sediments and strata are deposited.

1667 English poet John Milton (1608–1674) writes his epic *Paradise Lost*, the most well-known poem about the Creation and Fall. Milton's poem, which includes a description of how the Garden of Eden was washed away by the Flood, became exceedingly popular late in the 18th century. Charles Darwin took a pocket version of the poem with him aboard the *Beagle*.

1668 Italian physician Francesco Redi's (1626–1697) *Esperienze intorno alla generazione delgi inseffi (Experiments concerning the generation of insects)* refutes spontaneous generation by describing how maggots form in meat exposed to open air but do not form in meat isolated from flies. This and subsequent work by Louis Pasteur and Rudolf Virchow were used by creationists as evidence that life did not create itself, but required the hand of God. Redi, however, continued to believe that some organisms, such as intestinal worms, *could* arise spontaneously.

1669 In his concise *De solido intra Solidum naturaliter Contento Dissertationis Prodromas*, Nicolas Steno establishes the foundation of stratigraphy with his Principle of Superposition, which states that within undisturbed geologic strata, the lower layers are older than higher layers. Steno later applied this principle to determine the relative sequence of geologic events in the Tuscany region of Italy; this description was the first application of stratigraphic principles to the geologic record. Six years after publishing *De solido*, Steno was ordained as a Catholic priest, after which he gave up science and died penniless. On October 23, 1988, Steno was beatified by Pope John Paul II.

1670 Italian painter Agostino Scilla's (1639–1700) influential *La Vana Speculazione Disingannata dal Senso* (*Vain Speculation Undeceived by Sense*) depicts fossils as having an organic origin.

1675 German Jesuit Athanasius Kircher's (1602–1680) handsomely illustrated *Arca Noë* uses natural philosophy to explain the details of Noah's story. According to Kircher, the ark carried 130 "beasts," 30 pairs of snakes, and 150 kinds of birds.

1676 Oxford University chemist Robert Plot (1640–1696) discovers a "real bone, now petrified" femur from a dinosaur, but suggests that it was from an elephant brought to Britain by the Romans (or part of a femur from a giant human who lived before the biblical Flood). Plot described the bone the following year in his opus *Natural History of Oxfordshire*, the first illustrated book to mention fossils in England. Plot viewed fossils as naturally formed rocks resembling bony parts of humans and other animals, all of which he attributed to the grandeur of God's design. The "enigmatic thighbone" discovered by Plot, which weighed nearly 20 pounds, has since been lost.

1681 Thomas Burnet's (1636–1715) *The Sacred Theory of the Earth: Containing an Account of Its Original Creation, and of All the General Changes Which It Hath Undergone, Or Is to Undergo, until the Consummation of All Things* (published originally in Latin, and expanded in English in 1684) founds scriptural geology by trying to reconcile Scripture with geology. Burnet, who relegated God to the part of a playwright instead of a direct actor, used an analogy involving a clockmaker to argue that God's role in nature was indirect: "We think him a better Artist that makes a Clock that strikes regularly every Hour from the Springs and Wheels which he puts in the Work, than he that hath so made his Clock that he must put his Finger to it every Hour to make it strike." Burnet warned readers about using the Bible as one's source of information about nature: "Tis a dangerous thing to engage the authority of scripture in disputes about the natural world, in opposition to reason; lest time, which brings all things to light, should discover that to be evidently false which we had made scripture to assert. . . . We are not to suppose that any truth concerning the natural world can be an enemy to religion; for truth cannot be an enemy to truth, God is not divided against himself." Burnet viewed the earth's crust as a natural ruin and argued that the earth's features are marks of fracture and ruin rather than cogs of a smooth-working machine. Burnet, who claimed that all events in the earth's history could be explained by natural processes, also advocated natural theology—namely, that God's plan for the earth and humans is understandable by studying nature. Noah's Flood was a foundation for Burnet's ideas. After concluding that 40 days and nights of rain could not have covered the entire earth with water, Burnet

argued that there must have been underground oceans that flooded the earth when its original crust cracked open (consistent with the claim in Genesis 8:2 that "fountains of the great deep" erupted). Burnet's book does not mention fossils, but it later influenced the debate about fossils. *Sacred Theory*, one of the most popular geologic works of the 17th century, was praised by Burnet's contemporaries, including Isaac Newton. Burnet's book also included physical and moral claims consistent with sacred teaching (the second half of *Sacred Theory* was devoted to prophecy, not history) and described the Flood as a major incident that punished the sins of humanity. However, Burnet's lack of a direct role for God cost him his job as chaplain and secretary to the cabinet of King William III. In the late 1700s, James Hutton claimed that Burnet's book "surely cannot be considered in any other light than as a dream, formed upon a poetic fiction of a golden age."

1686 John Ray—the "father of natural history"—proposes the first definition of a species and argues that species have not changed over time. As Ray noted, "the works created by God at first" had been "by Him conserved to this day in the same state and condition in which they were first made."

1687 Isaac Newton (1643–1727) claims in *Philosophiae Naturalis Principia Mathematica*—in which was developed the foundation of classical physics—that "this most beautiful system of the sun, planets, and comets, could only proceed from the counsel and dominion of an intelligent and powerful Being." Newton claimed that God's wisdom is evident in the stars being far from each other, lest the "stars should, by their gravity, fall on each other."

1688 Robert Hooke tells the Royal Society that earthquakes, and not a flood, are responsible for fossils buried atop mountains. In 1705, Hooke's controversial ideas were published in *Posthumous Works*.

1691 John Ray's sermons about the relationship between nature and God are published as *The Wisdom of God Manifested in the Works of Creation* (Figure 6). Setting a pattern for European science for the subsequent 200 years, Ray claimed that adaptations are permanent traits designed by God. Ray's sermons criticized Epicurus's materialism and cited the wonders of nature as testaments to God's power. Ray believed that organisms have no history; they have always been the same, lived in the same places, and done the same things as when they were first created. Whereas Cicero had

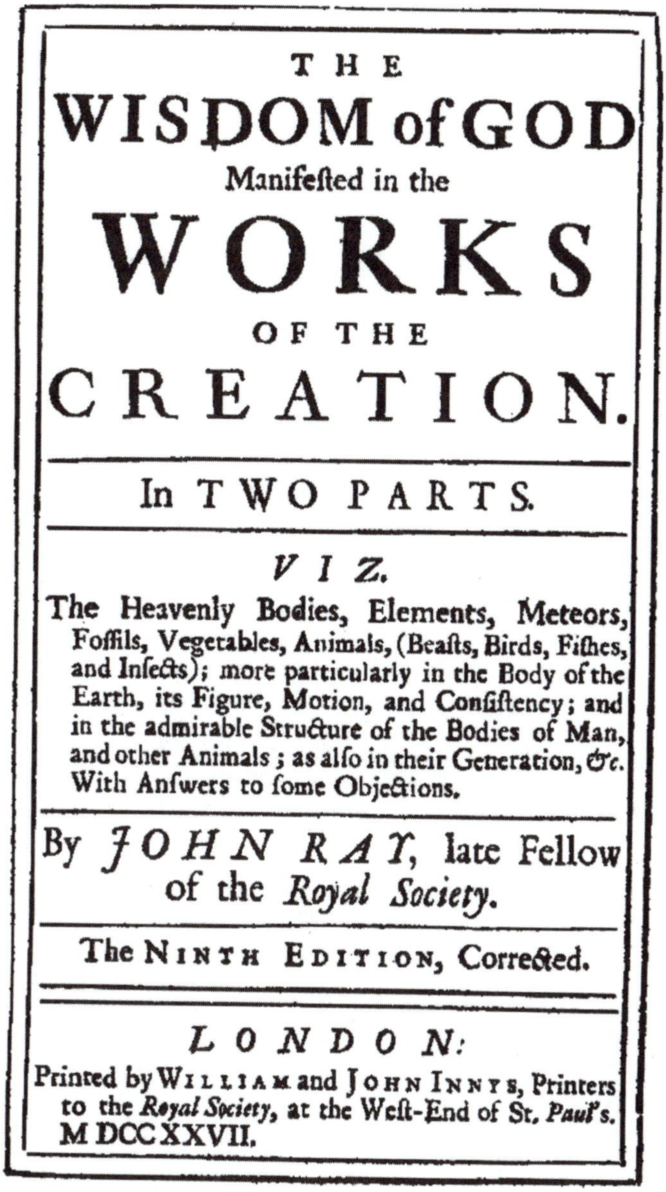

THE

WISDOM of GOD

Manifeſted in the

WORKS

OF THE

CREATION.

In TWO PARTS.

VIZ.

The Heavenly Bodies, Elements, Meteors, Foſſils, Vegetables, Animals, (Beaſts, Birds, Fiſhes, and Inſects); more particularly in the Body of the Earth, its Figure, Motion, and Conſiſtency; and in the admirable Structure of the Bodies of Man, and other Animals; as alſo in their Generation, &c. With Anſwers to ſome Objections.

By *JOHN RAY*, late Fellow of the *Royal Society*.

The NINTH EDITION, Corrected.

LONDON:

Printed by WILLIAM and JOHN INNYS, Printers to the *Royal Society*, at the Weſt-End of St. *Paul's*. M DCC XXVII.

Figure 6 John Ray's *The Wisdom of God* was the first English manifesto of natural theology. This book, which impeded the development of evolutionary thinking for many years, was Ray's most popular and influential work.

used an analogy involving a sundial to promote the argument from design, Ray argued that a clock shows evidence of a designer, and since nature is more perfect than a clock, then nature must also include a master designer.

1692 Thomas Burnet's *Archaeologiae Philosophicae* argues that biblical history is an allegory, not a literal history of the earth.

1695 Geologist and physician John Woodward's (1665–1728) *An Essay Toward a Natural History of the Earth* suggests a Flood-based model for earth's geologic features. Woodward, who invoked miracles (e.g., a providential suspension of gravity) and claimed that fossils are "the real Spoils of once living Animals," attributed biostratigraphy to animals settling out of the floodwaters according to their specific gravity. In contrast to Thomas Burnet, who viewed the earth as a "disorderly Pile of Ruines and Rubbish," Woodward viewed everything in nature as divinely designed by God and "in nearly the same *condition* that the *Universal Deluge* left it." Woodward—a leading critic of Thomas Burnet's *Sacred Theory*—was buried near Isaac Newton in Westminster Abbey (Figure 7).

1696  In *A New Theory of the Earth*, British theologian and mathematician William Whiston (1667–1752) argues that earth originated from "ancient Chaos, just before the beginning of the six days of Creation" and that the biblical Flood was caused by a comet nearly striking the earth. This near-miss allegedly provided water from above (i.e., rain from water in the comet's tail) and below as the earth's deformed crust allowed vast amounts of trapped groundwater to escape. (In 1694, English astronomer Edmond Halley [1656–1742] had suggested that the biblical Flood was caused by a comet's impact. However, facing ecclesiastical pressure, Halley recanted his claim a few days later.) Whiston argued that fossils atop mountains were the result of a flood, and that each day of creation was actually a year because in the beginning there was no daily rotation of the earth on its axis, only an annual revolution around the Sun (this moved creation from 4004 BCE to 4010 BCE; see also Appendix A). Although Whitson was later charged with heresy for questioning the doctrine of the Trinity, he was never defrocked by the Church. Whitson's ideas anticipated arguments that reappeared in the 18th century with day-age creationism.

1696 English clergyman and natural philosopher William Derham's (1657–1735) *The Artificial Clockmaker* presents a teleological argument for the existence of God. More than a century later, William Paley used these arguments and a similar analogy (i.e., based on the improved technology of a watch) in his monumental *Natural Theology*.

Figure 7 Westminster Abbey, one of the great churches in Christendom, enshrines the history of England. The famed abbey houses the graves of several people associated with the evolution-creationism controversy, including John Woodward, James Ussher, Charles Lyell, William Thomson (Lord Kelvin), Alfred Tennyson, and Charles Darwin. (*Randy Moore*)

1699 In *The Anatomy of a Pygmie*, English anatomist Edward Tyson (1650–1708) notes that humans and apes resemble each other in 47 key respects, whereas monkeys resemble apes in only 34 key respects. Tyson also noted that the brains of chimps and humans have a "surprising" resemblance.

CALLED

GENESIS.

Year before the common Year of CHRIST, 4004. – Julian Period, 0710. – Cycle of the Sun, 0010. – Dominical Letter, B.
Cycle of the Moon, 0007. – Indiction, 0005. – Creation from Tisri, 0001.

[1] CHAPTER 1

1 *The creation of heaven and earth.* 14 *Of the sun, moon, and stars.* 26 *Of man in the image of God.* 29 *Also the appointment of food.*

IN the [a]beginning [b]God created the heaven and the earth.

2 And the earth was without form, and void; and darkness *was* upon the face of the deep: [c]and the Spirit of God moved upon the face of the waters.

3 ¶ [d]And God said, [e]Let there be light: and there was light.

4 And God saw the light, that *it was* good: and God divided † the light from the darkness.

5 And God called the light [f]Day, and the dark-

Before CHRIST 4004.

a John 1. 1, 2. Heb. 1. 10. b Ps. 8. 3. & 33. 6. & 89. 11, 12. & 102. 25. & 136. 5. & 146. 6. Isa. 44. 24. Jer. 10. 12. & 51. 15. Zech. 12. 1. Acts 14. 15. & 17. 24. Col. 1. 16, 17. Heb. 11. 3. Rev. 4. 11.

Before CHRIST 4004.

s Jer. 31. 35. ‖ Or, *creeping.* † Heb. *soul.* † Heb. *let fowl fly.* † Heb. *face of the firmament of heaven.* u ch. 6. 20. & 7. 14. & 8. 19. Ps. 104. 26.

18 And to [s]rule over the day, and over the night, and to divide the light from the darkness: and God saw that *it was* good.

19 And the evening and the morning were the fourth day.

20 ¶ And God said, Let the waters bring forth abundantly the ‖ moving creature that hath † life, and † fowl *that* may fly above the earth in the † open firmament of heaven.

21 And [u]God created great whales, and every living creature that moveth, which the waters brought forth abundantly after their kind, and every winged fowl after his kind: and God saw that *it was* good.

Figure 8 James Ussher's claim that creation occurred in 4004 BCE became famous in 1701 when it was placed in Bibles produced by the Church of England. Shown here is a page from a Bible published in the mid-1800s; note the "Before CHRIST 4004" in the center columns of text. Ussher's dates also appeared in the bestselling *The Scofield Reference Bible* in 1909, but are in relatively few Bibles today.

1701 The Church of England begins including dates from James Ussher's biblical chronology in its Authorized King James Version of the Bible (Figure 8). Ussher's dates remained in these and many other Bibles for almost three centuries.

1705 The discovery of a giant tooth along the banks of the Hudson River prompts some theologians to speculate about the existence of giant humans in ancient America. Some theologians believed that the tooth belonged to the behemoth, a biblical creature mentioned in the Book of Job 40:15-24. During upcoming decades, debates about the giant tooth and other large bones involved luminaries such as Thomas Jefferson and Daniel Boone. Most colonists eventually assumed that the tooth belonged to an elephant-like animal they called a mammoth.

1709 A reviewer of Swiss naturalist Johann Jacob Scheuchzer's (1672–1733) *Herbarium of the Deluge* shows how fossils have been accepted in natural history when he writes that "[h]ere are new kinds of coins, the dates of which are incomparably more ancient, more important and more reliable than those of all the coins of Greece and Rome."

1715 Edmond Halley proposes that measurements of oceanic salinity could be used to determine the earth's age and show evidence "of the Sacred Writ [that] Mankind has dwelt about 6,000 years." Although Halley never used this technique, others—most notably John Joly—later used Halley's idea to estimate the earth's age (see also Appendix A).

1720 French scientist René-Antoine Ferchault de Réaumur (1683–1757) tells the Paris Academy of Sciences that a Noachian flood cannot account for the thick sediments (composed largely of broken shells) underlying the Tours area. Instead, Réaumur suggested that the area was once covered by seas.

1726 James Hutton (1726–1797) is born in Edinburgh, Scotland. Hutton received a medical degree in 1749, but soon abandoned medicine and began farming in northeastern Berwickshire on land that he inherited. Hutton became "very fond of studying the surface of the earth." Hutton's views about uniformitarianism later became a foundation of modern geology (Figure 2).

1726 Johann Scheuchzer (1672–1733), who was strongly influenced by John Woodward's advocacy of Flood geology, announces in *Lithographia Helvietica* the discovery of a fossilized human, *Homo diluvii testis* ("human witness of the deluge"). Scheuchzer claimed that the fossil was a human "whose sins brought upon the world the dire misfortune of the Deluge," but Georges Cuvier later identified the fossil as an extinct salamander. Today, the fossil resides in the Teylers Museum, the oldest museum in the Netherlands.

1732 French priest Noël-Antoine Pluche's (1688–1761) popular eight-volume *Spectacle of Nature* argues that the alleged perfection of nature and the human body proves God's goodness.

1735 Swedish biologist Carolus Linnaeus (1707–1778; known after 1761 as Carl von Linné when granted nobility by the King of Sweden) begins publishing *Species Plantarum,* which in 1753 establishes the Latin binomial system for naming plants and is considered the beginning of contemporary botanical nomenclature. That same year, Linnaeus's *Systema Naturae* (published in Latin, as was customary in Linnaeus' era) outlined his classification of life and established the foundation of systematic biology. *Systema Naturae,* which opened with an epigraph paraphrasing Psalm 103:24 ("O Lord, How manifold are your works! How wisely have you made them! How full is the earth with your riches!"), revolutionized classification by using common characteristics to categorize life. Linnaeus—the son of a clergyman—used a system of kingdoms, classes, orders, and genera of increasing specificity, each of which was a distinct "archetype." Although Linnaeus placed every plant and animal into its own unique and static place in the divinely created order of nature, his system was later adapted to evolutionary interpretations: the unifying theme in systematics, evolution, allows organisms to be classified based on their degree of shared ancestry. In the 1st edition of *Systema Naturae*, Linnaeus classified whales as fishes; in the 10th edition, whales were grouped with mammals. Linnaeus—whose motto was *Deus creavit, Linnaeus disposuit* ("God created, Linnaeus arranged")—had by the 10th edition of *Systema Naturae* (1758) included 7,700 plant species and 4,400 animal species in his system. Modern taxonomy traces its ancestry directly to this document.

1735 English poet Alexander Pope (1688–1744) pays homage to the Great Chain of Being (Figure 3) in his *Essay on Man*: "Who sees with equal eye, as God of all, The hero perish, a sparrow fall. . . . Where one step broken, the great scale's destroy'd, From Nature's chain whatever link you strike, Ten or ten thousand, breaks the chain alike."

1737 In *Genera Plantarum Eorumque Characteres Naturales*, Carolus Linnaeus advocates for the divine creation and constancy of species: "There are as many species as the Infinite Being produced different forms in the beginning; which forms afterwards produced more forms, but always similar to themselves according to inherent laws of generation, so that there are no more species now, than came into being in the beginning."

1736 Bishop Joseph Butler's (1692–1752) *Analogy of Religion, Natural and Revealed* claims that "the only distinct meaning of the word 'natural' is STATED, FIXED, or SETTLED; since what is natural as much requires and presupposes an intelligent agent to render it so, i.e., to effect it continually or at stated times, as what is supernatural or miraculous does to effect it for once." When Charles Darwin included this quotation as an epigraph in the 2nd edition of *On the Origin of Species*, some readers believed that Darwin was admitting a supernatural agency in the origin of species.

1739 In book III of *A Treatise of Human Nature*, Scottish philosopher David Hume (1711–1776) rejects claims that what *is* can justify what *ought* to be, a perspective that became known as *Hume's*

Guillotine. Despite Hume's argument, Social Darwinists later used what *is* (i.e., that natural selection *is* a fact of nature) to justify what *ought* to be (e.g., that "unfit" individuals *ought* to be sterilized).

1739 The first fossils are gathered in the United States by Baron Charles Lemoyne de Longueuil (1656–1729) at what is now Big Bone Lick State Park in Kentucky.

1744 Swiss biologist Albrecht von Haller (1708–1777) coins the term *evolution* (from the Latin *evolvere*, to unroll) to describe how embryos originate from preformed structures (*homunculi*) in eggs and sperm. Charles Darwin first used the word *evolution* in its contemporary sense in *The Descent of Man, and Selection in Relation to Sex* (1871).

1745 In *Vénus Physique*, French philosopher Pierre-Louis Moreau de Maupertuis (1698–1759) suggests that random events affecting heredity could create a new trait and that if these events were repeated across several generations, a new species could arise. Five years later, in *Essai de Cosmologie*, de Maupertuis glimpsed natural selection as a process that produces adaptation, but did not describe how it could produce new species.

1746 Carolus Linnaeus—who described the biblical Adam as "the first naturalist"—notes that "as a natural historian I have yet to find any characteristics which enable man to be distinguished on scientific principles from an ape." The following year, Linnaeus asked "the whole world for a generic difference between man and ape which conforms to the principles of natural history. I certainly know of none. . . . If I were to call man ape or vice versa, I should bring down all the theologians on my head. But perhaps I should still do it according to the rules of science." Linnaeus, who hoped to uncover God's plan for creation by classifying organisms, was the first person to include humans in a biological classification scheme.

1748 French anthropologist and well-traveled diplomat Benoît de Maillet's (1656–1738) posthumously published *Telliamed: Or, discourses between an Indian philosopher and a French missionary, on the dimunition of the sea, the formation of the Earth, the origin of men and animals, and other curious subjects, relating to natural history and philosophy* proposes a science-based estimate of the earth's age. De

Maillet, who did not try to reconcile geology with the Bible, used measurements of the decline of sea level to claim that the earth is more than two billion years old (see also Appendix A). To avoid problems with the church, de Maillet described his ideas in a fictitious series of discussions between a French missionary and an Indian philosopher named Telliamed (de Maillet spelled backward). *Telliamed* was advertised by a Baltimore publisher as "a very curious book."

1748 David Hume's *Enquiry Concerning Human Understanding* describes an imaginary speech by Epicurus to Athenians that rejects the "religious hypothesis" of "intelligence and design" in nature. Hume also proposed a rule for identifying miracles: "[t]hat no testimony is sufficient to establish a miracle, unless the testimony be of such a kind, that its falsehood would be more miraculous, than the fact, which it endeavours to establish." Hume concluded "it is easy to shew, that we have been a great deal too liberal in our concession, and that there never was a miraculous event on so full an evidence."

1749 Georges-Louis Leclerc, Comte de Buffon (1707–1788; Figure 9), publishes *L'Histoire Naturelle* (*Natural History)*, which will eventually include 43 volumes and become one of the most widely read science books of the 18th century. *L'Histoire Naturelle,* which made Buffon a celebrated writer, ridiculed Linnaeus's classification system while urging readers to leave the understanding of God to theologians. Buffon, like most scientists before him, believed that all organisms were created by a deity and, like Aristotle and predecessors, placed humans atop a hierarchical arrangement of life. However, Buffon marginalized divine action (without being openly atheistic) and rejected the proposition that God intervened in nature, a bold position that forced him to issue several apologies to theologians (e.g., "I abandon everything in my book . . . contrary to the narrative of Moses"). Buffon ascribed the origin of life to spontaneous generation (at the North Pole), viewed species as real, fixed entities ("Species are the only beings in Nature; they are perpetual beings, as old, as permanent as Nature itself"). He also believed that closely related species might share a common ancestor and accepted change within different populations of a species, but believed this change from the original form represented degeneration. Buffon argued against living creatures having been individually designed by an intelligent Creator; for example, a pig "does not appear to have been formed upon an original, special, and perfect plan, since it is a compound of other animals; it has evidently useless parts, or rather parts of which it cannot make any sense, toes all the bones of which are perfectly formed, and which, nevertheless, are of no service to it." Buffon noted that animals have "unlimited fecundity," that "each family, as well in animals as in vegetables, comes from the same origin," and that all animals come from one species that over time had produced "all the races of animals which now exist." Buffon—who speculated that humans and apes have a common ancestor—recognized that individuals can be grouped according to type, and he proposed using the ability to interbreed as the primary feature for such classification, anticipating the biological species concept developed early in the 20th century.

1751 In *Encyclopédie, ou dictionnaire raisonné des sciences, des arts et des métiers* (*Encyclopedia, or a systematic dictionary of the sciences, arts, and crafts*), French philosopher Denis Diderot (1713–1784) resoundingly supports the search for knowledge: "All things must be examined, debated, investigated without exception and without regard for anyone's feelings." In his "figurative system of human knowledge," Diderot placed theology under "philosophy" rather than recognize revelation as a discrete form of knowledge. *Encyclopédie* also questioned the scientific authority of the Catholic Church and threatened the power of the French aristocracy; in so doing, and by reaching so many of the literati, Diderot's books influenced the French Revolution of the late 18th century.

1751 Linnaeus produces *Philosophia Botanica*, the first textbook of descriptive systematic botany, in which he stressed biological continuity and the smooth transitions between taxa by claiming that "[n]ature does not make leaps. All plants show an affinity with those around them, according to their geographical location." Charles Darwin used the phrase *Natura non facit saltum* (*Nature makes no leap*) seven times in *On the Origin of Species* to emphasize gradualism.

1751 Self-confessed deist Benjamin Franklin's (1706–1790) *Observations Concerning the Increase of Mankind* accurately predicts the growth rates of the population of Great Britain and North America. This book, which lamented the "swarm" of German immigrants who threatened to make Pennsylvania "a colony of Aliens," strongly influenced English economist Thomas Malthus (whose work, in turn, influenced Charles Darwin and Alfred Russel Wallace).

Figure 9 Georges-Louis Leclerc, Comte de Buffon, helped establish a philosophical perspective of science based on explaining nature strictly through observable events. Although Buffon's view of organic change fit perfectly with the Great Chain of Being that disallowed the addition of intermediate forms to an already perfect progression, Buffon refused to accept that God originated this process because a deity would not concern himself with, for example, "the particular fold in a beetle's wing." This bronze statue of Buffon is on the grounds of Jardin des Plantes in Paris, where Buffon worked.

1753 Volume Four of Georges-Louis Buffon's (Figure 9) *L'Histoire Naturelle* notes that similarities among vertebrates suggest their common ancestry.

1754 German philosopher and deist Hermann Samuel Reimarus's (1694–1768) *Principal Truths of Natural Religion* rebuffs Epicurean criticisms of the argument from design. In this book, Reimarus transformed John Ray's metaphor involving a clock into one involving a watch, anticipating the proposals of William Paley.

1757 The Roman Catholic Church's Congregation of the Index withdraws its general objections to books teaching Copernican doctrine, but continues to forbid publication of Galileo's *Dialogue*.

1759 London's Natural History Museum (NHM) opens to the public. Except for parts of World Wars I and II, the Museum has been open ever since. NHM was the first national, public, and secular museum in the world, and today its more than 70 million items span every scientific discipline. It is famed for its skeletons of dinosaurs, especially the 84-foot-long *Diplodocus carnegii* (a gift from American steel magnate—and social Darwinist—Andrew Carnegie to King Edward VII) that dominates the Museum's main hall.

1760 In *Disquisitio de Sexu Plantarum*, Carolus Linnaeus expands upon the proposal (introduced in 1751 in the short *Plantae Hybridae*) that new species within genera can arise via hybridization ("[i]t is impossible to doubt that there are new species produced by hybrid generation"). Furthermore, Linnaeus claimed that this hybridization was guided by God. As was common in his era and since, Linnaeus accepted that God created life in the Garden of Eden and believed that classifying organisms would reveal the pattern of creation. Linnaeus accepted the biblical account of the Flood, but did not believe that it could have moved organisms very far inland and covered them in sediments in the time available. As he noted, "[h]e who attributes all this to the Flood, which suddenly came and as suddenly passed, is verily a stranger to science and himself blind, seeing only through the eyes of others, as far as he sees anything at all."

1761 French biologist Jean-Baptiste Robinet (1735–1820) argues in his five-volume *De la Nature* (*On Nature*) that all matter is alive, that the Creator made organisms on a hierarchical scale, and that nonhuman organisms can move on an evolutionary escalator toward the top of the scale, where humans are.

1762 English Romantic painter George Stubbs's surprisingly violent *Horse Attacked by a Lion* is one of a popular series of paintings that foreshadow the notion of a struggle for existence.

1763 The fossil discovered by Robert Plot in 1676 is illustrated in Richard Brookes's *The Natural History of Waters, Earths, Stones, Fossils, and Minerals*. On one of its ends, the fossil resembled a pair of human testicles, promoting Brookes to name it *Scrotum humanum*. It was later determined to be part of the dinosaur *Megalosaurus*. *Scrotum humanum* is the first name applied to a dinosaur, in conformity with the rules of biological nomenclature.

1764 In *Contemplation de la Nature*, Swiss naturalist Charles Bonnet (1720–1793) uses the term *scala naturae* to describe the Great Chain of Being (Figure 3). Bonnet's often-reproduced depiction of the Great Chain of Being showed the natural world as a ladder ascending from "ethereal matter" to "man." Bonnet, who became famous for discovering parthenogenesis of aphids, later argued that females contain small forms of all future generations and that catastrophic evolution causes these small forms to improve (i.e., evolve upward). Bonnet, who claimed that this kind of evolution enabled inorganic matter to come alive and humans to become angels, also believed that each species was an unchanging part of a divine plan: "The centuries transport from one to another this magnificent spectacle; and they transmit exactly what they receive. Nothing changes, nothing alters; perfect identity. Victorious over the elements, time and death, the species are conserved, and the period of their duration is unknown to us."

1766 Georges-Louis Buffon (Figure 9) again argues that similar species share a common ancestor and that their geographic distribution provides clues about the species' history. Buffon also claimed that the environment (and not biotic factors, such as competition) causes species to evolve, but he never proposed a convincing mechanism for how this could occur. In volume five of *L'Histoire Naturelle*, Buffon claimed that the cooler, more humid climate of North America (compared with Europe) had produced inferior animals and indigenous people. This so-called "Theory of American Degeneracy" was later refuted by Thomas Jefferson in *Columbian Magazine*. Jefferson's attack also appeared in the *Federalist Papers*

(1787–1788) when Alexander Hamilton (1755–1804) ridiculed the charges of American degeneracy.

1768 François-Marie Arouet (1694–1778), better known by the pen name Voltaire, ponders the earth's history when noting "I still confess that it is demonstrated to the eyes that it must have required a prodigious multitude of centuries to perform all the revolutions that have occurred to this globe." In *Candide*, Voltaire—appalled by an earthquake in Portugal that killed 100,000 people—ridiculed the idea that the world is designed for its inhabitants.

1768 Lazzaro Spallanzani (1729–1799) adds to the evidence against spontaneous generation by showing that no organisms grow in a nutrient broth if it is first heated and allowed to cool in a closed flask.

1770 Dutch miners unearth a giant skull, teeth, and jaws, and create a debate about the creature's origin. This discovery, which Georges Cuvier later named *Mosasaurus*, made headlines throughout Europe.

1770 As biologists and others continue to gather fossils of ancient organisms, most people continue to accept the Great Chain of Being (Figure 3) and its implication that extinction is impossible. For example, John Wesley (1703–1791) noted that "[d]eath . . . is never permitted to destroy the most inconsiderable species," and Thomas Jefferson agreed that "such is the economy of nature, that no instance can be produced of her having permitted any one race of her animals to become extinct; of her having formed any link in her great work, so weak as to be broken." Less than 40 years later, Georges Cuvier documented that extinction is real.

1770 Poet and physician Erasmus Darwin (1731–1802)—Charles's eccentric grandfather—promotes the idea of common descent by printing the allegorical motto *E conchis omnia* ("Everything from shells") on his carriage. Soon thereafter, when a public outcry ensued, he had the motto removed. Erasmus, who died before Charles Darwin was born, helped found the family fortune that later sustained Charles in a life of privileged comfort.

1770 Whereas the 1st edition (1735) of Carolus Linnaeus's *Systema Naturae* included only a title page, eleven pages of "observations" and taxonomic tables, and two one-page leaflets, the 13th (and last) edition comprises 3,000 pages.

1774 German pastor Johann Esper (1732–1781) writes in *A Detailed Account of Newly Discovered Zooliths* that he "dare not presume" that human remains that he discovered were contemporaneous with animal fossils among which they were found. Instead, Esper claimed that the fossils were brought together by the biblical Flood. This was the first discovery of human fossils.

1774 Anglo-Irish writer Oliver Goldsmith (1730–1774) notes that "[n]ature is varied by imperceptible gradations, so that no line can be drawn between any two classes of its productions, and no definition made to comprehend them all."

1776 German anatomist Johann Friedrich Blumenbach's (1752–1840) *On the Natural Varieties of Mankind* argues that changing environmental conditions caused Caucasians to "degenerate" into other races. Blumenbach—a critic of the Great Chain of Being—coined the term *Caucasian* and was the first person to apply the word *race* to humans.

1776 Scottish philosopher Adam Smith's (1723–1790) magnum opus *An Inquiry into the Nature and Causes of the Wealth of Nations*—the founding work of classical economics—provides a blueprint for how free enterprise can produce economic growth. Smith's *laissez faire* depicts an "invisible hand" guiding individuals to act for personal gain that produces economic growth and distributes resources. Just as Smith tried to understand "the nature and causes of wealth," Thomas Malthus later tried to determine "the nature and causes of poverty." Charles Darwin used both of these ideas to understand nature itself.

1778 Georges-Louis Buffon's (Figure 9) *Époques de la Nature* asserts that the earth existed long before the appearance of humans or any other life-form. Buffon, who was influenced by Benoît de Maillet's *Telliamed*, postulated two major episodes of spontaneous generation: The first episode formed organisms that eventually became extinct, and the second episode formed the ancestors of today's major groups of animals. Buffon also concluded—from his studies in the 1760s of the cooling rates of metal spheres—that the earth is approximately 74,832 years old (see also Appendix A). However, Buffon was dissatisfied with his modest timescale and suspected that he had overlooked some "hidden cause" (*cause cachée*); Buffon later extended the earth's age to seven million years to account for the fossil record. Buffon described seven *époques*, each of which was a "milestone on the eternal road of time." Buffon's reliance on a scientific approach that invoked only natural causes was revolutionary and influenced later generations of scientists, including Charles Darwin, who noted Buffon's contributions in *On the Origin of Species*. Buffon was the first scientist to advocate day-age creationism, noting that "what can we understand by the six days the sacred writer designates so precisely by counting them one after the other if not six spaces of time, six intervals of duration?"

1779 David Hume's posthumously published *Dialogues Concerning Natural Religion* attacks the argument from design for asserting the existence of a creator. In *Dialogues*, the protagonists Philo and Cleanthes discussed inferring the existence of a deity by observing an apparently complex and designed world. Hume rejected the design-based argument by invoking the existence of evil, by noting that human creations are subject to later improvements, and by questioning the assumption of complexity. Hume's arguments were included in many subsequent refutations of the design argument (including those of the modern ID movement).

1779 French scientist Jean-Sylvain Bailly (1736–1793) claims that after Noah's flood, humans moved from the North Pole to Central Asia, and then to the rest of the world.

1780 A giant skull is unearthed from a stone quarry in the Netherlands. In 1796, Georges Cuvier showed that the skull belonged to a huge, prehistoric reptile.

1785 Scottish geologist James Hutton studies rocks at Glen Tilt of the Scottish Highlands and finds granite extending into metamorphic schists. Hutton concluded that the granite formed from cooling magma and that it was therefore younger than the schists within which it was found. Hutton inferred that the Glen Tilt rock formation was evidence of intense subterranean heat, the engine that creates new rock. On March 7 and April 4, Hutton's ideas about the earth's history were reported to the recently formed Royal Society of Edinburgh in a 30-page dissertation titled *Concerning the System of the Earth, Its Duration, and Stability*. Hutton claimed that rocks now at the earth's surface were formed from pressurized sediments

under the ocean, raised to the surface by pressure and heat, and eroded to the sea, where the process began anew. During these upheavals, molten rock is injected into cracks of dislocated strata. Hutton stated what came to be known as *uniformitarianism*: "The past history of our globe must be explained by what can be seen to be happening now. No powers are to be employed that are not natural to the globe, no action to be admitted except those of which we know the principle." Hutton later expanded his ideas into *Theory of the Earth, with Proofs and Illustrations*. Hutton believed that all creatures except humans are ancient, and that all of the earth's changes have been purposeful (i.e., the world has been constantly replenished for man's use). Hutton, who claimed that geologic strata showed no evidence of historical direction, was vilified as an "infidel" by conservative critics.

1785 English archdeacon William Paley's (1743–1805) *Principles of Moral and Political Philosophy* argues that the world is designed in a particular way for good reasons and uses this claim to justify land ownership and social classes. This book, which went through 15 editions in Paley's lifetime, was one of the most influential books of the British Enlightenment. In 1802, Paley expanded his argument to include the intelligent design of nature in *Natural Theology*.

1785 Thomas Jefferson's *Notes on the State of Virginia* discusses discoveries of "mammoths" in Kentucky to refute Buffon's (Figure 9) "Theory of American Degeneracy." Jefferson believed that mammoths—"the largest of all terrestrial beings"—quashed Buffon's claim. Famed explorer Alexander von Humboldt (1769–1859) modeled his books after Jefferson's *Notes*.

1788 James Hutton visits Siccar Point on the rocky Berwickshire coast near Scotland's border with England (Figure 10). The Siccar Point unconformity—a point where two sets of rocks formed at different times come into contact, and described by Hutton as "a junction washed bare by the sea"—shows basal, vertical sediments that are overlain by horizontal layers. Hutton knew that sediments are generally laid down horizontally in water and concluded that the lower layers must therefore have been tilted and raised by pressure from below, such as during an earthquake. The tilted sediments were subsequently submerged and covered by additional sediments, after which seismic activity again raised

Figure 10 Siccar Point, where geologist James Hutton glimpsed the "abyss of time," is one of the most famous geological sites in the world. After Hutton visited Siccar Point in 1788, he was convinced of the earth's vast age. This view shows Siccar Point from the North, with the North Sea in the background. (*Randy Moore*)

the sediments. Hutton published his findings in the inaugural volume of *Transactions of the Royal Society of Edinburgh*. These observations helped Hutton appreciate the earth's vast age (Figure 2).

1788 The Linnean Society of London, the world's oldest extant biological society, is founded. The Linnean Society derived its name from famed Swedish naturalist Carl von Linné, whose biological collections and books are maintained by the Society. Those collections—which include 14,000 plants, 158 fish, 1,564 shells, 3,198 insects, 1,600 books, and more than 3,000 letters and documents—were bought in 1784 from Linné's widow by Sir James Edward Smith (1759–1828), the Society's first president. The Society's role remains as stated in its charter ("The cultivation of the Science of Natural History in all its branches") and the Society maintains historical biological collections and encourages debate, research, publications, and meetings. The Society also publishes several journals, including *Biological Journal of the*

Linnean Society, A Journal of Evolution, which is a direct descendant of the oldest biological journal in the world, *Journal of the Proceedings of the Linnean Society*. On the evening of July 1, 1858, the Linnean Society hosted the first public presentation of Charles Darwin's and Alfred Russel Wallace's theory of evolution by natural selection.

1789 American Presbyterian minister Samuel Stanhope Smith's (1751–1819) *Essay on the Cause of the Variety of Complexion and Figure in the Human Species*—the first American book on racial variation—argues that natural causes (and not separate creations) created the different races of humans.

1789 Charles Darwin's eclectic grandfather, Erasmus Darwin, publishes a book of poetry, *The Botanic Garden*, which speculates about cosmological theories and notes that "the ingenious theory of Dr Hutton" implies an eternal nature of the earth.

1791 The First Amendment to the U.S. Constitution (along with the Second through Tenth Amendments, collectively known as the Bill of Rights) is ratified and states "Congress shall make no law respecting an establishment of religion, or prohibiting the free exercise thereof; or abridging the freedom of speech, or of the press; or the right of the people peaceably to assemble, and to petition the government for a redress of grievances." Public institutions—including, eventually, public schools—were thenceforth restricted from establishing or favoring any religious perspective (via the Establishment Clause) and from limiting religious expression (via the Free Exercise Clause). The Establishment Clause (combined with the Fourteenth Amendment's restrictions on states' abilities to infringe upon individuals' rights) has formed the basis of contemporary court decisions against the prohibition of teaching evolution and the inclusion of creationism in public schools. In contrast, attempts to restrict the teaching of evolution have invoked the Free Exercise Clause by claiming that teaching evolution promotes one religion ("secular humanism") over another (e.g., Christianity).

1793 Jean-Baptiste Pierre Antoine Lamarck, Chevalier de Monet (1744–1829), who had earlier tutored the children of Georges-Louis Buffon, helps reorganize the Muséum National d'Histoire Naturelle in Paris and is appointed a professor there the following year. The museum was to be run by 12 professors in 12 scientific disciplines, and Lamarck was in charge of studying "insects, worms, and microscopic animals." Lamarck knew little about these organisms, but later coined the word *invertebrate* to describe them. In 1809, Lamarck proposed the first scientifically testable model for evolution—one based on the inheritance of acquired traits.

1793 French comparative anatomist Étienne Geoffroy Saint-Hilaire (1772–1844) is appointed to the newly founded Muséum National d'Histoire Naturelle in Paris, where he develops ideas about the transformation of organisms operating through modifications of a universally shared body plan. Geoffroy went so far as to propose that arthropods are built upon the same body plan as vertebrates, claiming that the exoskeleton of insects is merely a single vertebra used on the outside of the body. Geoffroy claimed that transformation follows a preordained plan, but his proposal of organic change was still a radical concept. In 1830, conflicts with colleague Georges Cuvier over the possibility of organic change led to their famous public debates.

1794 Scottish geologist James Hutton's *An Investigation into the Principles of Knowledge* describes the earth as a "beautiful machine" designed for an intended effect. Buried in the 2,138 pages was a chapter about variety in nature that anticipated Charles Darwin's theory of natural selection: "if an organised body is not in the situation and circumstances best adapted to its sustenance and propagation, then, in conceiving an indefinite variety among the individuals of that species, we must be assured, that, on the one hand, those which depart most from the best adapted constitution, will be the most liable to perish, while, on the other hand, those . . . which most approach to the best constitution for the present circumstances will be best adapted to continue, in preserving themselves and multiplying the individuals of their race. . . . If those organized bodies shall thus multiply, in varying conditions according to particular circumstances . . . we might expect to see . . . a variety in the species of things which we might term a race." Hutton supported his suggestion by citing his studies of animal breeding, noting that dogs survived by "swiftness of foot and quickness of sight . . . the most defective in respect of those necessary qualities, would be the most subject to perish, and that those who employed them in greatest perfection . . . would be those who would remain, to preserve themselves, and to continue the race." The same "principle of variation" would influence "every species of plant, whether growing in a forest or a meadow." However, Hutton provided no data to support his suggestion, and his idea about life's diversity received little attention.

1794 Erasmus Darwin's multivolume *Zoonomia, or, the Laws of Organic Life*—one of the earliest works to proclaim evolution—argues that biological urges such as lust and hunger produce the physiological assets that help satisfy those urges. Darwin believed that change was based on progress, development, and metamorphosis and wondered "in the great length of time since the earth began to exist, perhaps millions of ages before the commencement of the history of mankind, would it be too bold to imagine that all warm-blooded animals have arisen from one living filament."

1795 German scholar and savant Alexander von Humboldt discovers a layer of fossils in limestone from the Jura Mountains in Switzerland. He later named the period corresponding to these fossils the Jurassic period (see also Appendix B).

1795 In *The Age of Reason*, deist Thomas Paine (1737–1809) attacks the Bible in the name of the Enlightenment: "Take away from Genesis the belief that Moses was the author, on which only the strange belief that it is the word of God has stood, and there remains nothing of Genesis but an anonymous book of stories, fables, and traditionary or invented absurdities and downright lies. The story of Eve and the serpent, and of Noah and his ark, drops to a level with the Arabian tales, without the benefit of being entertaining; and the account of men living to eight or nine hundred years becomes as fabulous as the immortality of the giants of the Mythology." Paine—who dismissed Christianity as "too absurd for belief"—claimed to "detest" the Bible "as I detest everything that is cruel," but found in nature the revelation of God's benevolence and power.

1795 James Hutton publishes his most famous and influential work, the two-volume, 1,100-page *Theory of Earth*, in which he claims that ongoing, endless cycles of erosion, sedimentation, and uplift are constantly yet imperceptibly changing the earth, thereby challenging the claim that the earth is only 6,000 years old. Hutton argued that it is unnecessary to invoke processes other than ordinary, everyday events to explain geology. With *Theory of Earth*, Hutton opened geology to scientific observation, removed it from the influence of Bible-based chronologies (*Theory of Earth* does not mention Genesis), and gave geology its most transforming idea—namely, that the earth is ancient (Figure 2). Although Hutton's writings were full of deistic metaphysics and teleology ("The glove of this earth is evidently made for man"), his conclusions meant that vast periods of time were available for evolution to occur. To Hutton, the earth was a self-renewing "beautiful machine" whose history "must be explained by what can be seen to be happening now . . . no powers are to be employed that are not natural to the globe, no action to be admitted except those of which we know the principles." Hutton's geology was characterized by perpetual change ("rest exists not any where") instead of stasis and Hutton invoked no great catastrophes (e.g., a worldwide flood) to explain the earth's geology. Instead, Hutton claimed, "in examining things present we have data from which to reason with regard to what has been. . . . The ruins of an older world are visible in the present structure of our planet." Hutton believed that the earth's age was "indefinite" and beyond comprehension, as suggested in the poetic and now-famous sentence that concluded *Theory of Earth*, "we find no vestige of a beginning,—no prospect of an end." Hutton's ideas, which became a foundation of modern geology (Figure 11), were later developed by Charles Lyell and became the basis of uniformitarianism, a foundation of modern geology.

1796 Surveyor and canal-excavation engineer William Smith (1769–1839) observes that within layers of sedimentary rock, "each stratum contained organized fossils peculiar to itself, and might, in cases otherwise doubtful, be recognised and discriminated from others like it, but in a different part of the series, by examination of them." This is the first statement of the Principle of Faunal Succession, a scientific theory stating that fossils in undisturbed strata of sedimentary rocks succeed each other vertically in a reliable, specific order over vast areas of land. This principle enabled Smith to use fossils from one part of England to predict and identify fossils in particular strata of other parts of England. Smith did not publish his idea until 1816 in *Strata Identified by Organized Fossils*.

1796 In a paper about the extinct fossil mammal *Megatherium*, French anatomist Georges Cuvier (1769–1832; Figure 12)—the founder of comparative anatomy—confirms the reality of extinctions. Cuvier's documentation of extinctions, including mass extinctions, was nearly heretical because it "broke" the Great Chain of

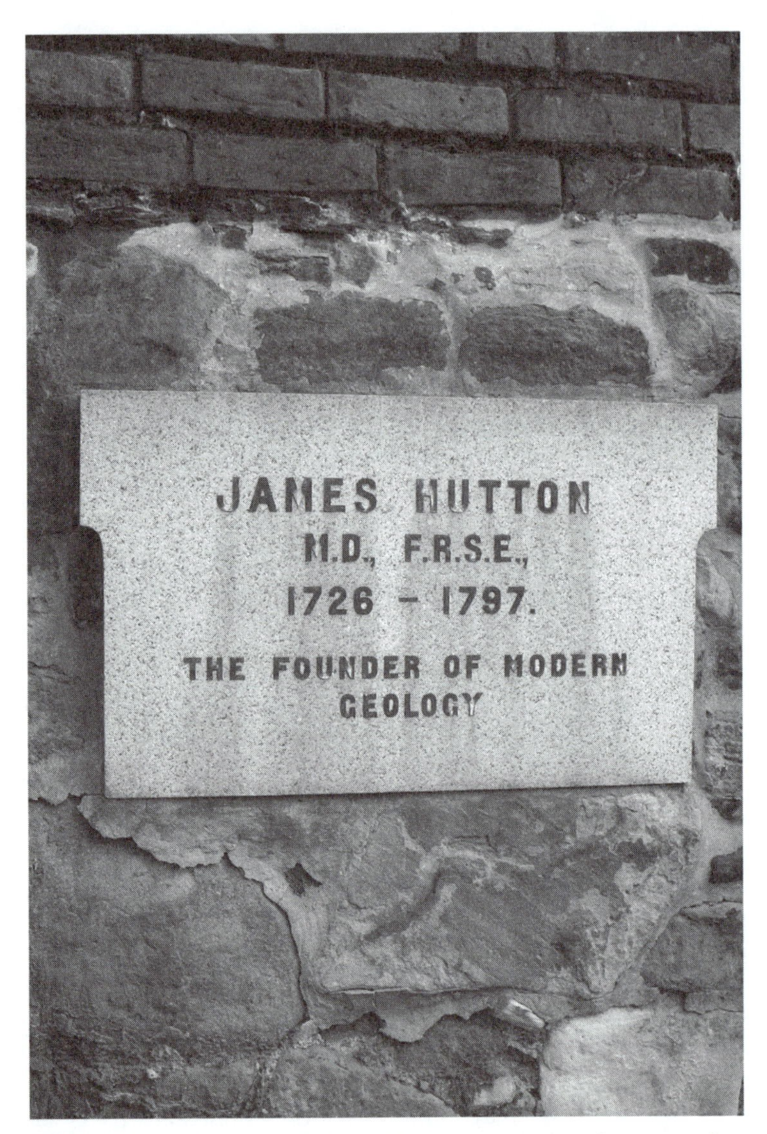

Figure 11 Scottish geologist James Hutton gave geology its most enduring idea—namely, the concept of "deep time" (see also Figure 2). This photo shows Hutton's tomb in a secluded area of Greyfriars Kirk churchyard in Edinburgh. (*Randy Moore*)

Being (Figure 3), in which each species had its own ideal structure and function. Cuvier knew that the history of life is recorded in layers of rocks containing fossils, that life on earth had changed over time, and that life on ancient earth was different from life today. To reconcile his findings with his religious beliefs, Cuvier claimed that the fossilized organisms in different layers of rocks

Figure 12 Georges Cuvier, who attributed animals' similarities to common functions rather than shared ancestry, established the discipline of vertebrate paleontology and documented extinctions. These extinctions, which Cuvier believed were caused by God-driven catastrophes ("revolutions"), "broke" the Great Chain of Being. Although Cuvier's discoveries supported Charles Darwin's theory of evolution by natural selection, Cuvier rejected evolution because it was "contrary to moral law, to the Bible, and to the progress of natural science itself." (*Library of Congress*)

had died in a series of God-directed catastrophes—that is, that there had been "a world previous to ours, destroyed by some kind of catastrophe" and that "life on earth has been disturbed by terrible events." Each of these catastrophes (Cuvier preferred the less theologically loaded term "revolutions," which were analogous to the traumatic political changes in France) eliminated some species, and each catastrophe was followed by the separate creations or immigrations of new species. Cuvier, who established the discipline of paleontology, realized that an alliance between stratigraphy and paleontology could help explain the earth's history. However, he rejected evolution, claiming that it was "contrary to moral law, to the Bible, and to the progress of natural science itself." No human fossils were known to Cuvier, who believed that humans came into existence after the biblical Flood. Cuvier's catastrophism was popular for several decades.

1796 Thomas Jefferson, who in 1801 will become the third president of the United States, hires American explorer William Clark (later of Lewis and Clark fame) to search for remains of extinct mastodons in Kentucky. When Congress authorized the Lewis and Clark expedition of 1803, Jefferson helped finance the expedition and asked the explorers to send bones of large animals back to Washington. (At one point, the unfinished East Room of the White House was filled with bones of ancient animals and was called the "Mastodon Room.") In 1797, while en route to his inauguration as Vice President, Jefferson delivered a large box of fossils to the American Philosophical Society.

1797 Charles Lyell (1797–1874) is born at his family's estate near Forfarshire, Scotland. James Hutton died the same year at age 71. Lyell and Hutton, two of the most influential geologists of all time, strongly influenced the thinking of Charles Darwin. Indeed, Darwin later described himself as a "zealous disciple" of Lyell.

1798 English economist and demographer Thomas Malthus (1766–1834) anonymously publishes *An Essay on the Principle of Population, as it Affects the Future Improvement of Society with Remarks on the Speculation of Mr. Godwin, M. Condorcet, and Other Writers*. Malthus made a simple prediction—namely, that the human population would inevitably outstrip its supply of food: "Population, when unchecked, increases in a geometrical [exponential] ratio. Subsistence increases only in an arithmetic [linear] ratio. . . . As many more individuals of each species are born than can possibly survive . . . it follows that any being, if it vary ever so slightly in a manner profitable to itself . . . will have a better chance of survival, and thus be naturally selected." Malthus claimed that "extermination, sickly seasons, epidemics, pestilence, and plague" were inevitable, and if diseases and wars were not enough, population would "fairly be resolved into misery and vice" by "gigantic, inevitable" famines. Malthus argued that starvation was designed to teach the values of hard work and moral behavior. In later editions of *Essay*, Malthus expanded his discussion of population growth to claim that lower social classes were responsible for a variety of societal ills; he suggested that these people should practice "moral restraint" by "delaying the gratification of passion from a sense of duty" so that they would not bring into the world children that they could not feed. Malthus, who rejected the claim that social reform can bring happiness to all, viewed high fertility rates in a new and negative way, claiming that even though increased fertility might increase gross output, it also tends to reduce output per capita. In a discussion of how tribal groups compete for limited resources, Malthus also coined the phrase *struggle for existence*, a perspective that was a key part of the theory of evolution by natural selection formulated by Charles Darwin and Alfred Russel Wallace. Malthus focused on humans (some have argued that Britain's inaction during the Irish famine in 1845 was due to the impact of Malthus's ideas), but Darwin extended Malthus's idea "with manifold force to the whole animal and vegetable kingdoms." Although Malthus's *Essay* was later associated with repressive measures against the poor (Charles Dickens modeled Ebenezer Scrooge of *A Christmas Carol* after Malthus), Wallace praised *Essay* as "the most important book I read," and Darwin referred to his theory of evolution by natural selection as an application of Malthus's ideas. In 2001, evolutionary biologist Ernst Mayr judged Malthus's "population thinking" as "the foundation of modern evolutionary theory."

1798 Austrian composer Joseph Haydn (1732–1809) writes his great oratorio *The Creation* (*Die Schöpfung*), which is based on Genesis.

1799 Alexander von Humboldt and French botanist Aimé Bonpland (1773–1858) begin their five-year New World voyage, collecting over 60,000 specimens. In one of the 30 books (containing 1,425 maps and illustrations) describing this journey, von Humboldt—who Charles Darwin described as "the greatest traveling scientist who ever lived"—was among the first to suggest that the land masses of the Atlantic were once joined. Although publishing costs bankrupted von Humboldt, his work later informed Darwin's understanding of biogeography.

1799 United States Vice President Thomas Jefferson publishes a paper in *Transactions of the American Philosophical Society* describing *Megalonyx*, a North American fossil ground-sloth similar to the extant South American sloth. The fossil was later renamed *M. jeffersonii* in his honor. Jefferson's paper was one of the first American paleontology publications.

1799 Workers discover a massive femur, along with other bones, in the Hudson River Valley; this discovery begins "Mammoth Fever," which sweeps the United States.

1800 English antiquarian John Frere (1740–1807) claims that ancient tools, "weapons of war," and human remains found in Suffolk "may tempt us to refer them to a very remote period, even beyond that of the present world."

1801 Philadelphia painter Charles Willson Peale (1741–1827) and his son Rembrandt Peale (1778–1860) excavate the first complete skeleton of an American mastodon, an event captured in the elder Peale's painting *Exhuming the Mastodon*. By this time, the mastodon—one of the first organisms to be declared extinct—had become a symbol of American antiquity.

1801 William Smith, who two years earlier had published the first geologic map (a circular map of the geology around Bath), sketches the first geologic map of an entire country. This 8.5-foot by 6-foot map, which later became known as "The Map That Changed the World," was based on fossils and the mineral composition of rocks.

1802 In a letter to the Danbury Baptist Association (a religious minority in Connecticut), Thomas Jefferson uses the phrase "wall of separation between church and state." This phrase led to the shorthand "separation of church and state" used today for the First Amendment's Establishment Clause. This phrase has often been used in legal decisions specifying that the teaching of creationism in science classes of public schools is unconstitutional.

1802 James Hutton's friend John Playfair (1748–1819)—a famed mathematician and ordained minister of the Church of Scotland—publishes *Illustrations of the Huttonian Theory of the Earth*, a book in which Hutton's principle of uniformitarianism first reaches a wide audience. Playfair noted "the theory of Dr Hutton stands here precisely on the same footing with the system of Copernicus." Playfair's book, which urged full separation of religion and science, accompanied Charles Darwin aboard the *Beagle*. Ironically, *Illustrations* contained no illustrations.

1802 The term *biology* in its modern sense is used independently by Gottfried Treviranus (1776–1837) in *Biologie oder Philosophie der lebenden Natur* and by Jean-Baptiste Lamarck in *Hydrogéologie*.

1802 William Paley publishes his influential bestseller *Natural Theology, Or, Evidence of the Existence and Attributes of the Deity Collected from* *the Appearances of Nature*, which states his version of the argument from design. *Natural Theology* was a cornerstone of science at Cambridge University in the 1800s; Charles Darwin read it while a student there. Paley, who claimed that God created organisms perfectly adapted to the environment, opened *Natural Theology* by introducing the "watchmaker" metaphor to provide evidence of a creator—that is, if a person finds a watch in a field, he or she would assume the watch had a designer and maker. (John Ray had earlier used an analogy based on a clock to argue for natural theology, much like Cicero had used an analogy involving a sundial to make the same argument.) According to Paley, "every indication of contrivance, every manifestation of design, which existed in the watch, exists in the works of nature." Paley, who argued that life was unfailingly harmonious and happy ("It is a happy world after all"), also invoked what Adam Smith had called an "invisible hand" to explain what he considered evidence of design (Paley assumed that it was the hand of God). As Paley noted, "[d]esign must have had a designer. That designer must have been a person. That person is GOD." The watchmaker metaphor became a pervasive icon in the modern ID movement, and references to this metaphor formed the basis of Richard Dawkins's *The Blind Watchmaker* (1986). Paley's book, which mentions neither miracles nor the Bible, remains the best-known exposition of natural theology.

1802 Massachusetts farm boy Pliny Moody discovers giant fossilized tracks that appear to have been made by a giant bird. These tracks were later shown to have been made by dinosaurs and were the first known dinosaur remains discovered in North America.

1803 Erasmus Darwin's posthumously published *The Temple of Nature* uses rhymed couplets to describe a gradual progress of life toward higher levels of complexity and greater mental powers. Darwin, one of the most famous people in England, never proposed natural selection, but he did suggest that all species have a common ancestor. Darwin also suggested that species' survival was governed by "laws of nature" rather than divine authority and that new species arise because of competition and sexual selection. When Charles Darwin—Erasmus's grandson—was 70 years old, he wrote Erasmus's biography, *The Life of Erasmus Darwin.* When Charles's daughter Henrietta edited the book, she removed parts of the text (about 16% of the total) that she considered too salacious for the Victorian audience.

1803 Rembrandt Peale's (1778–1860) book *An Historical Disquisition on the Mammoth* references Georges Cuvier's work while dramatizing catastrophic extinctions for the public.

1803 President Thomas Jefferson appoints Meriwether Lewis (1774–1809) and William Clark (1770–1838) to explore the uncharted American West and search for "the remains and accounts of any [minerals and animals] which may be deemed rare or extinct." Lewis and Clark expected to find mountains of salt, unicorns, giant beavers, and living mastodons. They found none of these, but they did find fossils that challenged contemporary ideas about the earth's history.

1804 Georges Cuvier (Figure 12), a devout Christian who considers Genesis as history, ponders the earth's age when he claims that fossils found near Paris are "thousands of centuries" old. He later asked "would there not also be some glory for man to know how to burst the limits of time, and . . . to recover the history of the world, and the succession of events that preceded the birth of the human species." In a separate paper, Cuvier again confirmed the reality of extinction by noting that fossilized animals are unlike anything still living. Cuvier believed that animals' similarities resulted from common functions, not common ancestry, and that an animal's function was determined by its structure. Cuvier, who claimed that nature had a "repugnance" for changes in design, was convinced that every species needs all of its parts just the way they are, and that any change would cause the functioning of the organism to collapse.

1804 In the first of a three-volume work titled *Organic Remains of a Former World*, James Parkinson (1755–1824) describes fossils as the remains of Noah's Flood. Parkinson later referred to fossils as the remains of a prehuman world. Parkinson, who treated the days of creation as vast periods of time, united Genesis with Georges Cuvier's catastrophism.

1807 The Geological Society of London is established. In 1859, the Society awarded Charles Darwin its highest honor.

1809 Charles Robert Darwin (1809–1882; Figure 13) is born into a life of privilege in the small country town of Shrewsbury, Shropshire, England (about 160 miles northwest of London). Darwin shared his birth date (February 12) with Abraham Lincoln, both of whom

Figure 13 In 2000, Charles Darwin replaced fellow Victorian Charles Dickens on the British £10 note. Although Darwin was not the first person to propose evolution, his *On the Origin of Species* (1859) documented a testable mechanism for evolution and, in the process, changed the course of biology. (*Randy Moore*)

later appeared on the currency of their countries. In 1859, Darwin's *On the Origin of Species* famously proposed evolution by natural selection, which philosopher Daniel Dennett (b. 1942) later called "the single best idea anyone ever had."

1809 Jean-Baptiste Lamarck (Figure 14) publishes his most famous book, *Philosophie Zoologique*, in which he shocks Georges Cuvier and much of Europe by proposing life's "tendency to progression" and "tendency to perfection." Lamarck claimed that the fixity of life is an illusion, that life is instead in a constant state of advancement and improvement—"there is no species which is absolutely constant"—driven by spontaneous generation, and that this advancement is too slow to be perceived except with the fossil record. Lamarck argued for the inheritance of acquired traits and claimed that evolution is driven by organisms' needs as they strive to fulfill their way of life. According to Lamarck, species are little more than arbitrary points along a continuum, and environmental changes alter the needs of organisms living in those environments. In turn, these altered needs changed the organisms' behaviors, and these altered behaviors then led to the greater or lesser use of structures (this change in structural utility occurred through the involvement of a "nervous fluid"). The more an organism used a part of its body, the more developed that part became (similarly, disuse of a part resulted in its decay). Lamarck was his era's most renowned advocate of evolution, and his godless model for evolution was the first testable hypothesis to explain how a species could change over time. Lamarck was exceedingly confident about his idea, referring to it as a "permanent truth, which can only be doubted by those who have never observed or followed the operations of nature." However, Lamarck's conclusions were not accompanied by a persuasive mechanism, which prompted many of the leading scientists of Lamarck's time to reject (and sometimes ridicule) his ideas. Nevertheless, Lamarck's proposals were popular with the public (they reinforced the self-congratulatory Victorian emphasis on progress produced by diligence and hard work), so much so that Charles Darwin alluded to them in later editions of *On the Origin of Species*. After the publication of *Origin*, there was a revival of Lamarckism, but the claims associated with this resurgence were based on flawed experiments and fraudulent data (e.g., the "midwife toad" fraud in 1919). Although Darwin privately described *Philosophie Zoologique* as "veritable rubbish," he explicitly admitted the possibility of the inheritance of acquired traits in Chapter VIII of *On the Origin of Species*. Darwin later acknowledged that Lamarck was a "justly celebrated naturalist" who "was the first man whose conclusions on the subject excited much attention." Lamarck's idea is often depicted in textbooks as giraffes stretching to reach leaves atop trees, with the trait of lengthened necks then being passed to their offspring. However, *Philosophie Zoologique* included relatively little about giraffes.

1811

Mary Anning (1799–1847) unearths the first complete *Ichthyosaurus* from the limestone (Mesozoic) cliffs near her home in the southern England coastal town of Lyme Regis. The self-taught Anning became one of the most famous fossil-hunters in history, finding an almost perfect skeleton of *Plesiosaurus* (which became the basis for one of the "reconstructions" of the Loch Ness Monster) and the first *Pterodactylus* (later renamed *Dimorphodon marconyx* by Richard Owen). Although Anning published only one paper ("Note on the Supposed Frontal Spine in the Genus Hybodus," which appeared in 1839 in *Magazine of Natural History*), she became a legendary figure in paleontology. Anning's many discoveries, some of which continue to be displayed at NHM, helped convince people of the reality and extent of past extinction events.

Figure 14 Jean-Baptiste Lamarck made many important contributions to biology, but he is usually remembered for his misguided theory of evolution based on the inheritance of acquired traits. The monument to Lamarck shown here was erected in 1908 near the entrance of Paris' Jardin des Plantes. The relief, which depicts an old, blind Lamarck being cared for by his daughter Aménaïde Cornélie, is accompanied by the inscription "Posterity will admire you. Posterity will avenge you, father."

1812 Georges Cuvier's (Figure 12) four-volume *Researches on the Fossil Bones of Quadrupeds* includes an appendix by French chemist Alexandre Brongniart (1770–1847) reporting that some strata in the Paris Basin contain fossils of marine mammals, others contain fossils of freshwater animals, and others bear no fossils at all. This appendix was one of the first published stratigraphic analyses.

Cuvier concluded that these ancient forms of life had appeared and disappeared suddenly due to the effects of catastrophes such as the biblical Flood. After each catastrophe, there were new creations and immigrations of organisms that had avoided the catastrophes because they lived elsewhere. In *Discourse on the Revolutionary Upheavals on the Surface of the Earth and on the Changes Which They Have Produced in the Animal Kingdom*, Cuvier noted "man, to whom a mere instant has been granted on earth, would have the glory of reconstructing the history of thousands of centuries which predated his existence, and thousands of beings which have never been his contemporaries. . . . Would it not be glorious for geologists to burst the limits of time?" Cuvier concluded, "lands once laid dry have been reinundated several times. . . . Life in those times was often disturbed by these frightful events. Numberless living things were victims of such catastrophes: some inhabitants of the dry land, were engulfed in deluges; others, living in the heart of the seas, were left stranded when the ocean floor was suddenly raised up again; and whole races were destroyed forever, leaving only a few relics which the naturalist can scarcely recognize." Cuvier also correctly identified the first known fossil remains of a pterodactyl and claimed that pterosaurs were flying reptiles. In 1812, Cuvier's *Inquiry into Fossil Remains*—the first major work of paleontology—described extinct species and classified them with Linnaeus's system, but claimed that they were the products of earlier, separate creations triggered by catastrophes instead of the ancestors of living organisms. Cuvier's view became known as *catastrophism*. In 1813, Cuvier's *Essay on the Theory of the Earth* used the fossil record to confirm extinctions and popularize catastrophism.

1812 Swiss botanist Augustin-Pyrame de Candolle (1778–1841) is the first person to use the word *taxonomy* in reference to the classification of species.

1813 British Romantic painter Joseph Turner (1775–1851) draws crowds to the Royal Academy with his exhibit of *Deluge*. Turner's famous painting, which tapped a centuries-long fascination of artists and poets with the biblical Flood, was accompanied by lines from Milton's *Paradise Lost*: "Down rush'd the rain, Impetuous, and continued till the earth no more was seen."

1813 Scottish naturalist Robert Jameson's (1774–1854) "Mineralogical Notes, and an Account of Cuvier's Geological Discoveries" (published in *Essay on the Theory of the Earth*) adds design-based language to Cuvier's work while presenting it as a scientific verification of Scripture.

1814 In a review of Georges Cuvier's *Essay on the Theory of the Earth* published in *Christian Instructor*, Thomas Chalmers (1870–1847) proposes gap creationism as "one way of saving the credit of the literal history" described in Genesis. Chalmers suggested that there was a gap between Genesis 1:1 (i.e., "In the beginning God created the heavens and the earth") and Genesis 1:2 (i.e., "And the earth was without form"). This gap was caused by some cataclysm, after which God recreated life on earth in the six days recorded in Genesis 1 from the chaos that had befallen the initial creation. Gap creationists interpret Genesis 1:1 as referring to a *prior* creation event and Genesis 1:2 to the most recent creation, thereby leaving an undefined period between the two verses into which virtually anything can be inserted. Gap creationism became famous when it was included in *The Scofield Reference Bible* in 1909.

1815 Jean-Baptiste Lamarck (Figure 14) restates his theory of evolution in the seven-volume *Histoire Naturelle des Animaux sans Vertèbres* (*Natural History of the Invertebrates*). These volumes, the last of which did not appear until 1822, introduced the terms *vertebrate* and *invertebrate* and founded modern invertebrate zoology. During upcoming years, Lamarck—who had claimed that blind fish in caves lost their eyesight as a result of not using their eyes—began to lose his eyesight. This prompted Georges Cuvier (Figure 12) to tell Lamarck that "perhaps your own refusal to use your eyes to look at nature properly has caused them to stop working."

1815 With *A Delineation of the Strata of England and Wales with part of Scotland*, William "Strata" Smith founds modern stratigraphy, the scientific discipline of describing and interpreting rock successions and their characteristic fossils. Although Smith had little interest in determining the ages of the rocks he examined (he accepted the biblical creation story), his work transformed the work of Steno and Hutton into practical geology while providing a key insight to modern evolutionary studies. Smith's map was a follow-up to his 1801 map and is considered by the British Geological Society to be the "Map That Changed the World" (Figure 15). Smith's *Strata Identified by Organized Fossils* (1816–1819) announced his famous Principle of Faunal Succession—that is, that there are

fossils "peculiar to each stratum," and that fossil assemblages succeed each other in a predictable manner. Smith, who claimed that his method of tracing strata by the fossils they contained was a "science not difficult to learn," noted "each layer of these fossil organized bodies must be considered as a separate creation, or is an undiscovered part of an older creation." Smith later spent two

Figure 15 In 1815, William Smith produced the first geological map of an entire country. This so-called "Map That Changed The World" was based on fossils as well as the mineral composition of rocks, and showed that fossils are not distributed randomly as in a flood, but instead are arranged in a definite order. (*IPR/110-82-CT British Geological Survey. © NERC 2008. All rights reserved*)

months in a debtors' prison before being honored by geologists and given a pension by the king.

1815 Alexander von Humboldt begins publishing his seven-volume *Personal Narrative of Travels to the Equinoctial Region of the New Continent*. John Henslow gave Charles Darwin a copy of *Personal Narrative* to take aboard the *Beagle*. Von Humboldt's books inspired many naturalists, including Darwin, Charles Lyell, and Louis Agassiz.

1816 In *Treatise on the Records of Creation and the Moral Attributes of the Creator*, Anglican John Bird Sumner (1780–1862) defends gap creationism by claiming that there may have been previous worlds from whose remains our own was organized. Sumner later became Archbishop of Canterbury.

1817 While doing research for his medical degree, Christian Pander (1794–1865) identifies three germ layers—ectoderm, endoderm, and mesoderm—in chick embryos. In 1828, Karl Ernst von Baer (1792–1876) extended Pander's work by showing that all vertebrate embryos possess the same three germ layers. Von Baer eventually developed the foundations of embryology and proposed that anatomically similar structures in different species can be traced to homologous structures found in embryos.

1819 Robert FitzRoy (1805–1865) joins the Royal Navy. In 1824, FitzRoy passed his examination for promotion to lieutenant "with full numbers," a score never before achieved. FitzRoy later captained the voyage that carried Charles Darwin around the world.

1820 Drawing on his observations of rock striations far removed from existing glaciers, Ignaz Venetz (1788–1859) proposes, in *Mémoire sur les Variations de la température dans les Alpes de la Suisse,* that much of Europe was once covered by ice.

1820 Anglican clergyman William Buckland (1784–1856) returns from an extensive trip to Europe convinced that similar successions of geologic strata in continental Europe and in Britain contain similar fauna. That same year, Buckland's *Vindiciae Geologiae; or the Connexion of Geology with Religion Explained* used gap creationism to reconcile geology with Noah's Flood and biblical creation. The genial Buckland claimed that the word *beginning* in Genesis referred to an unspecified period between the formation of the earth and

the creation of today's organisms, during which there were many extinctions and successive creations. Buckland's so-called "diluvium theory," which was based on Georges Cuvier's "successive catastrophes theory," was embraced by theologians confident of evidence of a worldwide flood.

1820 In *Essai Élémentaire de Géographie Botanique* (which Charles Lyell described as "luminous" in *Principles of Geology*), Augustin Pyrame de Candolle famously depicts the struggle for survival by noting that "[a]ll the plants of a given country are at war with another." Darwin's *On the Origin of Species* (1859) gave scientific authority to the metaphor of warring species: "From the war of nature, from famine and death, the most exalted object which we are capable of conceiving, namely the production of the higher animals, directly follows." Darwin also referred to "De Candolle's war of nature." *Essai Élémentaire de Géographie Botanique* became a founding document of biogeography.

1820 HMS (His Majesty's Ship) *Beagle*—a 90-foot-long, double-masted, 235-ton "Cherokee Class" warship built at a cost of £7,803—is launched from the Royal Naval Dockyard in Woolwich, England. In 1831, the *Beagle*—one of several ships that have had that name—began a voyage that helped Charles Darwin formulate his ideas about evolution by natural selection.

1821 In his speculative poetic drama *Cain: A Mystery,* Lord Byron (1788–1824) invokes the ideas of Georges Cuvier (Figure 12) to assault orthodox Christianity.

1822 English surgeon Gideon Mantell (1790–1852) and his wife Mary Ann discover several large fossil bones and teeth in a quarry near Whiteman's Green in Cuckfield, England. Renowned anatomists Richard Owen (1804–1892) and Georges Cuvier dismissed the finds as belonging to known mammals. However, while examining collections in London's Hunterian Museum (at the Royal College of Surgeons), Mantell met curator Samuel Stutchbury (1798–1859), who showed Mantell a recently prepared skeleton of an iguana. Mantell believed that the teeth he had found were similar to those of an iguana—but 20 times larger—and named his creature *Iguanosaurus.* Later, on the advice of geologist William Conybeare (1757–1857), Mantell changed the name to *Iguanodon* ("iguana-tooth"). In 1825, Mantell announced his discovery at a meeting of the Royal Society of London. By this time, Cuvier

(Figure 12)—admitting "I am quite convinced of my mistake"—agreed that Mantell's fossils were from an unknown monster. Mantell's find was the first discovery of a dinosaur, although the word *dinosaur* would not be coined by Owen until 1842. Mantell—who proclaimed that he would "ride on the back of my *Iguanodon* into the temple of immortality!"—became obsessed with dinosaurs, and in 1839, his wife left him and his medical practice failed. Facing mounting debt, Mantell sold his fossils to NHM. Mantell was involved in a carriage accident in 1841 that damaged his spine and left him in constant pain. Near the end of his life, Queen Victoria gave Mantell a civil gratuity of £100 per year. When Mantell died in 1852 of an opium overdose, Owen removed part of Mantell's spine, which he kept at the Royal College of Surgeons. Dinosaurs have played a prominent role in the evolution-creationism controversy ever since the Mantells's discovery. As Polish paleontologist Zofia Kielan-Jaworowska (b. 1925) noted in *Hunting for Dinosaurs* (1969), "the study of animals that lived on earth millions of years ago is not merely a study of their anatomy, but first and foremost a study of the course of evolution on earth and the laws that govern it."

1822 Johann Mendel (1822–1884) is born in Heinzendorf of Austria-Hungary. Mendel adopted the first name Gregor when he entered the monastery in 1843. Mendel's studies of inheritance in the garden pea later formed the basis of modern genetics.

1822 Belgian statesman and amateur geologist Jean-Baptiste-Julien d'Halloy (1783–1875) names the so-called "French Chalk" in Belgium the Terrain Cretacé, thus providing the source of the name for the Cretaceous period (see also Appendix B).

1822 The Congregation of the Holy Office approves books describing movement of the earth. The first *Index of Forbidden Books* not listing Galileo's *Dialogue* appeared 13 years later.

1823 English naturalist Adam Sedgwick (1785–1873)—who would later be one of Charles Darwin's professors at Cambridge—studies some of the oldest known rocks of Wales and later names the era corresponding to these fossils the Cambrian period (after the Roman name for the province, Cambria; see also Appendix B).

1823 William Buckland's *Reliquiae Diluvianae* argues that a recent worldwide flood marks the boundary of today's human world and

the prehuman world of the remote past. Buckland—Oxford University's first geologist—later noted that it is "impossible to ascribe the formation of [geologic] strata to . . . the single year occupied by the Mosaic deluge. . . . The strata [document] periods of much greater antiquity." Buckland referred to evidence of a flood as *diluvium*. More than 1,000 copies of *Reliquiae Diluvianae* were sold in the first six months after its publication.

1823 Alfred Russel Wallace (1823–1913) is born in the town of Usk in what is now Wales. Wallace later codiscovered, with Charles Darwin, the theory of evolution by natural selection.

1824 In a pioneering move that led to the recognition of dinosaurs as extinct reptiles, William Buckland publishes the first scientific description of a dinosaur in a paper titled "Notice of the Megalosaurus or great Fossil Lizard of Stonesfield," which appears in *Transactions of the Geological Society of London*. Buckland based his conclusions on part of a lower jaw (with teeth) and some postcranial bones. After consulting with his friend William Conybeare, Buckland named the fossil *Megalosaurus*, or "Great Lizard," a large meat-eating theropod of the Middle Jurassic (see also Appendix B). (Robert Plot had described a similar bone in 1677 but did not recognize it as belonging to a dinosaur.) In 1826, Cuvier (Figure 12) gave the dinosaur a complete binomial: *Megalosaurus bucklandi*, or "Buckland's Big Lizard." When Buckland later discovered coprolites (fossilized excrement), he suggested that carnivores had existed since the Creation, but later realized that this contradicted a consequence of Original Sin. Buckland rationalized this by claiming that carnivores are an example of God's benevolence because carnivores reduced the suffering of old and sick individuals by killing them.

1825 Sixteen-year-old Charles Darwin enrolls in the University of Edinburgh to study medicine. Darwin dropped out of college at the end of his second year, when watching gruesome surgeries on children became more than he could bear.

1825 Thomas Henry Huxley (1825–1895) is born in Ealing, just outside London. Huxley later became a towering figure in the history of evolutionary biology and a fierce advocate of Charles Darwin's theory of evolution.

1826 Embryologist Karl von Baer discovers that mammals produce eggs, and his subsequent two-volume *History of the Development of*

Animals becomes a founding work of modern embryology. By providing a framework for studying similarities between taxa, von Baer became a pioneer of comparative anatomy, a discipline that has contributed greatly to understanding evolution.

1826 George Bugg's (1769–1851) *Scriptural Geology; or, Geological Phenomena Consistent Only with Literal Interpretations of Sacred Scriptures* attacks the antediluvian ideas of Georges Cuvier and William Buckland: Bugg argues, "The SCRIPTURAL ACCOUNT of the DELUGE, will alone account for the phenomena of the fossil strata."

1826 Scottish minister John Fleming (1785–1857) promotes the "tranquil flood theory," which had been suggested earlier by Carolus Linnaeus and William Buckland. According to this theory, the worldwide Flood was short, gradual, and left few visible effects.

1827 Scottish geologist Charles Lyell writes in *Quarterly Review*, "all discoveries which extend indefinitely the bounds of time must cause the generations of man to shrink into insignificance."

1827 At a March meeting of the Plinian Society (a student group dedicated to natural history) at Edinburgh University, Charles Darwin's friend William Browne (who later nominated Darwin for Society membership) reads a paper that presents a materialistic interpretation of the brain. The paper upset many people, and all references to the paper (including the minutes from the previous meeting announcing Browne's intention to deliver it) were removed from the minutes. Darwin learned from Browne's experience, noting "to avoid stating how far, I believe, in Materialism, say only that emotions, instincts, degrees of talent, which are hereditary, are so because the brain of a child resembles parent stock." That same month, Darwin read his first scientific paper, a study of fertilization in the seaweed-like creature *Flustra*, to the Plinian Society. During his career, Darwin's varied interests included insects, barnacles, plants, and humans. He produced several theoretical works that included incorrect (pangenesis and his explanation of the Parallel Roads of Glen Roy) and extensively validated (evolution by natural selection and his explanation of the formation of coral reefs) proposals.

1827 Inspired by the writings of economist Adam Smith, French zoologist Henry Milne-Edwards (1800–1885) argues that a "physiological division of labor" would increase organisms' efficiency and ability to survive. In 1859, Charles Darwin invoked Milne-Edwards's idea in Chapter IV of *On the Origin of Species*.

1827 The frontispiece of Gideon Mantell's *Illustrations of the Geology of Sussex* shows Mantell, Charles Lyell, and William Buckland—all wearing top hats and gentlemanly clothes—in 1825 at the Cuckfield quarry that yielded Mantell's iguanodon tooth and many other fossils. In the distance is the spire of Cuckfield Church. Today, the quarry has been filled in, and an athletic field occupies the space.

1827 Joseph Jackson Lister (1786–1869), the father of the more famous Joseph Lister (1827–1912; the founder of antiseptic surgery), develops achromatic lenses for microscopes. This contribution, combined with the accidental discovery of histological stains by Joseph von Gerlach (1820–1896) in the 1850s, enabled biologists to make discoveries that eventually transformed genetics, evolution, and cell biology.

1827 After being scolded by his father that "you care for nothing but shooting, dogs, and rat-catching, and you will be a disgrace to yourself and your family," Charles Darwin enrolls at Christ's College, Cambridge, and the following year begins studying for the clergy. During his three years at Cambridge, Darwin held conventional religious beliefs, but exhibited a notable lack of religious zeal. Darwin attended botany lectures and often went on plant-collecting expeditions led by John Henslow (1796–1861), who later helped Darwin secure passage aboard the *Beagle* (Darwin became known on campus as "the man who walks with Henslow"). Darwin remained interested in collecting insects, and it was at Cambridge that he began to appreciate the vast diversity of species. Darwin studied William Paley's *Natural Theology*, which argued that nature's complexity was evidence of design, and that design required a designer; however, Darwin later rejected Paley's arguments "now that the law of natural selection has been discovered." While at Cambridge, Darwin lived in the same dormitory rooms that had housed Paley 70 years earlier; these rooms were the most expensive available to undergraduates of his rank. In 1831, Darwin graduated 10th in a class of 178 students from Christ's College. To get his degree, Darwin pledged his adherence to the "Thirty-Nine Articles" (established in 1563) outlining the basic doctrine of the Church of England. Today, portraits of Darwin and Paley hang side-by-side in the dining hall of Christ's College.

1828 German chemist Friedrich Wöhler (1800–1882) synthesizes urea from inorganic compounds: "I can make urea without the use of kidneys, either man or dog." By demystifying basic physiological processes, Wöhler's work began the use of science to reject vitalism—that life involves a fundamental spirit that cannot be isolated.

1828 The temperamental Robert FitzRoy assumes command of the *Beagle* after then-Captain Pringle Stokes kills himself off the coast of Tierra del Fuego.

1828 French naturalist Étienne Geoffroy Saint-Hilaire (1772–1844), curator of vertebrates at the Muséum d'Histoire Naturelle in Paris, cautiously argues for extinction produced by *monde ambient* ("conditions of life"), but does not claim that existing species are evolving.

1828 Charles Lyell—a student of William Buckland's—visits the Temple of Serapis near Pozzuoli, Italy, where marine bivalves on the temple's columns document changes in the earth's surface (Figure 16). The temple became an icon of uniformitarianism when Lyell included a sketch of the temple's pillars in the frontispiece of his *Principles of Geology* (1830).

Figure 16 The Temple of Serapis near Pozzuoli, Italy, was made famous when Charles Lyell included its image in his influential *Principles of Geology*. The dark bands on the upright marble pillars were formed by mollusks that drilled into them after the columns were submerged in the sea. Whereas scriptural geologists had used geology to support their views of the Bible, Lyell used a religious relic to support his claims about geology. (*Randy Moore*)

1829 Jean-Baptiste Lamarck dies blind, destitute, and ostracized in Paris and is buried in an unmarked limestone pit (Figure 14). His books and meager possessions were sold at auction. Georges Cuvier's (Figure 12) eulogy denounced Lamarck as a speculative philosopher who courted heretical ideas about spontaneous generation.

1829 The eccentric Francis Henry Egerton (1756–1829), the eighth Earl of Bridgewater, dies and leaves £8,000 to further William Paley's work in natural theology. The resulting eight books, which became known as the *Bridgewater Treatises*, were the century's most thorough attempt at establishing natural theology.

1829 Philippe-Charles Schmerling (1790–1836) discovers a partial cranium of a Neanderthal fossil in Engis, Belgium. Charles Lyell subsequently illustrated this fossil in his *Geological Evidences of the Antiquity of Man* (1863), but the fossil was not accurately identified as a Neanderthal for a century (see also Appendix C).

1830 Charles Lyell (1797–1875) publishes his monumental and ambitiously titled *Principles of Geology: being an Attempt to Explain the Former Changes of the Earth's Surface by Reference to Causes now in Operation*. In *Principles*, a name chosen to echo Isaac Newton's *Principia*, Lyell used the work of James Hutton and John Playfair as a starting point to describe an old earth whose features had been formed by "the slow agency of existing causes" operating for long periods of time. Although Lyell did not dispute the occurrence of local catastrophes, he popularized and refined Hutton's ideas about uniformitarianism while showing that the earth is something far more than the natural home for humans; rather, humans have been on a ceaselessly changing earth for only a tiny portion of its history. Lyell acknowledged Hutton as the first scholar to study geology as its own subject, but noted that "although [Hutton's idea of an ancient Earth] was vehemently opposed at first, and although it has gradually gained ground, and will ultimately prevail, it is yet far from being established." Although Hutton had dealt only with the physical aspects of geology, Lyell realized that geology would be incomplete if it ignored fossils, which record the changes encountered by life on earth. Lyell reasoned that the periodicity in climate would produce cycles of species through time, so that "huge *Iguanodon* might reappear in the woods, and the ichthyosaurs in the sea." *Principles* was the standard reference for geologists for several decades, and Charles Darwin read the book while aboard the *Beagle*. Lyell's approach to science strongly influenced Darwin's thinking as Darwin started developing his ideas about biological evolution. Although Lyell encouraged Darwin to publish *On the Origin of Species*, he initially rejected Darwin's claims about life's history and was the most reluctant of Darwin's confidants to publicly support evolution (early editions of *Principles* argued that organisms had been created perfectly adapted for local conditions). However, Lyell's own discoveries validated Darwin's ideas, prompting Thomas Huxley to note that Lyell was "doomed to help the cause he hated." By the 10th edition of *Principles* (1867), the religiously conservative Lyell endorsed Darwin's theory, noting that there is "a strong presumption in favour of the truth" of Darwin's theory. Lyell linked an incomprehensibly old earth with natural theology, but not with biblical cosmology: "We discover everywhere the clear proofs of a Creative Intelligence, and of His foresight, wisdom, and power." However, by the end of his life, Lyell—who believed that scriptural geologists were "wholly destitute of geological knowledge"—questioned traditional theism. *Principles of Geology*—the most influential geology book in history—forever raised geology's public image to a science by masterfully synthesizing vast amounts of data with elegant and compelling rhetoric and established geology as a discipline founded fully on scientific inquiry.

1830 After reading a review of Charles Lyell's *Principles of Geology* in *Quarterly Review*, Alfred Tennyson (1809–1892) writes his first poem about geology. The poem noted Lyell's uniformity of causes in time ("Old principles still working new results"). However, Tennyson's most famous poem involving geology—*In Memoriam*—would not be published for another 20 years.

1830 French antiquarian Jacques Boucher de Crèvecœur de Perthes (1788–1868) recognizes the Ice Age handaxes found near the Somme River as products of ancient humans. Today, the handaxes are accepted as being at least 500,000 years old.

1830 Georges Cuvier (Figure 12) and Étienne Geoffroy Saint-Hilaire, colleagues at the Muséum d'Histoire Naturelle in Paris, publicly debate the possibility of organic change of species. Geoffroy claimed that organisms share a common body-plan that has been altered in a deterministic manner through time to produce the biological diversity now apparent. Cuvier, however, was an ardent proponent of the fixity of species. Although Cuvier—a more skilled speaker and a world-renowned scientist—"won" the debates, his proposals were eventually overturned by more sophisticated evolutionary perspectives. The history of these debates is recorded by the street names surrounding the Muséum: *rue Cuvier* and *rue Geoffroy-St. Hilaire* intersect near the museum's botanical gardens. In a bit of scientific politics, *rue Lamarck*—which intersects *rue Becquerel*—is tucked away across town.

1830 HMS *Beagle*, with Robert FitzRoy as captain, returns to England with four Fuegians on board. FitzRoy captured the Fuegians to encourage the return of property stolen by the residents of Tierra del Fuego. However, FitzRoy soon decided to "improve" the captives (who FitzRoy described as "now scarcely superior to brute creation") via "a suitable education" and then "after two or three years, . . . send them back to their country." One of the Fuegians, Boat Memory, succumbed to smallpox soon after arriving in England, but the remaining three—given the names Jemmy Button, Fuegia Basket, and York Minster by the *Beagle*'s crew—became celebrities before returning to South America.

1830 English geologist Henry de la Beche's (1796–1855) watercolor *A More Ancient Dorset* famously depicts prehistoric life based on fossils discovered by Mary Anning. De la Beche was one of the few scientists who regularly and gratefully acknowledged Mary Anning's contributions to paleontology. That same year, de la Beche's cartoon titled *Awful Changes* lampooned Charles Lyell's claim in *Principles of Geology* that extinct plants and animals could return if suitable environments arose. In the cartoon, Professor Ichthyosaurus lectures about a fossilized human skull: "You will perceive that the skull before us belonged to some of the lower order of animals."

1831 Reverend Adam Sedgwick and others object to some of the claims in Charles Lyell's *Principles of Geology*, especially Lyell's denial of progression in the earth's history and his refusal to allow past causes of greater magnitude than those of today. Others disagreed; the following year, a reviewer in the *Spectator* noted that "[t]here are sermons in stones . . . but they want an interpreter: that interpreter is the enlightened geologist. Such a man is Mr. Lyell."

1831 English veterinarian William Youatt (1776–1847) publishes *The Horse*. Charles Darwin's first use of the term *natural selection* appears as a handwritten marginal note in Darwin's copy of this book.

1831 After noticing that Mesozoic rocks contained no fossilized mammals, Gideon Mantell describes the Mesozoic as the "Age of Reptiles" (see also Appendix B).

1831 The 23-year-old Charles Darwin earns a bachelor's degree from Cambridge, but is drawn to the prospect of travel. Earlier in the year, Cambridge botanist John Henslow received a letter from Cambridge astronomer George Peacock (1791–1858) describing an opening for a naturalist on the upcoming voyage of the *Beagle*. After considering the offer himself, Henslow approached his brother-in-law, Leonard Jenyns (1800–1893; like Henslow, a vicar with a background in natural history) about his interest in the position. Jenyns declined, after which Henslow contacted his former student, Charles Darwin, noting that although Darwin was not "a finished naturalist," he was certainly "qualified for collecting, observing, and noting anything worthy to be noted in Natural History. . . . I think you are the very man they are in search of."

1831 In scattered pages (most in an appendix) of *On Naval Timber and Arboriculture*, Scottish farmer Patrick Matthew (1790–1874) briefly describes what Charles Darwin will later call natural selection: "As nature, in all her modifications of life, has a power of increase far beyond what is needed to supply the place of what falls by Time's decay, those individuals who possess not the requisite strength, swiftness, hardihood, or cunning, fall prematurely without reproducing—either a prey to their natural devourers, or sinking under disease, generally induced by want of nourishment, their place being occupied by the more perfect of their own kind, who are pressing on the means of subsistence. . . . There is more beauty and unity of design in this continual balancing of life to circumstance, and greater conformity to those dispositions of nature which are manifest to us, than in total destruction and new creation. . . . [The] progeny of the same parents, under great differences of circumstance, might, in several generations, even become distinct species, incapable of co-reproduction." Matthew, who referred to his idea as "diverging ramification," suggested that if humans could produce new varieties of animals in just a few years via artificial selection, then nature might be able to produce new species over longer periods. However, because Matthew considered evolution by natural selection "a self-evident fact" to people with "unprejudiced minds with sufficient grasp," he did not provide convincing evidence for his idea, nor did he appreciate its possible implications. Matthew's idea remained unnoticed until Matthew drew attention to it in an article in 1860 in *Gardener's Chronicle*. When Charles Darwin learned of Matthew's book in 1860, he confessed to Lyell that Matthew's idea was "certainly, I think, a complete but not developed anticipation" of natural selection. Darwin later acknowledged Matthew in a reprint of *On the Origin of Species*: "[Matthew] clearly saw . . . the full force of the principle of natural selection." Darwin also sent a letter to *Gardener's Chronicle* apologizing "to Mr. Matthew for my entire ignorance of his publication," after which Matthew printed business cards describing himself as the "Discoverer of the Principle of Natural Selection." Others who described natural selection before Darwin include French freethinker Jean-Jacques Rousseau (1712–1778), German philosopher Johann Herder (1744–1803), and James Hutton.

1831 On December 27, the *Beagle*, with Robert FitzRoy as Captain, sails from Devonport to survey the Cape Verde Islands, the South American Coast, the Strait of Magellan, the Galápagos Islands, Tahiti, New Zealand, Australia, the Maldives, and Mauritius. The *Beagle*'s voyage was among the most ambitious and best-financed survey voyages sponsored by the Admiralty Board in more than a decade. Passengers included Charles Darwin and three Fuegians who had been brought by FitzRoy to England the previous year. The *Beagle*'s official naturalist (and surgeon) was Robert McCormick (1800–1890). However, FitzRoy could have little social contact with the ship's crew and needed a companion for his voyage, and FitzRoy wanted this person to be a naturalist. Although FitzRoy initially had reservations about Darwin serving in this role (he believed the shape of a person's skull indicated the person's character and concluded that Darwin's nose meant that Darwin was lazy and hesitant), Darwin was eventually accepted. While at sea, Darwin—who subsequently referred to December 27 as "my real birthday"—was often sick; he later noted, "I loathe, I abhor the sea." Darwin's 40,000-mile voyage aboard the *Beagle*—at a cost of about £1,000, paid by his father—lasted 58 months, 43 of which were spent in South America. Although the *Beagle*'s trip is often described as a sea voyage, Darwin spent over three of the years on land. Darwin's book about his voyage, retitled *The Voyage of the Beagle* at its third printing, remains one of the world's great travel books. Darwin later supported FitzRoy's election to the Royal Society and named a species of dolphin after him, although the two differed vehemently about Darwin's eventual conclusions based on the observations and collections made during the voyage.

1831 The British Association for the Advancement of Science (BAAS) is founded.

1831 In *A Preliminary Discourse on the Study of Natural Philosophy*, English scientist John Frederick William Herschel (1792–1871) describes a model of the interactions between observation and theory in science. Herschel claimed that nature is governed by laws and that understanding these laws—and "not insulated, independent facts"—was the goal of natural philosophy. In *On the Origin of Species* (1859), Charles Darwin tried to adhere to Herschel's model and noted in the opening pages that he was trying "to throw some light on the origin of species—that mystery of mysteries, as it has been called by one of our greatest philosophers." That philosopher was Herschel, who later dismissed natural selection as "the law of higgledy-piggledy." Herschel advocated day-age creationism; as he noted in a letter to Charles Lyell, "Time! Time! Time! We must not impugn the Scripture Chronology, but we *must* interpret it in accordance with *whatever* shall appear on fair enquiry to be the *truth* for there cannot be two truths. And really there is scope enough: for the lives of the Patriarchs may as reasonably be extended to 5,000 or 50,000 years apiece as the days of Creation to as many thousand millions of years."

1832 Georges Cuvier (Figure 12) dies six days after publicly denouncing Étienne Geoffroy Saint-Hilaire's "pantheism" and "useless scientific theories." Although Cuvier is often remembered for his staunch adherence to the immutability of species, he made tremendous contributions to comparative anatomy and is considered the father of paleontology. Cuvier's work, which forced the acceptance of extinction as a natural occurrence, figured prominently in the development of Darwin's ideas on evolution.

1832 During the *Beagle*'s stop at Montevideo, Uruguay, Charles Darwin receives the second volume of Charles Lyell's *Principles of Geology*.

1832 William Smith is awarded the Wollaston Medal; this is the first formal recognition of Smith's groundbreaking geologic work (Figure 15). Smith was later given an annual civic pension of £100 by King William IV (1765–1837).

Figure 17 The Galápagos archipelago, a part of Ecuador, was made famous by Charles Darwin's visit in 1835 aboard the *Beagle*. Today, the islands host more than 100,000 visitors annually. (*Randy Moore*)

1832 Ecuador annexes the Galápagos Islands, which are renamed Archipiélago de Colón in 1892 (Figure 17).

1832 In a review of the second volume of Lyell's *Principles of Geology*, geologist and theologian William Whewell (1794–1866) introduces the geologic terms *uniformitarianism* and *catastrophism*. Whewell later criticized Lyell for claiming that there is no direction to the earth's history and for assuming that the present is a representative period in the earth's history, noting that we must not "select arbitrarily the period in which we live as the standard for all other epochs." Whewell rejected evolution: "Species have a real existence in nature and a transmutation from one to another does not exist." In the 1960s, uniformitarianism was also attacked vehemently by Flood geologists such as Henry Morris and John Whitcomb, Jr., who claimed in *The Genesis Flood* that uniformitarianism is a false, unbiblical, unchristian view that is not even "a *possible* explanation of the earth's geologic formations, as any candid examination of the facts ought to reveal."

1833 Yale University's Benjamin Silliman (1779–1864), in a supplement to the second American edition of Robert Blakewell's

(1768–1843) *Introduction to Geology*, reconciles "the consistency of geology with sacred history" by interpreting each of the "days" of creation as long, indefinite periods. This is the first major published elaboration of day-age creationism. Famous 19th-century geologists who advocated day-age creationism included Arnold Guyot (1807–1884), Sir John William Dawson (1820–1899), Hugh Miller (1802–1856), and Silliman's son-in-law James Dwight Dana (1813–1895), the premier American geologist of his era. In the 21st century, day-age creationism is most famously advocated by astronomer Hugh Ross (b. 1945), who directs the Christian apologetics ministry Reasons to Believe.

1833 *The Bridgewater Treatises* are commissioned to show "the power, wisdom, and goodness of God as manifested in the Creation." In Treatise I, *On the Adaptation of External Nature to the Moral and Intellectual Constitution of Man*, Thomas Chalmers—who believed that both Scripture and nature led equally to God—attacked materialists who "reason exclusively on the laws of matter." In the volume *On Astronomy and General Physics Considered with Reference to Natural Theology* (Treatise III), William Whewell argued that the only way to understand nature is as the product of special creation: "If there be, in the administration of the universe, intelligence and benevolence, superintendence and foresight, grounds for love and hope, such qualities may be expected to appear in the . . . fundamental regulations by which the course of nature is . . . made to be what it is." This contrasted with Charles Darwin's view in 1859 in *On the Origin of Species*: "Thus, from the war of nature, from famine and death, the most exalted object which we are capable of conceiving, namely, the production of the higher animals, directly follows." Scottish anatomist and natural theologian Sir Charles Bell (1774–1842)—for whom Bell's palsy is named—authored the fourth *Bridgewater Treatise*, *The hand, its Mechanism and Vital Endowments as evincing Design*. Bell's descriptions reflected spiritual inspiration: "If we select any object from the whole extent of animated nature, and contemplate it fully and in all its bearings, we shall certainly come to this conclusion: that there is Design in the mechanical construction, Benevolence in the endowment of the living properties, and that Good on the whole is the result." Opposite the title page of *On the Origin of Species* (1859), Charles Darwin quoted Whewell's *Bridgewater Treatise*: "But with regard to the material world, we can at least go so far as this—we can perceive that events are brought about not by insulated interpositions of Divine power, exerted in each particular case, but by the establishment of general

laws." Whewell accepted that physics was governed by "general laws," but rejected such a claim about biology.

1833 Famed Harvard biologist Louis Agassiz (1807–1873), who viewed the fossil record as the orderly genesis of species by the Creator, concludes that there is a direct correspondence between sequences in the fossil record and the embryonic development of individual organisms. Agassiz, a protégé of Georges Cuvier (Figure 12) and one of the last prominent scientists to reject evolution, predicted he would "outlive this mania" about evolution.

1833 In volume three of *Principles of Geology*, Lyell notes that "a study of present processes is the 'alphabet and grammar of geology.'" Lyell identified the Pliocene, Miocene, and Eocene epochs of the earth's history (see also Appendix B). The brief, last period of geologic history in which humans appeared was later termed the Pleistocene (see also Appendix B), Lyell's term for Louis Agassiz's "Ice Age." Lyell spent most of the rest of his life revising *Principles* (the 12th edition appeared posthumously in 1875) and *Elements of Geology* (the 6th edition appeared in 1865), which provided most of his income. The complete edition of *Principles* cost 24 shillings, which was then two weeks' wages for a typical worker. For comparison, Robert Chambers's *Vestiges* (1844) and Charles Darwin's *Journal of Researches* (1845) each cost 7 shillings.

1833 Scriptural geologist George Fairholme (1789–1846) argues in *General View of the Geology of Scripture* that "all the appearances of the surface of the earth" can be accounted for by "an *attentive*, an *unprejudiced*, and, above all, a *docile* consideration" of creation and the biblical Flood.

1833 The three surviving Fuegians, "improved" by Captain Robert FitzRoy with a year's stay in England, disembark from the *Beagle* in Tierra del Fuego. Charles Darwin described the native Fuegians as resembling "the representations of devils on the stage" while noting "I could not have believed how wide was the difference between savage and civilised man; it is greater than between a wild and domesticated animal." He also noted of the three Fuegians that "three years has been sufficient to change savages into, as far as habits go, complete and voluntary Europeans."

1834 Thomas Malthus, whose *Essay* would strongly influence Charles Darwin, Alfred Russel Wallace, and others, dies while Darwin is at sea aboard the *Beagle*.

1834 English Romantic painter John Martin (1789–1854), who had earlier been commissioned to illustrate *Paradise Lost*, finishes his second version of the haunting *Deluge*, the most famous rendering of the biblical Flood.

1834 American minister Henry Cole's *Popular Geology Subversive of Divine Revelation*, published as "a letter to Rev. Adam Sedgwick," denounces Sedgwick's claims about geology and urges a return to diluvial catastrophism.

1834 In a letter to English poet Samuel Coleridge (1772–1834), William Whewell coins the term *scientist.*

1835 Well into the fourth year of the *Beagle*'s voyage, Charles Darwin lands at Chatham Island in Ecuador's Galápagos archipelago on September 15 and describes the archipelago as "a little world within itself; the greater number of its inhabitants, both vegetable and animal, being found nowhere else" (Figure 17). During his five weeks in the Galápagos, Darwin spent 19 days ashore, making collections and observations at four islands. Darwin remarked "I never dreamed that islands 50 or 60 miles apart, and most of them in sight of each other, formed of precisely the same rocks, placed under a quite similar climate, rising to nearly equal height, would have been differently tenanted." But he later noted, "[b]y far the most remarkable feature in the natural history of this archipelago [is] that the different islands to a considerable extent are inhabited by a different set of beings." Nicolas Lawson, vice governor of the prison colony on Floreana Island in the Galápagos Islands, told Darwin that he could "pronounce with certainty from which island [in the Galápagos] any tortoise had been brought from the shape of its shell." At the time, Darwin did not grasp the potential significance of Lawson's claim, thus possibly explaining why Darwin did not collect shells from all of the islands. Subsequent research in the Galápagos by David Lack and Peter and Rosemary Grant documented Darwin's concept of evolution by natural selection. Although many biology textbooks cite the Galápagos as central to Darwin's theory, *On the Origin of Species* mentions the Galápagos only six times.

1835 The Cambridge Philosophical Society publishes Charles Darwin's *Letters on Geology*, which includes some of Darwin's correspondence with John Henslow. From the *Beagle*, Darwin wrote that he was "a good deal" horrified that these letters were published. The Society reprinted the letters in 1960.

1836 While at the Cape of Good Hope, Charles Darwin goes ashore with Captain FitzRoy to meet astronomer Sir John Herschel. Herschel rejected Lyell's claim in *Principles* that species were created "in succession at such times and in such places as to enable them to multiply and endure for an appointed period, and occupy an appointed space on the globe." Lyell, who at this point rejected evolution in favor of unchanging species that were specially created, tried to explain the succession of species he knew from the fossil record as a series of creations; he claimed that groups of species particular to any area were "*centres* or *foci* of creation . . . as if there were favorite points where the creative energy has been in greater action than others."

1836 Congregational minister and Amherst College chemistry professor Edward Hitchcock (1793–1864) discovers curious fossilized footprints in Connecticut. In 1848, in a paper titled "An Attempt to Discriminate and Describe the Animals that Made the Fossil Footprints of the United States," Hitchcock concluded that the prints—which he called ornithichnites ("stony bird tracks")—were made by three-toed birds that were larger than ostriches (local folklore described the birds as "Noah's Ravens"). These prints, which are today displayed at the Amherst College Museum of Natural History, were later shown to be footprints of bipedal dinosaurs. They were the first known discovery of dinosaur remains in North America.

1836 In his two-volume *Geology and Mineralogy, Considered with Reference to Natural Theology* (Treatise VI of *Bridgewater Treatises on the Power, Wisdom and Goodness of God, as Manifested in the Creation*), William Buckland invokes gap creationism to reconcile geology and Scripture. According to Buckland, who rejected the occurrence of a worldwide flood, it was no longer possible to claim that the earth is only 6,000 years old, nor could anyone attribute all rock strata to a flood. Buckland's popular book, the only *Bridgewater Treatise* devoted to geology, went through four editions between 1836 and 1869.

1836 On October 2, the *Beagle* returns to England. The projected two-year voyage had lasted four years, nine months, and five days. Soon after his return, Charles Darwin met Charles Lyell at a dinner at Lyell's home. That year Darwin was also elected to the Geological Society of London (he served as its secretary from 1838 to 1841), which in 1859, honored him with its highest award—the Wollaston Medal—for his geologic research. After disembarking from the *Beagle*, Darwin never again left England.

1837 Charles Darwin begins describing his "dangerous" idea in a secret notebook labeled "Transmutation of Species." Page 36 of this notebook contained Darwin's first "tree of life," in which—beneath the words "I think"—he depicted life not as a hierarchical and philosophical ranking of "higher" and "lower" forms (as Aristotle and other naturalists had claimed), but instead as a branching tree showing shared origins. Instead of marching up a chain as Lamarck and other naturalists had suggested, Darwin's tree showed that species evolved; in some cases, one species gave rise to many species (as had occurred on the Galápagos Islands). At this point, Darwin knew that "one species does change into another," but could not explain how. A few pages later, he wrote "Heaven knows whether this agrees with Nature: Cuidado" ("cuidado" is Spanish for "careful"). Although Darwin did not publish his theory for 22 years, his "tree of life" became a metaphor for his view of how species evolve.

1837 Charles Babbage's (1791–1871) unofficial *Ninth Bridgewater Treatise* claims that nature is governed by predictable and intelligible laws, but that unusual events could result from natural processes. Babbage's views influenced Robert Chambers's *Vestiges of the Natural History of Creation* (1844) as well as Charles Darwin, as Darwin considered a mechanism for "descent with modification."

1837 Charles Darwin writes in his *Notebook B* that extinction "is a consequence . . . of non adaptation of circumstances." In Chapter X of his *On the Origin of Species*, Darwin later noted that extinction of old species is intimately linked to the production of new species.

1837 Charles Darwin gives ornithologist John Gould (1804–1881), Curator and Preserver at the museum of the Zoological Society of London, the birds he and others aboard the *Beagle* had collected from the Galápagos Islands. (Darwin did not label his specimens carefully, so he gathered specimens from Captain FitzRoy and others who had been aboard the *Beagle*.) Darwin believed the birds were blackbirds, "gross-bills," and finches, but one week later Gould reported that the birds were "a series of ground Finches which are so peculiar" because they formed "an entirely new group, containing 12 species." Two months later, Darwin met with Gould, who reported that the Galápagos "wren" was

another species of finch, and the "mocking-thrushes" (i.e., mockingbirds) that Darwin had collected on different islands were also separate species. Gould described Darwin's birds in 1837 in the *Proceedings of the Zoological Society of London*. Thomas Bell (1792–1880), who had been identifying the reptiles collected by Darwin in the Galápagos, gave Darwin a parallel conclusion—namely, that each island of the Galápagos chain had produced its own distinct species of iguana. Darwin later noted in his *Journal of Researches into the Geology and Natural History of the Various Countries Visited by the H.M.S. Beagle Under the Command of Captain FitzRoy, R.N. from 1832 to 1836* (later given the simpler title *The Voyage of the Beagle*) that finches of the Galápagos appeared as if "one species had been taken and modified for different ends," but he was not ready to say how this might have happened.

1838 Charles Darwin reads "for amusement" the 6th edition of Thomas Malthus's *An Essay on the Principles of Population*. Malthus argued that populations can grow exponentially (e.g., 2, 4, 8, 16, 32, 64, 128, 256, etc.), but that resources such as food can increase only linearly (e.g., 1, 2, 3, 4, 5, 6, 7, etc.), thus leading to competition and a struggle for existence. Darwin realized that Malthus's struggle, throughout the history of Lyell's ancient earth, might explain the great diversity of plants and animals that he encountered on his travels; as he observed: "Being well prepared to appreciate the struggle for existence . . . it at once struck me that under these circumstances favourable variations would tend to be preserved and unfavourable ones destroyed. The result of this would be the formation of a new species. Here then, I had at last got a theory by which to work." Darwin noted in his *Notebook N* that "[w]e can allow satellites, planets, suns, universe, nay whole systems of universes, to be governed by laws, but the smallest insect, we wish to be created at once by special act." To Darwin, natural selection was the force that constantly adjusts the traits of future generations, and he wrote that "[i]t is absurd to talk of one animal being higher than another." At this time, Darwin also experienced some of his earliest attacks of chronic stomach illness. Darwin's health plagued him for the rest of his life, and speculation about the nature of his illnesses continues to abound.

1838 Pastor and scriptural geologist George Young (1777–1848) publishes his 78-page *Scriptural Geology*. Young, a young-earth creationist, argued that the biblical Flood produced most of the earth's geologic features.

1838 Gideon Mantell's popular *Wonders of Geology*, which went through six editions in 10 years, includes John Martin's famous rendering of fighting prehistoric reptiles as its frontispiece.

1838 After failed careers in law and medicine, German botanist Matthais Jakob Schleiden (1804–1881)—whose brain housed a bullet from a failed suicide attempt—speculates that plants are made of cells. The following year, his friend, German zoologist Theodor Schwann (1810–1882), made a similar claim about animal tissues. Together, these biologists formulated the idea that all organisms are made of cells. However, Schleiden and Schwann suggested that cells arise from a chaotic fluid or from within cells as enlargement of nuclei (which had been discovered in 1831 by Robert Brown) or smaller granules (Schleiden and Schwann believed that nuclei are crystalline condensates in the process of becoming cells). This "free formation of cells" by crystallization was abandoned in 1858 when Prussian scientist Rudolf Virchow (1821–1902) showed that all cells come from preexisting cells. The so-called "cell theory" documented a trait shared by all living organisms and provided evidence for common ancestry.

1838 English inventor Samuel Rowbotham (1816–1884) performs the first of several "Bedford level experiments," intended to prove that the earth is flat. Among the people who challenged Rowbotham's claims was Alfred Russel Wallace in 1870.

1839 Botanist Joseph Hooker (1817–1911) joins the four-year expedition of the *Erebus* as it leaves England to explore the southern hemisphere. Hooker was eager for adventure and keen on duplicating the success his hero Charles Darwin had achieved from his travels aboard the *Beagle*. Hooker's expedition visited Australia, the southern tip of South America, New Zealand, and Antarctica, and Hooker amassed a large collection of plants from these locales. Fascinated by the regional patterns he observed, Hooker hoped to elucidate the overarching laws governing plant distribution. Soon after Hooker's return to England in 1843, Darwin confided to Hooker his ideas about evolution, and Hooker applied Darwin's ideas to the biogeographical patterns he studied.

1839 Just before turning 30, and five days after he paid the £70 fee for election to the Royal Society, Charles Darwin marries his first cousin Emma Wedgwood (1808–1896) at a ceremony officiated by Rev. Allen Wedgwood (Emma's cousin) at the Church of St. Peter near the Wedgwood mansion. For the rest of his life, Charles referred to Emma as "my greatest blessing." The newlyweds moved into a house near Charles's brother Erasmus in London to start their family; they eventually had 10 children, but only 7 reached adulthood (the infant mortality rate in Victorian England was 15%). The Darwins had no financial worries and never had to work.

1839 Robert FitzRoy publishes his day-by-day narratives of the *Beagle*'s voyage as part of a three-volume set of books. FitzRoy's work was ridiculed by many naturalists for adhering to biblical literalism, and Charles Lyell claimed that "[i]t beats all the other nonsense I have ever read on the subject." FitzRoy, who later dissociated himself from Charles Darwin's published account of the *Beagle*'s voyage, felt guilty because he believed that Darwin's experiences during the *Beagle*'s voyage had been used to undermine the Bible.

1839 Roderick Impey Murchison's (1792–1871) *The Silurian System* describes and names the Silurian System for an ancient Welsh tribe that opposed the Roman occupiers of the land (see also Appendix B). In the same year, geologist Charles Lyell identified the Pleistocene period (which in 1863 he estimated to have lasted 800,000 years; see also Appendix B). After studying fossils in a stratum of slate in Devonshire, England, Murchison and Adam Sedgwick named the Devonian period (for the Devonshire area in which they were working).

1840 Louis Agassiz's book *Études sur les Glaciers* supports Georges Cuvier's catastrophism and describes a universal ice age at the end of the Tertiary; he considered glaciers to be "God's great plough," replacing the Noachian flood as the most recent global catastrophe, and believed that the Creator had repeatedly destroyed and recreated life. Agassiz believed that life's history was directional ("the end and aim of this development is the appearance of man") and that the "history of the Earth proclaims its Creator," and he also proposed a separate origin for all human races. Agassiz—who interpreted biogeography as "the direct intervention of a Supreme Intelligence in the plan of the Creation"—believed that species came into existence when God thought of them and disappeared when God stopped thinking of them. As a result,

Agassiz believed that natural history was destined to become an analysis of the thoughts of God.

1840 Edward Hitchcock's popular *Elementary Geology*, which went through 30 editions in 20 years, typifies textbooks that used Charles Lyell's ideas to show that natural history is evidence of God's power.

1841 Hugh Miller (1802–1856), a self-taught Scottish geologist, publishes *The Old Red Sandstone,* which becomes one of the most popular books written by a scientist for a general audience (Old Red Sandstone is a prominent formation in Devonian rocks). Miller urged young people to abandon political agitation, become geologists, and read Charles Lyell's *Elements of Geology*, starting with "the description of the upper Silurian rocks." Miller's book was exceedingly popular and was reissued more than 25 times. Miller believed that an old earth proclaimed God's glory much better than did a recently created earth.

1841 Twenty-two-year-old Herman Melville (1819–1891) visits the Galápagos Islands (Figure 17) while aboard the New England whaler *Acushnet*. Melville, who published 10 sketches of the archipelago, rejected—and at times parodied—Charles Darwin's conclusions about the islands. In *Moby-Dick* (1851), Melville quoted from Darwin's *Journal of Researches* (later *The Voyage of the Beagle*).

1841 William Smith's nephew John Phillips (1800–1874) proposes the geologic Mesozoic and Kainozoic (later changing the spelling to Cainozoic and now referred to as Cenozoic) eras (see also Appendix B). The boundaries of these geologic eras, like those of Adam Sedgwick's Paleozoic, are marked by mass extinctions that produced sharp breaks in the fossil record. Phillips defended catastrophists' claims that fauna often appeared and disappeared from the earth, after which they were replaced by divine creation. After Charles Darwin published *On the Origin of Species* in 1859, Phillips claimed that his study of marine deposits "showed no evidence of evolution having occurred."

1842 Just before moving out of "dirty, odious London," Charles Darwin develops the ideas in his "Transmutation" notebook into a 35-page outline of "descent with modification" (as evolution was called in

Figure 18 Down House is an 18-acre estate in Downe, Kent (16 miles southwest of London) where Charles Darwin wrote most of his books. Down House is now a public museum. (*Randy Moore*)

Darwin's day). Soon thereafter, Charles and Emma Darwin bought a large house on an 18-acre estate near the village of Downe, where they lived for the rest of their lives (Figures 18 and 19).

1842 Charles Darwin publishes *The Structure and Distribution of Coral Reefs*, his first scientific book and the first volume of his geologic trilogy (other volumes include *Volcanic Islands* in 1844 and *Geological Observations on South America* in 1846). Darwin argued that small changes can, over vast periods, create major geologic features. Darwin's last book—*Formation of Vegetable Mould through the Action of Worms* (1881)—made the same point. Through 1846, Darwin—an accomplished geologist—published 19 papers (or notices) about geologic topics, after which his day-to-day interests shifted to biology.

1842 In an article published in the *Report of the British Association for the Advancement of Science,* British anatomist Richard Owen recognizes that *Iguanodon, Megalosaurus*, and other similar creatures share several important traits. Owen grouped them into "a distinct tribe or suborder of Saurian Reptiles, for which I would propose the name *Dinosauria*," thereby introducing the word *dinosaur* ("terrible

Figure 19 Charles Darwin and his family lived in Down House (Figure 18) for more than 40 years. (*Randy Moore*)

lizard"). For Owen, dinosaurs were the "height of reptilian achievement" whose demise was caused by air pollution.

1843 American geologist James Hall (1811–1898) documents the similarities between the Lower Paleozoic fossils of New York and those of England. Although Hall led a movement to establish a separate nomenclature for the geologic periods of North America, by the time of his death in 1898, Hall's and others' work clearly showed the same basic sequences in the fossil record in North America as in Europe. By the time Charles Darwin and Alfred Russel Wallace announced their theory of evolution in 1858, hundreds of geologists had established the validity and practical use of William Smith's Principle of Faunal Succession—namely, that a similar pattern of fossil distribution can be observed over broad areas of the earth's surface. Most of these geologists were devout Christians who were indifferent or hostile to the concept of evolution. Since the 1800s, hundreds of biostratigraphers have documented a well-defined order of fossils in geologic strata.

1844 Acknowledging Jean-Baptiste Lamarck (Figure 14) and Charles Lyell as his inspirations, publisher and amateur scientist Robert Chambers (1802–1871; Figure 20) publishes his controversial, popular, flawed—and influential—*Vestiges of the Natural History of*

Creation, a 390-page book that is "the first attempt to connect the natural sciences into a history of creation." Chambers proposed an evolution of the universe as well as of living organisms, contending that organisms arise by spontaneous generation and are transformed in a steadily upward progression into ever more complex forms. Chambers claimed that the earth was not specifically created by God, but instead by laws that expressed God's will; "How can we suppose that the august Being . . . was to interfere personally and on every occasion when a new shellfish or reptile was ushered into existence? . . . The idea is too ridiculous for a moment to be entertained." Chambers argued that there is "a system of Mercy and Grace behind the screen of nature" but inferred that nature "has the fairness of a lottery, in which every one has the like chance of drawing the prize." Chambers suggested that humans originated in Asia (Thomas Huxley and Ernst Haeckel later made similar suggestions). Chambers published his book anonymously (his wife transcribed the entire manuscript so as to disguise the handwriting) to protect his business interests from the backlash likely to follow introduction of his controversial ideas. *Vestiges*—a *cause celébre* and bestseller—was denounced by scientists and theologians alike, and it showed Charles Darwin some of the obstacles he had to overcome if his theory of evolution was to gain acceptance. Despite its critics, the 1st edition of *Vestiges* sold out in a few days, and by 1860 it had sold almost 24,000 copies (for comparison, *On the Origin of Species* sold only about 10,000 copies in the decade after its release, and annual sales of *Origin* did not consistently overtake those of *Vestiges* until the 20th century). When Darwin read *Vestiges* at NHM in November 1844, he noted "I must allude to this" and "never use the word [sic] higher and lower." The reaction to *Vestiges*, the most controversial book of its time, ushered in an era of public controversy about evolution.

1844 Charles Darwin tells Leonard Horner (1785–1864, Charles Lyell's father-in-law) that "the great merit of *Principles of Geology*, was that it altered the whole tone of one's mind and therefore that when seeing a thing never seen by Lyell, one yet saw it partially through his eyes."

1844 Gideon Mantell's *Medals of Creation* explains fossils as medallions struck by the Creator to commemorate the success of epochs.

1844 Charles Darwin expands his earlier outline of natural selection into a 231-page essay that rejects the fixity of species. The table of

Figure 20 Publisher and amateur scientist Robert Chambers was one of the most successful businessmen in Victorian England. His sensational bestseller, *Vestiges of the Natural History of Creation*, began the public controversy about evolution. (*Special Collections, Tutt Library, Colorado College, Colorado Springs, Colorado*)

contents of this essay, as well as much of the essay's text, was similar to that of *On the Origin of Species*, which did not appear for another 15 years. After confiding to his friend, botanist Joseph Hooker, that he had discovered "the simple way which species become exquisitely adapted to various ends," Darwin likened his idea to "confessing a murder." This was Darwin's first discussion with a colleague about his ideas on evolution. Soon thereafter, Darwin wrote his wife a note: "I have just finished my sketch of my species theory. If, as I believe, my theory in time [will] be accepted even by one competent judge, it will be a considerable step for science. . . . I therefore write this in case of my sudden death, as my most solemn and last request . . . that you will devote £400 to its publication, and further will yourself, or through Hensleigh [Emma's brother], take trouble promoting it."

1844 Ellen G. White (1827–1915), a co-founder of Seventh-Day Adventism (in 1863) and a follower of apocalyptic preacher

William Miller (1782–1849), predicts that Jesus will return on October 22. Soon after the so-called "Great Disappointment," in which Jesus failed to reappear, White began having visions from God revealing creation to have occurred in six days as well as a global flood that buried life and created fossils. White's teachings profoundly influenced the 20th-century young-earth creationism movement, primarily through Adventist George McCready Price.

1845 Biblical scholar John William Burgon's (1813–1888) poem *Petra* describes Petra, the inaccessible city in present-day Jordan. The poem is remembered for its final lines: "A rose-red city—half as old as time." According to most biblical chronologies, Petra is approximately 3,000 years old.

1845 Charles Darwin publishes his account of his five-year (1831–1836) journey aboard the *Beagle*. Darwin's book, which eventually was retitled *The Voyage of the Beagle*, became a bestseller and one of the world's best-known travel books. That same year, Darwin wrote to Joseph Hooker that "geographical [distribution] . . . will be the key which will unlock the mystery of species." By this time, Darwin was "an unbeliever in every thing beyond [my] own reason."

1845 Robert Chambers (Figure 20) responds to his critics by publishing *Explanations: A Sequel to "Vestiges of the Natural History of Creation" by the Author of that Work*. The final edition of *Vestiges* appeared in 1860, a year after publication of Charles Darwin's *On the Origin of Species*.

1845 Refusing to abandon their support of slavery, Baptists in the southeastern United States meet in Augusta, Georgia, and form the Southern Baptist Convention. Southern Baptists later became vocal supporters of the antievolution movement, and many of their leaders (e.g., Frank Norris) used creationism to support racial discrimination.

1846 Mary Anning, whose reputation bordered on legendary even during her short lifetime, is made an honorary member of the all-male Geological Society of London. Late in her life, William Buckland helped Anning secure an annual "Civil List" pension of £25, thereby making Anning the first woman to receive government support for her scientific research. The following year, Anning—by that time a national celebrity—died of breast cancer and was buried at St. Michael the Archangel Church in Lyme Regis. Her obituary was published in the *Quarterly Journal of the Geological*

Society, an organization that barred women from membership until 1904. After Anning's death, The Royal Society donated funds for a stained-glass window honoring Anning's memory in the parish church at Lyme Regis. Anning, the subject of the tongue twister "she sells seashells by the seashore," remains one of the most influential and interesting people in the history of paleontology. As lamented by English novelist John Fowles (1926–2005) in *The French Lieutenant's Woman* (1969), Mary Anning never had a new species named after her until long after her death.

1846 Thomas Huxley, bereft of a scholarship for continuing his medical studies, joins the Royal Navy and becomes assistant surgeon on the *Rattlesnake*, and then starts a four-year expedition to explore northern Australia and New Guinea. Huxley attended to day-to-day medical matters, assisted the ship's naturalist, and later published several papers on the anatomy and classification of organisms collected during the voyage. Huxley returned to England in 1850 but was dismissed from the Royal Navy in 1854 for failing to appear for active duty. Soon thereafter, he secured a professorship and started giving public lectures on scientific topics. Huxley's vocal and eloquent defense of Charles Darwin's theory of evolution by natural selection earned him the nickname "Darwin's Bulldog."

1847 Joseph Hooker publishes "Enumeration of the Plants of the Galápagos Archipelago," based on Charles Darwin's collections of plants from the islands.

1847 French physician Prosper Lucas's (1805–1885) *Traité Philosophique et Physiologique de L'Hérédité Naturalle* catalogues the number and diversity of heritable traits. Charles Darwin cited Lucas's book in Chapter I of *On the Origin of Species*.

1847 The last mooring of the *Beagle* is recorded in a Photographic Office Survey Chart. The entry positions the *Beagle* in the River Roach.

1847 Louis Agassiz claims that God, and not evolution, is responsible for the fact that "the brain of the Negro is that of the imperfect brain of a seven months' infant in the womb of a White." Agassiz was the last prominent biologist to reject evolution; as Charles Darwin noted in a letter to American botanist Asa Gray, "Agassiz's name no doubt is a heavy weight against us."

1848 Self-taught biologists and entrepreneurs Henry Bates (1825–1892) and Alfred Russel Wallace, both of whom dropped out of school at the age of 13, depart for the Amazon to collect exotic plants and animals to sell to museums and collectors. Before he returned more than a decade later, Bates collected 14,712 animal species, more than 8,000 of which were newly described species. Bates's discoveries—especially those about coevolution and mimicry—later provided strong support for Darwin's theory of evolution by natural selection. Unlike Darwin's voyage aboard the *Beagle*, in which Darwin was the captain's paying guest, Wallace and Bates had to work for a living.

1848 In *Archetype and Homologies of the Vertebrate Skeleton*, Richard Owen—a student of Georges Cuvier—presents an updated argument from design while explaining homologies as variations of vertebrate archetypes. Owen famously depicted the vertebrate archetype—an idealized form of all vertebrates—as a skeletal zeppelin hovering above humans, birds, fishes, and other vertebrates. Owen, who tutored Queen Elizabeth's children, noted the similar organization of human hands, bat wings, and whale flippers. Owen's explorations of homology, along with the establishment of NHM, became his most lasting legacy to modern biology.

1848 The death of Robert Waring Darwin (1766–1848) provides an inheritance of more than £50,000 for his son Charles.

1849 Hugh Miller's *Footprints of the Creator*—a response to Robert Chambers's *Vestiges of the Natural History of Creation* (1844)—states a lingering concern for evangelical Christians: If there was no Fall, and Adam merely took the first human step in an ongoing development, then there is no need for salvation and therefore no need for Christ (the "second Adam") to die for the sins of the world.

1849 In *On the Nature of Limbs*, Richard Owen notes that some structures seem to be "made in vain," but holds that they somehow characterize the archetype. Such traits, which reflect the evolutionary history of a lineage, are described by modern biologists as *vestigial traits*.

1849 In an article titled "Lyell's Second Visit to America," American philosopher and educator Francis Bowen (1811–1890)—the editor of *North American Review*—notes that "the literal interpretation of the first chapter of the book of Genesis has come to be regarded by nearly all educated Christians in the same light with the Papal opposition to the doctrine that the earth revolves around the sun."

1850 Alcide d'Orbigny (1802–1857), the most famous French paleontologist of his generation following Cuvier, publishes *Prodrome de Paléontologique Stratigraphique Universelle*, which presents his "universal" (i.e., global) history of life. D'Orbigny, like Georges Cuvier, was a catastrophist, and used characteristic fossils in 27 strata to claim that there had been 27 separate floods and creations.

1850 Following the death of William Wordsworth (1770–1850), Alfred Tennyson (later Alfred, Lord Tennyson)—one of the most famed poets of the Victorian age—becomes Poet Laureate of the United Kingdom. The previous year he had published *In Memoriam, A.H.H*, a tribute (on which he had worked 17 years) to his friend and Cambridge classmate Arthur Henry Hallam. Tennyson's poem described Charles Lyell's view of uniformitarianism and a growing sense that nature is characterized by pitiless suffering and death. *In Memoriam* introduced the famous phrase "Nature, red in tooth and claw."

1850 In two articles published in *The Unitarian Christian Examiner*, Louis Agassiz argues that human races originated not from a single pair of people, but from separate groups of people created by God in areas where the races live today. Agassiz claimed that differences among human races were as "primitive" as those separating animals in one area from those of another.

1851 British philosopher and economist Herbert Spencer's (1820–1903) *Social Statics* argues for eliminating the less fit. Spencer, like Charles Darwin, was strongly influenced by Thomas Malthus.

1851 In *The Religion of Geology and Its Connected Sciences*, influential geologist Edward Hitchcock stresses that religion has nothing to fear from science because "scientific truth is religious truth." Hitchcock argued that the Bible is not meant to explain nature, that the Flood may have been regional, and that geology—more than any other science—proves that God has guided history.

1851 The death of Anne Elizabeth "Annie" Darwin (1841–1851) removes the last vestiges of her father Charles Darwin's religious faith. Charles did not attend Annie's funeral, but remembered

Annie by writing a 12-page memorial a week after she died. The memorial ends poignantly: "We have lost the joy of the household, and the solace of our old age: she must have known how we loved her; oh that she could now know how deeply, how tenderly we do still and shall ever love her dear joyous face. Blessings on her."

1852 After four years of exploring and collecting, Alfred Russel Wallace's return from Brazil aboard the *Helen* ends in disaster when the ship catches fire and sinks in the mid-Atlantic (Wallace was rescued 10 days later by the *Jordeson*). As a result, Wallace had virtually nothing to show for his years in Brazil (fortunately, Wallace's agent—Samuel Stevens—had insured the collection for £200), and his hopes for joining the scientific elite were derailed. Two years later, Wallace left for Southeast Asia to try again. This time his luck was better.

1852 British artist Benjamin Waterhouse Hawkins (1807–1889) is commissioned to "illustrate and realize—the revivifying of the ancient world—to call up from the abyss of time and from the depths of the earth, those vast forms of gigantic beasts which the Almighty Creator designed with fitness to inhabit and precede us in possession of this part of the earth called Great Britain." Two years later, public displays of Hawkins's models of dinosaurs—some of which were made of several tons of clay—triggered dinosaur-mania and forever changed the way people viewed ancient life on earth (Figure 21).

1853 First Baptist Church of Minneapolis, Minnesota, is founded. In the early 1900s, this church became the national headquarters of the fundamentalist movement in the United States. Led by William Bell Riley (1861–1947) from 1897 to 1947, First Baptist Church housed the World's Christian Fundamentals Association (WCFA) and regularly hosted antievolution crusaders such as William Jennings Bryan, Frank Norris, and Billy Sunday.

1853 German anatomist Hermann Schaaffhausen (1816–1893) notes in a paper titled "On the Constancy and Transformation of Species" that "the immutability of species . . . is not proved." Schaaffhausen later described the first Neanderthal fossils.

1853 French aristocrat Arthur de Gobineau's (1816–1882) *Inequality of the Races* and *Moral and Intellectual Diversity of the Races* (1856) claim that Aryans originated in Central Asia and are the world's

Figure 21 Benjamin Waterhouse Hawkins created the world's first sculptures of dinosaurs in this "Model-Room" on the grounds of the Crystal Palace in Sydenham, London. This illustration of Hawkins's workshop covered an entire page of *The Illustrated London News* on December 31, 1853. That night, Hawkins hosted a dinner for 21 guests in the belly of one of his *Iguanodon* models.

superior race. Gobineau, who believed that Middle Eastern men were inferior and resulted from a separate creation, was the first person to argue that history is a racial struggle and that its study is a type of biology.

1853 German linguist August Schleicher (1821–1868) publishes a genealogical classification of languages based on a branching-tree model similar to what Charles Darwin would later describe for organisms in *On the Origin of Species*. Schleicher subsequently noted that he "set up family trees of languages known to us precisely the same way as Darwin has attempted to do for plant and animal species." Darwin cited Schleicher's work in *The Descent of Man* (1871).

1854 Alfred Russel Wallace arrives on the island of Borneo, which serves as the locus for his collections and conclusions regarding the relationship between biogeography and the origin of species. Wallace later became one of the first prominent scientists to raise concerns about humans' impacts on the environment.

1854 English naturalist Edward Forbes's (1815–1854) article titled "On the Manifestation of Polarity in the Distribution of Organized Beings in Time" claims that there were two major periods of creation. Although this quasi-supernatural "polarity theory" had little impact, it provoked Alfred Russel Wallace to write his monumental "Sarawak paper" the following year. Also in 1854, Charles Darwin wrote to his wife Emma that Charles Lyell and Forbes were his top choices to publish his 1844 essay in the event of his own untimely death.

1854 German scientist Heinrich Ernst von Beyrich (1815–1896) describes fossils in northern Germany from an era that he names the Oligocene period (see also Appendix B).

1854 American geologist Ferdinand Vandeveer Hayden (1829–1887) finds dinosaur teeth in the Rocky Mountains.

1854 Joseph Hooker's *Himalayan Journal*, which Hooker dedicated to Charles Darwin, describes the geology and botany of the Himalayas, as well as numerous adventures during his 1847–1850 expedition (including how he and his party were taken prisoner for several weeks by the Rajah of Sikkim). Darwin cited Hooker's work in Chapter XI of *On the Origin of Species.*

1854 In the garden at the St. Thomas Monastery in Brünn (today Brno in the Czech Republic), Augustinian monk Gregor Mendel begins studying inheritance of seven traits in the garden pea (*Pisum sativum*). The area surrounding Brünn was agricultural, and because the monastery was a regional center of scientific study, Mendel justified his studies of hybridization as providing potential benefit to local farmers. A meticulous experimentalist and record keeper, Mendel generated large sets of data (frequently based on observations of hundreds of individuals from each generation) that allowed him to identify the critical frequencies of the alternate forms of the traits he studied. Mendel's work, which would be ignored for 35 years, established the foundation of modern genetics.

1854 The world's first life-size models of dinosaurs, sculpted at a cost of £13,729 by Benjamin Waterhouse Hawkins, greet Queen Victoria and 40,000 other people attending the opening ceremonies for the Crystal Palace Exhibition in London's Hyde Park, providing the public with some of the first opportunities to experience dinosaurs. (Until Hawkins created his models, dinosaurs were poorly understood and of little interest to anyone except a few paleontologists.) The sculptures, some of which contained several tons of rocks and clay, included *Iguanodon* and *Megalosaurus*. To produce the exhibit, Hawkins worked closely with Richard Owen, the first director of NHM and originator of the word *dinosaur*. The Crystal Palace began the exhibit with a New Year's Eve dinner for 21 guests in the belly of one of Hawkins's 30-foot-long *Iguanodon* sculptures (Figure 21). More than one million people visited the exhibit during each of the next 50 years as dinosaurs moved to the forefront of the public's imagination. In 2002 and 2007, Hawkins's antediluvian monsters were restored and once again attracted throngs of visitors.

1855 While in Sarawak, Malaysia, Alfred Russel Wallace writes "On the Law Which Has Regulated the Introduction of New Species," which is published later the same year in *The Annals and Magazine of Natural History*. Wallace stated his conclusion in italics near the beginning and end of the paper: "*Every species has come into existence coincident both in space and time with a pre-existing closely allied species*." Wallace's "Sarawak paper," which later became a landmark in evolutionary biology, initially failed to generate the interest he had hoped for. However, Edward Blyth (1810–1873), an English biologist working in India, wrote to Charles Darwin that "Wallace has, I think, put the matter well; and according to his theory the various domestic races of animals have been fairly developed into species." Wallace's paper inspired Charles Lyell to start his journals about the species question. Lyell told Darwin about Wallace's paper, adding that Darwin might be scooped, but Darwin was not overly worried. Darwin summarized Wallace's

paper with a marginal note that read "nothing very new." Within a year, however, Darwin wrote in his pocket diary that he "began by Lyell's advice writing Species Sketch."

1856 Richard Owen, famed anatomist and foe of Charles Darwin, becomes Superintendent of the Natural History Departments at NHM.

1856 Alfred Russel Wallace travels to Indonesia, where he notes the existence of two distinct zoogeographic regions, divided by an east-west boundary that Thomas Huxley later called "Wallace's Line." Wallace had identified the edge of the Indo-Australian continental plate.

1856 Charles Darwin begins work on his "big book," which ultimately becomes the monumental *On the Origin of Species*. That same year, Darwin began corresponding with Alfred Russel Wallace. Darwin praised Wallace's 1855 "Sarawak paper" and dismissed the fact that it had generated little interest among scientists by claiming that "very few naturalists care for anything beyond the mere description of species."

1856 Sixteen human-like bones are discovered by limestone workers in Feldhofer Cave in the Neander Valley (Neander Thal in Old German) near Düsseldorf, Germany. These remains, which were characterized by a low brow, thick jaw, and thick bones, were passed by local teacher Johann Carl Fuhlrott (1803–1877) to Hermann Schaaffhausen, who in 1858 published a detailed description of the bones; this description noted that the skull had a shape similar to that of a large ape. Schaaffhausen's work was translated into English in 1861 by surgeon George Busk (1807–1886). In 1864, these 41,000-year-old fossils became the type specimen of *Homo neanderthalensis* (*Neanderthal 1*), now known to be an extinct cousin of *Homo sapiens* (see also Appendix C).

1856 In a letter to his friend Joseph Hooker, Charles Darwin notes that "[w]hat a book a Devil's Chaplain might write on the clumsy, wasteful, blundering low and horridly cruel works of nature." In 2003, biologist Richard Dawkins (b. 1941) published a best-selling book titled *A Devil's Chaplain: Reflections on Hope, Lies, Science, and Love*.

1857 Charles Darwin tells American biologist Asa Gray (1810–1888) about his theory of evolution: "I have come to the heterodox conclusion that there are no such things as independently created species." Before now, Darwin had disclosed his idea only to Charles Lyell and Joseph Hooker.

1857 Scottish missionary David Livingstone's (1813–1873) *Missionary Travels and Researches in South Africa* becomes a sensational bestseller. Charles Darwin later described *Travels* as "[t]he best travels I ever read."

1857 English marine biologist Philip Gosse (1810–1888) publishes *Omphalos: An Attempt to Untie the Geological Knot* (later titled *Creation* in hopes of boosting sales), in which he claims that God used fossils and geology to suggest a history that does not actually exist. Gosse, who corresponded with Charles Darwin, struggled to reconcile science and religion. Gosse claimed that God created the world *as if* scientific claims about the earth's history are true—that is, trees were created with tree rings, animals were created with excrement already in them, geologic strata were created with fossils already in them, and Adam and Eve had navels (*omphalos* is Greek for *navel*). Gosse argued that all geologic evidence for an ancient earth had been created about 6,000 years ago when God created the universe. (Gosse called these events that did not occur "prochronic," meaning "outside time.") Gosse's claim that God had created evidence of a past that never existed was not popular; critics ridiculed it while equating it with divine fraud because it implied that humans cannot trust what God created. Despite these criticisms, Gosse's ideas—which became known as *ideal-time creationism*—were famously resurrected in 1961 by Henry Morris and John Whitcomb, Jr. (without mentioning Gosse by name) in their influential book *The Genesis Flood.*

1857 In the popular *The Testimony of the Rocks; or, Geology in Its Bearings on the Two Theologies, Natural and Revealed*, which was completed just before his suicide, Scottish geologist Hugh Miller promotes day-age creationism. Miller, who edited *The Witness* (the ecclesiastical journal of the Free Church of Scotland), described an old earth and a localized Noachian Flood and predicted that the ideas of "anti-geologists" (i.e., young-earth creationists who demanded a literal reading of the Bible) would soon be as obsolete as the ideas of those who had earlier promoted Ptolemy's geocentric view of the solar system. Miller, apparently a better geologist than prophet, was wrong, because the second half of the 20th century witnessed a revival of "anti-geology" by young-earth creationists.

The Testimony of the Rocks was one of the most popular geology books of the 19th century.

1858 Amateur naturalist William Parker Foulke (1816–1865) and a crew from Philadelphia's Academy of Natural Sciences excavate from a ravine in Haddonfield, New Jersey, the most complete (49 individual bones and teeth) dinosaur skeleton ever found. Joseph Leidy (1823–1891), who became known as the "father of North American vertebrate paleontology," studied the skeleton (later named *Hadrosaurus foulkii*) and, in doing so, became the first person to fully describe the anatomy of a dinosaur. A second *Hadrosaurus* skeleton was later excavated at Haddonfield and taken to Philadelphia, where it was displayed at the American Philosophical Society. When biologist Othniel Marsh (1831–1899) later viewed the Haddonfield site with Edward Drinker Cope (1840–1897; Figure 22), Marsh offered the miners money if they would send the bones to him instead of Cope; this triggered the bitter, public, and spectacular "Bone War" between Marsh and Cope. In 1994, the Haddonfield site was made a national monument.

1858 Charles Darwin receives a letter from Alfred Russel Wallace, who was then halfway through eight years of collecting specimens across the Malay Archipelago (a vast group of islands between Australia and southeastern Asia). Wallace included a 3,764-word handwritten manuscript titled "On the Tendency of Varieties to Depart Indefinitely from the Original Type" (written over two evenings in February 1858) that described Wallace's ideas about natural selection. In a cover letter accompanying the manuscript, Wallace told Darwin that he hoped the idea would be as new to Darwin as it was to him. It was not, for Wallace's manuscript described the same concepts that Darwin had been secretly developing for years. As Darwin told Lyell after reading Wallace's letter, "I never saw a more striking coincidence. All my originality, whatever it may amount to, will be smashed. . . . Even his terms now stand as heads of my chapters. . . . I rather hate the idea of writing for priority, yet I certainly should be vexed if any one were to publish my doctrines before me." Darwin was shocked by Wallace's letter; Darwin's journal for June 18—the day that Darwin received Wallace's letter—tersely recorded "interrupted by letter from A R Wallace." As per Wallace's request (if Darwin "thought [the paper] sufficiently important"), Darwin sent Wallace's manuscript to Charles Lyell. Distraught at the prospect of losing credit

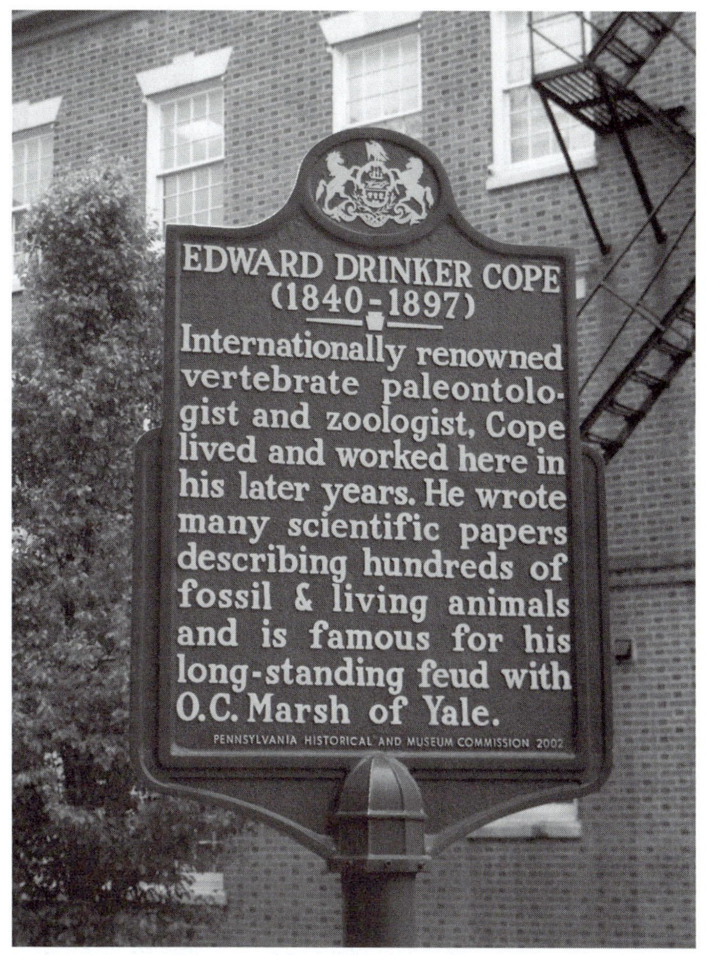

Figure 22 The bitter rivalry between paleontologists Othniel Marsh and Edward Drinker Cope, which was covered by newspapers throughout the country, fascinated the public. This monument on Philadelphia's Pine Street marks the site of Cope's residence, which was filled with fossils when Cope died. (*Randy Moore*)

for an idea he had developed many years earlier, yet having to contend with a seriously ill child, Darwin put the matter into the hands of Joseph Hooker and Lyell. Hooker and Lyell proposed that at a July 1 meeting of the Linnean Society of London, three documents would be read: extracts from Darwin's 1844 essay that outlined his conceptualization of natural selection; a letter from Darwin to Asa Gray that established Darwin as the original discoverer of natural selection; and Wallace's paper. Darwin did not attend the meeting—he was mourning the death of his son

Charles Waring Darwin (1856–1858), who had died two days earlier of scarlet fever—and Wallace (who was sick at Dorey, now Manokwari in New Guinea) did not even know about the meeting. The papers were read by Thomas Huxley's friend, George Busk. There was little response to the presentation, possibly because of the announcement of the death of botanist Robert Brown (1773–1858; a former president of the Linnean Society and the discoverer of the cell nucleus, cytoplasmic streaming, and Brownian motion), but Joseph Hooker expressed relief at no longer having to keep secret the ideas Darwin had confided to him in 1844. The paper was published in August, 1858 under the impressive title "On the tendency of species to form varieties; and on the perpetuation of varieties and species by natural means of selection by Charles Darwin Esq., FRS, FLS, & FGS and Alfred Wallace Esq., communicated by Sir Charles Lyell, FRS, FLS, and J. D. Hooker Esq., MD, VPRS, FLS, &c." Darwin began writing a book describing his idea, and he finished the final chapter of *On the Origin of Species* on March 19, 1859. At the end of 1858, Thomas Bell (1792–1880)—the president of the Linnean Society—noted in his annual report that "the year which has passed has not, indeed, been marked by any of those striking discoveries which at once revolutionize, so to speak, the department of science on which they bear." However, Darwin understood the importance and potential impact of his idea: "It is no doubt the chief work of my life."

1858 Rudolf Virchow shows that diseased cells arise from other diseased cells. Based on this work and its implications (e.g., his belief that tumors arise from single malignant cells), Virchow proposed that cells can arise only from preexisting cells ("*Omnis cellula e cellula*") and do not crystallize, as had been proposed by Schleiden and Schwann. Virchow's discovery later led to the discovery of a mechanism for cell division. Virchow's claim, which became part of what is now known as the cell theory, rejected spontaneous generation. To distinguish it from the claims of Schleiden and Schwann, Virchow's conclusion is sometimes referred to as the *cell doctrine*.

1858 Thomas Huxley (Figure 23) writes to Joseph Hooker that "Wallace's impetus seems to have set Darwin going in earnest, and I am rejoiced to hear we shall learn his views in full, at last. I look forward to a great revolution being effected." When Charles Darwin announced his theory the following year in *On the Origin of*

Species, Huxley lamented "[h]ow extremely stupid of me not to have thought of that."

1859 In *Essay on Classification*, Louis Agassiz proclaims that the living world is characterized by "premeditation, wisdom, greatness, prescience, omniscience, providence . . . all these facts . . . proclaim aloud the One God whom man may know, and natural history must, in good time become the analysis of the thoughts of the Creator of the Universe." Agassiz advocated multiple creations, thereby contradicting both evolution and a literal interpretation of Genesis. This same year, Agassiz organized the Museum of Comparative Zoology at Harvard University along three themes—zoology, geography, and embryology. Alfred Russel Wallace later noted that "[i]t is surely an anomaly that the naturalist most opposed to the theory of evolution should be the first to arrange the museum in such a way as best to illustrate that theory." Throughout his life, Agassiz continued to reject evolution, claiming "[w]e are children of God, not of monkeys."

1859 When Charles Darwin submits his manuscript to publisher John Murray III (1808–1892), Murray has difficulty with the text and asks Reverend Whitwell Elwin (1816–1900)—the editor of *Quarterly Review*—for advice. Elwin advises Darwin to write a book about pigeons because "[e]verybody is interested in pigeons." Although Darwin gave Murray the option of not publishing his manuscript, Murray followed Charles Lyell's recommendation to publish Darwin's "important new work." On November 24—22 years after Darwin had opened his secret "Transmutation of Species" notebook—John Murray Publishing of London (which had published all of Lyell's books) released Darwin's 502-page book *On the Origin of Species by Means of Natural Selection, or The Preservation of Favoured Races in the Struggle for Life*. Murray printed 1,250 copies of *On the Origin of Species* (a modest print-run, even by mid-19th-century standards), 139 of which were distributed as promotional copies. Booksellers bought all of the 1,111 remaining copies (for 15 shillings—about £35 in 2009—apiece) on the first day they were for sale. Darwin's five subsequent revisions of *On the Origin of Species* were translated into several languages that took Darwin's idea throughout the world. Unlike Lamarck's (Figure 14) *Philosophie Zoologique*, which was purely a theoretical book, Darwin's *Origin* was an overwhelming compendium of facts. It includes only one sentence about human evolution, but that sentence could be the understatement of the 19th century:

"Light will be thrown on the origin of man and his history." Whereas Lamarck had spoken of the "march of nature," Darwin wrote of "transformation." When Darwin sent a copy of the book to Alfred Russel Wallace, he enclosed a note: "God knows what the public will think." In *Origin*, Darwin (1) replaced the notion of a perfectly designed and benign world with one based on an unending, amoral struggle for existence; (2) challenged prevailing Victorian ideas about progress and perfectibility with the notion that evolution produces change and adaptation, but not necessarily progress, and never perfection; (3) offered no larger purpose in nature for humanity other than the production of fertile offspring; and (4) liberated readers from the conceit of providentially supervised special creation of each species with the argument that all life—humans included—are not a special product of creation, but of evolution acting according to principles that act on other species. For Darwin, natural selection replaced divine benevolence as an explanation for adaptation. Darwin's theory was based on an ancient earth, and he wrote that anyone not grasping earth's antiquity "may at once close this book." *Origin* shifted thought from a foundation of untestable awe of special creation to a science-based examination of the natural world based on natural mechanisms and historical patterns. Whereas scientists before Darwin had often invoked purpose to explain biology (e.g., that a particular structure was present because it was pleasing to a deity), Darwin's idea replaced purpose with function and history; a structure was there because it was (or had been) an adaptation. Darwin had many defenders, most notably Harvard scientist (and evangelical Christian) Asa Gray in the United States and Thomas Huxley (Figure 23; "Darwin's Bulldog") in England. As Huxley wrote to Darwin in 1859, "I trust you will not allow yourself to be in any way disgusted or annoyed by the considerable abuse and misrepresentation which unless I greatly mistake is in store for you. Depend upon it you have earned the lasting gratitude of all thoughtful men. . . . Some of your friends at any rate are endowed with an amount of combativeness which . . . may stand you in good stead—I am sharpening up my beak and claws in readiness." Darwin appreciated their work; as he noted in 1860, "I see daily more and more plainly that my unaided book would have done absolutely nothing." Throughout the uproar that followed publication of his book, Darwin stayed at Down House (Figures 18 and 19); he was interested in what was happening, but stayed out of the fray. Not surprisingly, *Origin* was condemned by many religious leaders, and William Whewell, Master of Trinity

College at Cambridge, refused to allow it into the college library (despite the fact that Darwin quoted Whewell prominently opposite the book's title page). John Henslow damned Darwin's book with faint praise, calling it "a stumble in the right direction." The following year, Darwin wrote to Charles Lyell that if he were starting anew, he would use the phrase *natural preservation* rather than *natural selection*. Darwin, who believed that "with a good book as with a fine day, one likes it to end with a glorious sunset," closed his book with this famous paragraph: "It is interesting to contemplate an entangled bank, clothed with many plants of many kinds . . . so different from each other, and dependent on each other in so complex a manner, have all been produced by laws acting around us. . . . Thus, from the war of nature, from famine and death, the most exalted object which we are capable of conceiving, namely, the production of the higher animals, directly follows. There is grandeur in this view of life, with its several powers, having been originally breathed into a few forms or into one; and that, whilst this planet has gone cycling on according to the fixed law of gravity, from so simple a beginning endless forms most beautiful and most wonderful have been, and are being, evolved." Although Darwin's model for evolution by natural selection became synonymous with evolution, the last word in *Origin* is the only use of the word *evolve* or its cognates in the book.

1859 Cambridge geologist Adam Sedgwick—who coined the term *Paleozoic* to encompass the life forms of the Cambrian through Permian Periods (see also Appendix B)—tells Charles Darwin "I have read your book with more pain than pleasure." Sedgwick criticized Darwin for not linking science with morality, noting that without that link, humanity "would suffer a damage that might brutalize it." Sedgwick, one of Darwin's former professors at Cambridge, believed that geologic evidence documented widespread changes of species under providential guidance.

1859 Henry Bates returns from the Amazon, admitting "the contemplation of Nature alone is not sufficient to fill the human heart and mind." Bates's descriptions of his discoveries later provided strong evidence for Darwin's theory of evolution by natural selection.

1859 In his *Introductory Essay to the Flora Tasmaniae*, released one month after *On the Origin of Species*, Joseph Hooker becomes the first biologist to state in print his support for evolution by natural selection.

1859 After reading *On the Origin of Species*, Alfred Russel Wallace describes it to his friend George Silk: "It is the 'Principia' of Natural History. . . . Mr. Darwin has given the world a new science . . . his name should stand above that of every philosopher of ancient or modern times." While Darwin later concentrated on producing evidence supporting his theory, Wallace published on topics such as miracles, vaccines, spirituality, the true identity of Shakespeare, the advisability of labor strikes, and social problems due to the private ownership of land, as well as biology.

1859 Asa Gray hosts the American premiere of Charles Darwin's ideas at a meeting of the Cambridge Scientific Club in his home in Cambridge, Massachusetts: "What are termed closely related species may in many cases be lineal descendants from a pristine stock, just as domesticated races are." At the meeting was Louis Agassiz, a staunch opponent of evolution.

1860 A small, isolated impression of a feather is found in Upper Jurassic (see also Appendix B) limestone near Solnhofen, Germany. The next year, this feather was linked to *Archaeopteryx*, the earliest known bird.

1860 William Jennings Bryan (1860–1925) is born in Salem, Illinois, the same town in which John Scopes would later attend high school and learn biology. Bryan later ran unsuccessfully for President of the United States three times (1896, 1900, and 1908) and became a leader of the fundamentalist movement in the early 1900s. In 1925, Bryan participated in the Scopes Trial, the most famous event in the history of the evolution-creationism controversy.

1860 Louis Agassiz writes in *American Journal of Science* that Darwin's ideas "have not made the slightest impression on my mind, nor modified in any way the views I have already propounded."

1860 In *Essays and Reviews*, Victorian theologian Benjamin Jowett (1817–1893) attacks Philip Gosse's *Omphalos* by noting that "it is ridiculous to suppose that the world appears to have existed, but has not existed during the vast epochs of which geology speaks to us." Jowett is famous for saying "[d]oubt comes in at the window, when Inquiry is denied at the door."

1860 An anonymous author writes, in the periodical *All the Year Round*, of Charles Darwin's *Origin*: "It is well for Mr. Charles Darwin, and a comfort to his friends, that he is living now, instead of having lived in the sixteenth century. . . . But we have come upon more tolerant times. If a man can calmly support his heresy by reasons, the heresy will be listened to."

1860 Charles Darwin writes to Asa Gray that "I cannot see, as plainly as others do, and as I should wish to do, evidence of design and beneficence on all sides of us. There seems to me too much misery in the world." Gray argued "a fortuitous cosmos is simply inconceivable. The alternative is a designed cosmos."

1860 Baden Powell (1796–1860), Frederick Temple (1821–1902)—a future Archbishop of Canterbury—and five other writers publish the bestselling *Essays and Reviews*, which praises Darwin's "masterly" *Origin*, argues that people do not require revelation of things that they can discover for themselves, and declares belief in miracles to be atheistic. The seven authors were labeled the "seven against Christ" and threatened with indictment for heresy. Powell's death before he could face the courts was described as his having been "removed to a higher tribunal."

1860 Cleric-geologist Edward Hitchcock (who was president of Amherst College from 1845 to 1854) updates *Elementary Geology,* a bestseller that uses gap creationism to reconcile geology with Genesis. Although earlier editions of *Elementary Geology* had disparaged the evolutionary claims of Lamarck, the 1860 edition questioned Darwin's theory. Hitchcock proclaimed that "scientific truth is religious truth," accepted an ancient earth, and argued that the Bible and science must be related. Hitchcock collected more than 20,000 fossilized footprints of dinosaurs and argued that the prints were left by huge, extinct birds. Hitchcock, a charter member of NAS, was America's leading advocate of catastrophism-based gap creationism.

1860 

Asa Gray becomes the first person to publish a review of *On the Origin of Species*. Gray's review—which began with "[t]his book is already exciting much attention"—appeared in the March edition of *American Journal of Science and Arts*, the leading science periodical in the United States. Gray—an orthodox Christian—described Darwin's views about variation as "general, and even universal" and considered Darwin's argument as "fair and natural" and not necessarily atheistic. Although Darwin later rated the review as "by far the best which I have read," Gray told readers that nature is filled with "unmistakable and irresistible indications of design." Gray rejected the purposelessness he felt Darwin's theory implied, yet he became America's foremost Darwinist of his time and corresponded regularly with Darwin. Gray published several widely read articles that discussed how evolution by natural selection is consistent with a religious perspective, a position now referred to as *theistic evolution*. Gray's advocacy of Darwinism culminated in his book *Darwiniana* (1876), and later editions of Gray's influential textbook *First Lessons in Botany and Vegetable Physiology* replaced "the Creator established a definite number of species at the beginning, which have continued by propagation, each after its kind" with "nearly related species probably came from a common stock in earlier times." Although Darwin noted in 1876 that Gray "knows my book as well as I do myself," Gray's Harvard colleague, Louis Agassiz, remained one of Darwin's biggest scientific foes. In 1860 in the *American Journal of Science and Arts,* Agassiz dismissed Darwin's theory as a collection of "mere guesses" that is "a scientific mistake, untrue in its facts, unscientific in its methods, and mischievous in its tendency."

1860 John Phillips, an advocate of the argument from design, becomes the first person to use the formation of geologic strata as an hourglass to measure the earth's age (see also Appendix A). In *Life on the Earth*, Phillips gave the earth an age of 96 million years and cited God as the ultimate creator. Charles Darwin referred to this work as "unreadably dull," a response similar in tone to Phillips's claim that Darwin's estimate of the earth's age was an "abuse of arithmetic" (in Chapter IX of *On the Origin of Species*, Darwin had estimated that it took "306,662,400 years; or say three hundred million years" to carve the Weald, a region of weathered ridges and valleys in southeast England stretching from Kent to Surrey). Others who subsequently used sedimentation rates of geologic strata to estimate the earth's age included Thomas Huxley (100 million years in 1869), Alfred Russel Wallace (28 million years in 1880), and Alexander Winchell (3 million years in 1883).

1860 Patrick Matthew (1790–1874) claims in *Gardener's Chronicle* that he discovered natural selection. Although the eccentric Matthew had summarized natural selection decades earlier in an appendix of *Naval Timber and Arboriculture*, Matthew failed to discuss the importance of natural selection and provided little evidence for his claims. Consequently, although Darwin was not the first to discover natural selection, and later editions of *On the Origin of Species* acknowledged Matthew and others, Darwin was the first to develop and grasp its significance.

1860 During the meeting of the BAAS at Oxford, Thomas Huxley (Figure 23) and Anglican Bishop Samuel Wilberforce (1805–1873) clash over the topic of evolution. Huxley—who had implored other scientists to defend Darwin's idea "if we are to maintain our position as the heirs of Bacon and the acquitters of Galileo"—had pledged to the publicity-shy Charles Darwin that "as for your doctrines I am prepared to go to the stake if requisite," and the Oxford meeting provided that opportunity. In a long speech that Joseph Hooker described as brimming with "ugliness and emptiness and unfairness," Wilberforce denounced Darwin's theory. When Wilberforce had finished, Huxley purportedly whispered "the Lord hath delivered him into mine hands" and rose to proclaim *On the Origin of Species* to be "the most potent instrument for the extension of the realm of knowledge which has come into man's hands since Newton's *Principia*." The clash between Huxley and Wilberforce—

Figure 23 Thomas Henry Huxley—also known as "Darwin's Bulldog"—was a towering figure in the history of biology, and the most vocal publicist for evolution by natural selection. Darwin claimed that Huxley often wrote with "vitriol rather than ink."

later parodied in Charles Kingsley's *Water-Babies*—was attended by Robert FitzRoy, the man who provided Darwin with the vehicle for his conclusions. FitzRoy, dressed in a rear admiral's uniform, stood and waved a Bible while telling the crowd that he regretted the publication of Darwin's book. Although not an actual debate, the legendary exchange between Wilberforce and Huxley was later hailed as a turning point in the acceptance of evolution by the scientific community and represented a serious challenge to the church's authority. At this same meeting, English academic Frederick Temple criticized the church's claim that apparent gaps in history are evidence of divine intervention. In the 1880s, Temple openly advocated natural selection as the mechanism for evolution.

1860 The American Academy of Arts and Sciences holds a special meeting to discuss Darwin's ideas. This meeting became famous for its clash between prominent Harvard naturalists Asa Gray and Louis Agassiz.

1860 *The Edinburgh Review* prints Richard Owen's anonymous—and somewhat confusing—review of *On the Origin of Species*. Owen asked "[b]ut do the facts of actual organic nature square with the Darwinian hypothesis? Are all the recognised organic forms of the present date, so differentiated, so complex, so superior to conceivable primordial simplicity of form and structure, as to testify to the effects of Natural Selection continuously operating through untold time? Unquestionably not." Owen described Darwin's observations as "few indeed and far apart, and leaving the determination of the origin of species very nearly where [Darwin] found it." Darwin described Owen's review as "extremely malignant, clever, & . . . very damaging." Today, Owen's copy of *On the Origin of Species* is at Shrewsbury School in Shropshire, the school that Darwin attended as a child.

1860 The 2nd edition of *On the Origin of Species* is published only a few weeks after the appearance of the 1st edition. In the United States, the 2nd edition sold for $1.25. Darwin was frustrated by the revision; he wrote to publisher John Murray, "I hope never again to have to make so many corrections or rather additions which I have made in hopes of making my many rather stupid reviewers at least understand what is meant." Many changes in the 2nd edition of *Origin* were minor corrections of printing errors, but others were more substantive; for example, Darwin deleted his famous "wedge" metaphor from Chapter III (i.e., to be part of "the face of Nature," a new wedge can exist only by forcing another out). Possibly to appease concerns from his family and the public, Darwin added "by the Creator" to the last line of *Origin*: "life . . . having been originally breathed by the Creator into a few forms or into one." Darwin later regretted this addition. That same year, the German translation of *Origin* (by paleontologist Heinrich Georg Bronn [1800–1862]) omitted any hint that evolution is applicable to humans, and includes an appendix critical of Darwin. In many respects, Darwin's message became obscured as he repeatedly tried to defend, clarify, and qualify his ideas in subsequent editions of *Origin*.

1860 In *Westminster Review*, Thomas Huxley (Figure 23) praises *On the Origin of Species* as "the most compendious statement of well-sifted facts bearing on the doctrine of species that has ever appeared" and denounces special creation as "verbal hocus-pocus" that has been "a mere specious mask for our ignorance." Huxley argued that Charles Darwin's work "does not so much prove that natural selection does occur, as that it must occur."

1861 *Archaeopteryx lithographica* ("ancient wing of the printing stone"; Figure 24) fossils are discovered in Upper Jurassic limestone near Solnhofen, Germany, and are named by paleontologist Hermann von Meyer: "I consider *Archaeopteryx lithographica* a suitable name for this animal." (The name *lithographica* came from the fine-grained limestone that surrounded the fossil and that was used in the printing industry.) The 150-million-year-old fossil was an unusual feathered, bird-like creature that had dinosaur-like traits. Famed anatomist Richard Owen bought the specimen (today known as BMNH 37001 or the "London specimen"), along with more than 1,000 other fossils, for £700 for NHM (for comparison, a typical house at the time sold for £100, and £700 was double the Museum's annual acquisitions budget). Owen bought the fossil from physician Karl Häberlein, who—needing to raise a dowry for his daughter—had acquired the fossil from quarrymen in return for medical services. The following year, Charles Darwin told a colleague that the *Archaeopteryx* fossil is "by far the greatest prodigy of recent time. It is a grand case for me, as no group was so isolated as birds." Darwin referenced the famous fossils in later editions of *On the Origin of Species*.

1861 French explorer Paul du Chaillu (1835–1903) exhibits stuffed gorillas in London. This was the first time gorillas had been seen outside of Africa. Chaillu's exhibit generated much excitement

Figure 24 *Archaeopteryx* (in German, *Urvogel,* meaning *original bird*) was a feathered birdlike creature that had several dinosaur-like traits that provided evidence for birds having evolved from non-avian dinosaurs. Specimens of *Archaeopteryx* are named for the cities having museums that house them. Shown here is the famous "Berlin specimen," which was discovered in 1877 and which today is housed in Berlin's Museum of Natural History. This museum, which is more commonly known as the Humboldt Museum, also houses the largest mounted dinosaur skeleton in the world—*Brachiosaurus*, which weighed approximately 55 tons when it roamed Earth. (*Carnegie Museum of Natural History*)

but also caused many people to recoil at the notion of human evolution.

1861 Six weeks before his death, German zoologist Johann Andreas Wagner (1797–1861) begins the controversy about *Archaeopteryx* (which he referred to as *Griphosaurus*) by challenging "the

Darwinians . . . insofar as they wish to promulgate the *Griphosaurus* as a creature in transition, to show me the intermediate stages that mediated the transition of any living or fossil animal from one class to another. If they cannot—and they can't—their views should be rejected as fantastic delusions, that have nothing to do with exact science." The next year, Charles Darwin noted that *Archaeopteryx* and evolution had "killed poor Wagner, but on his death-bed he took consolation in denouncing it as a phantasia."

1861 In *Physical Geography of the Globe*, astronomer John Herschel claims that Charles Darwin's theory "gave no indication of the Creator's foresight" and argues "an intelligence, guided by a purpose, must be continually in action to bias the directions of the steps of change." Darwin responded that "the point which you raise on intelligent Design has perplexed me beyond measure. . . . One cannot look at this Universe with all living productions & man without believing that all has been intelligently designed; yet when I look to each individual organism, I can see no evidence of this. For I am not prepared to admit that God designed the feathers in the tail of the rock-pigeon to vary in a highly peculiar manner in order that man might select variations & make a Fan-tail." This was the first use of the phrase *intelligent design* in its modern sense. Darwin was later buried beside Herschel in Westminster Abbey (Figure 7).

1861 In his presidential address to the Geological Society of London, Leonard Horner cites geologic evidence as a basis for proposing that James Ussher's "creation date" of 4004 BCE be removed from the English Bible (Figure 8).

1861 The death of John Henslow—to whom Charles Darwin owed much of his attachment to natural history—prompts Darwin to note that Henslow had a bigger influence on his career "than any other. . . . I fully believe a better man never walked the earth."

1861 Alfred Russel Wallace returns to England as the 3rd edition of *On the Origin of Species* is published. During his Malay expedition, Wallace collected more than 125,000 specimens and traveled more than 14,000 miles within the archipelago. That same year, in a confession to his brother-in-law Thomas Sims, Wallace claimed to be "an utter disbeliever in almost all that you consider the most sacred truths" and that "whatever may be our state after death, I can have no fear of having to suffer for the study of nature and the search for truth." Wallace later embedded natural selection in theism.

1861 The most obvious change in the 3rd edition of *On the Origin of Species* is the addition of a prefatory "Historical Sketch of the Recent Progress of Opinion on the Origin of Species." Charles Darwin also deleted his calculation of the time required for the erosion of the Weald in southern England, and expanded his claim in Chapter IV that natural selection has no need for a "necessary and universal law of advancement or development."

1862 Henry Bates's study of coloration in butterflies establishes Batesian mimicry (Bates called it "mimetic analogy"), in which some good-tasting, nonpoisonous animals mimic the warning colors of bad-tasting, poisonous animals. Bates—whose work transformed the oddities of natural history collectors into powerful evidence for evolution—argued that this mimicry is "a most beautiful proof of natural selection" and wrote to Darwin "I think I have got a glimpse into the laboratory where Nature manufactures her new species." Whereas *On the Origin of Species* relied heavily on arguments involving artificial selection, Bates's work provided independent evidence from the field supporting Darwin's ideas. Despite its drab title ("Contribution to an Insect Fauna of the Amazon Valley, Lepidoptera: Heliconidae"), Darwin praised Bates's mimicry paper as "one of the most remarkable and admirable papers I ever read in my life. . . . It will have lasting value."

1862 In *On the Various Contrivances by Which British and Foreign Orchids are Fertilised by Insects, and on the Good Effects of Intercrossing*, Charles Darwin argues that species having different origins have evolved mutual ecological relations that have affected important aspects of their morphology. When Darwin argued that the beauty of orchids results from adaptations for attracting pollinators, he was attacked by advocates of natural theology (e.g., George Campbell, the Duke of Argyll) who believed that beauty was designed for humans' benefit by the creator. Darwin also predicted the existence of a long-snouted moth that pollinates *Anagraecum sesquipedale*, a Madagascar orchid having an 11-inch floral tube, at the bottom of which is nectar. In 1903—41 years after Darwin's prediction and 21 years after his death—entomologists in Madagascar discovered just such a moth: *Xanthopan morgani praedicta*. The *praedicta* was added to honor Darwin's prediction.

1862 Famed Scottish scientist William Thomson (1824–1907; Figure 25), later known as Lord Kelvin, publishes his first major paper about the age of the earth, claiming that the Sun had illuminated the earth for 100 million years (see also Appendix A). In a separate paper, Thomson studied the cooling rates of the earth, thereby beginning his first attack on the "extreme quietist, or 'uniformitarian' school." Thomson, who charged that uniformitarianists had overlooked

Figure 25 Famed Scottish inventor William Thomson, later known as Lord Kelvin, used thermodynamic-based arguments to claim that Earth is 20–400 million years old. Thomson's stature lent credibility to his claims and was a significant problem for Charles Darwin's theory of evolution by natural selection, which required a much older Earth. (*Library of Congress*)

essential tenets of thermodynamics, used these calculations to estimate that the earth is 20–400 million years old. Thomson's subsequent estimates of the earth's age included 100 million (1868), 50 million (1876), 20–50 million (1881), and 24 million years (1893), all too short for evolution by natural selection to produce life's diversity. Thomson did not know that as the earth cools, new heat is being generated internally by radioactive decay.

1862 French chemist and devout Catholic Louis Pasteur (1822–1895), best known today for his contribution to pasteurization, refutes the idea of spontaneous generation.

1863 British philosopher and economist Herbert Spencer coins the phrase *survival of the fittest* to describe Charles Darwin's model of evolution by natural selection. Although neither this phrase nor the word *evolution* appeared in the 1st edition of *On the Origin of Species*, Darwin liked Spencer's phrase, noting in the 5th edition of *Origin* that "the expression often used by Mr. Herbert Spencer of the Survival of the Fittest is more accurate, and is sometimes equally convenient." Alfred Russel Wallace also liked Spencer's phrase—so much so that he crossed out "natural selection" through much of his copy of *Origin* and wrote over it "survival of the fittest." Darwin included the phrase "survival of the fittest" in the title of Chapter IV ("Natural Selection, or the Survival of the Fittest") of the 5th edition of *Origin*. Although Spencer was primarily responsible for popularizing the term *evolution*, his view of evolution was neo-Lamarckian.

1863 Louis Agassiz's *Methods of Study in Natural History* compares evolution to medieval alchemy: "The philosopher's stone is no more to be found in the organic than the inorganic world, and we shall seek as vainly to transform the lower animal types into the higher ones by any of our theories, as did the alchemists of old to change the base metals into gold."

1863 Charles Kingsley (1819–1875), a respected English zoologist and theologian who was chaplain to Queen Victoria, publishes *The Water-Babies*, a popular children's novel that was denounced by creationists as evolutionary propaganda for children. Kingsley's book portrayed nature as Mother Cary, who used natural laws to create new animals ("I sit here and make them make themselves"), thereby replacing Paley's watchmaker with the creative forces of nature that reward the good and punish the bad. This

Spencerian self-help version of Lamarckism (i.e., that struggle improves individuals and therefore improves the species) was a typical and popular response by liberal Christians to Darwinism. *The Water-Babies* mythologized the various aspects of Darwin's theory of evolution by natural selection. Earlier, Charles Darwin had suspected that Kingsley would support his theory of evolution and sent Kingsley an advance copy of *On the Origin of Species*. After reading *Origin*, Kingsley wrote to Darwin that, "[a]ll that I have seen of it awes me . . . if you [are] right, I must give up much that I have believed." *The Water-Babies*, which was reprinted for more than a century, described Richard Owen, Thomas Huxley, and Darwin as "very wise men . . . you should listen respectfully to all they say." Also in 1863, Kingsley wrote in a letter to English theologian Frederick Maurice (1805–1872) that he was "busy working out points of natural theology, by the strange light of Huxley, Darwin, and Lyell," and that "Darwin is conquering everywhere, and rushing in like a flood, by the mere force of truth and fact. The one or two who hold out [against Darwin] are forced to try all sorts of subterfuges as to fact, or else by evoking the *odium theologicum*."

1863 Henry Bates documents 11 years of work in *The Naturalist on the River Amazons*, which Charles Darwin (who wrote the preface) praised as "the best work of Natural History Travels ever published in England." Bates, who encouraged Alfred Russel Wallace to develop his theories of organic evolution, discovered that closely related species were often separated geographically by rivers, and he later realized that this was evidence of geographical speciation and evolution by natural selection. Bates described butterfly wings as "expanded membranes [on which] nature writes, as on a tablet, the story of the modification of species." In homage to his exploratory spirit, Bates's grave in London's East Finchley Cemetery is crowned by a globe. *Naturalist*, which is still in print, became a standard against which biologists have measured the ecological impact of the past century.

1863 In a paper published in *Proceedings of the Royal Geographical Society of London*, Alfred Russel Wallace describes what would become known as Wallace's Line, which separates the islands of Bali and Lombok and divides the Indo-Malayan (tigers, rhinoceri, orangutans) and Austro-Malayan (kangaroos, cuscus, and other marsupials) fauna. Although these islands are separated by a strait only 20 miles wide, each has a unique fauna—

for example, woodpeckers, starlings and thrushes dominate Bali, and cockatoos and honey-suckers dominate Lombok. Geologists later discovered that Bali was once connected to the Asiatic continental shelf, but never to Lombok.

1863 Ellen White and others establish the Seventh-Day Adventist movement. White's writings—more than 5,000 articles and 40 books—continue to be regarded by her followers as prophetic. The Adventist church currently has more than 14 million members in 200 countries. Early in the 20th century, Adventists were among the most consistent voices for young-earth creationism.

1863 Charles Lyell's *Geological Evidences of the Antiquity of Man* discusses three popular issues: the age of the human race, the existence of ice ages, and Charles Darwin's theory of evolution by natural selection. Lyell argued that humans originated at least 100,000 years ago. Lyell's book, which helped to establish prehistoric archaeology in Britain, was a bestseller; it went through three editions in less than a year. However, Darwin was disappointed in the book's "want of originality" and Lyell's "excessive caution" about evolution.

1863 In *Evidence as to Man's Place in Nature*—the first book to take a scientific approach to human evolution—Thomas Huxley (Figure 23) uses comparative anatomy to conclude that "whatever system of organs be studied, the structural differences that separate Man from the Gorilla and the Chimpanzee are not so great as those which separate the Gorilla from the lower apes . . . man is, in substance and in structure, one with the brutes" (Figure 26). Huxley defined an issue that continues to inspire anthropologists and others: "The question of questions for mankind—the problem which underlies all others, and is more deeply interesting than any other—is the ascertainment of the place which Man occupies in nature and of his relations to the universe of things." Although Huxley had chided Charles Darwin by noting that Darwin had "loaded [himself] with an unnecessary difficulty in adapting *Natura non facit saltum*—Nature does not make leaps so unreservedly," *Evidence* promoted gradualism. (Huxley later noted "Nature does make jumps now and then," an idea that was popularized in the early 1970s by Niles Eldredge and Stephen Jay Gould as *punctuated equilibrium*.) Huxley emphasized the importance of Darwin's ideas, noting that "if any process of physical causation can be discovered by which the genera and families of ordinary animals have been produced, that process of causation is amply sufficient to account for the origin of Man." For Huxley, there was only one process that could do this—"that propounded by Mr. Darwin."

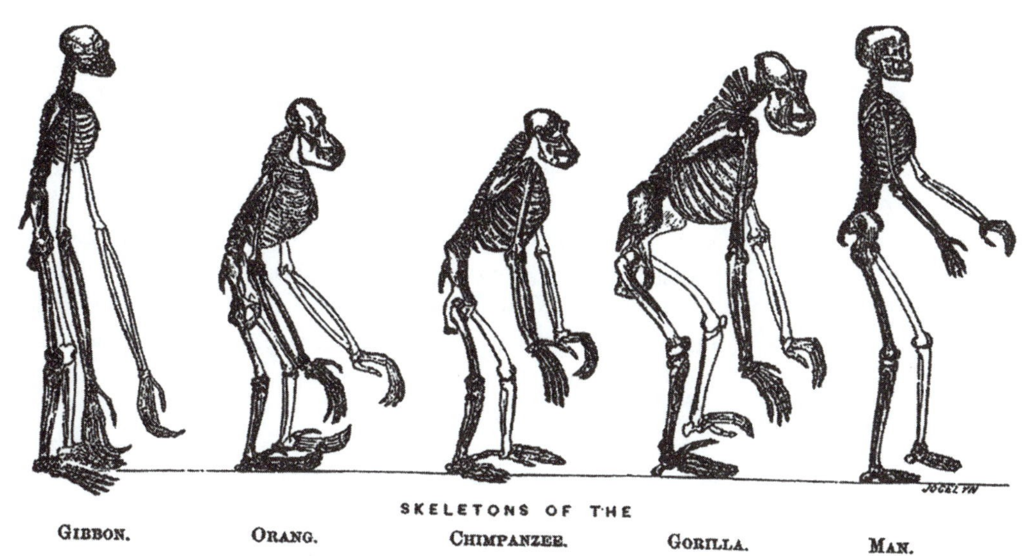

Figure 26 Thomas Huxley's 159-page *Evidence as to Man's Place in Nature*, which featured this famous frontispiece drawn by Benjamin Waterhouse Hawkins, argued for the simian ancestry of humans.

1863 The National Academy of Sciences (NAS) is established by an Act of Incorporation signed by President Abraham Lincoln (1809–1865). NAS established the National Research Council (NRC) in 1916, and later became a leading advocate for the teaching of evolution, producing supporting books such as *Teaching About Evolution and the Nature of Science* (1998) and *Science, Evolution, and Creationism* (2008).

1863 The term *Neanderthal Man* is coined by Irish anatomist William King (1809–1886). The following year, King assigned the fossil *Neanderthal 1* to *Homo neanderthalensis*.

1864 In *Religion and Chemistry*, Harvard University chemist Josiah Parsons Cooke (1827–1894) claims "the existence of an intelligent Author of nature, infinite in wisdom and absolute in power, may be proved from the phenomena of the material world with as much certainty as can be any theory of science."

1864 Carl Akeley (1864–1926) is born in Clarendon, New York. Akeley's revolutionary advances in taxidermy brought distant, unimagined animals to life for the viewing public. Today, Akeley's work is displayed in the Akeley Hall of African Mammals at the American Museum of Natural History (AMNH), an exhibit considered to be among the world's greatest museum displays.

1864

Charles Darwin is awarded the Copley Medal by the Royal Society, an organization whose original purpose was to glorify God and improve man's estate. Edward Sabine (1788–1883), President of the Royal Society, noted that "speaking generally and collectively we have not included [*On the Origin of Species*] in our award."

1864

After a "vision" from God, Ellen White claims that she was "carried back to the creation and was shown that the first week, in which God performed the work of creation in six days and rested on the seventh day, was just like every other week." This claim linked Adventism with young-earth creationism and established the Seventh-Day Adventists' opposition to evolution: "When men leave the word of God in regard to the history of creation, and seek to account for God's creative works upon natural principles, they are upon a boundless ocean of uncertainty. . . . The genealogy of our race, as given by inspiration, traces back its origins, not to a line of developing germs, mollusks, and quadrupeds, but to the great Creator." White's descriptions of her many "visions" profoundly influenced George McCready Price, who recast White's visions in scientific terms to produce *flood geology*, a precursor of creation science. Price, in turn, strongly influenced Henry Morris and John Whitcomb, Jr., whose *The Genesis Flood* started the modern creationism movement. Like young-earth creationists who followed him, Price rejected modern geology, claiming that Noah's Flood produced the geologic record.

1864

Benjamin Disraeli (1804–1881), later Britain's Prime Minister, decries the idea that humans evolved from earlier species: "Is man an ape or an angel? I, my lord, I am on the side of the angels. I repudiate with indignation and abhorrence those newfangled theories."

1864

Jemmy Button (ca. 1815–1864), the most famous of the Fuegians that Robert FitzRoy had brought to England for a year, dies of infection in his native land.

1864

In *Journey to the Center of the Earth*, French writer Jules Verne (1828–1905) describes travelers who, while descending through geologic epochs, witness the days of biblical creation, including a battle between *Ichthyosaurus* and *Plesiosaurus*. Verne's book, which converted prehistoric discoveries into high drama, brought painter John Martin's gothic images to life while exploring the dissonance between Genesis and the scientific claims about evolution and an old earth.

1865 Gregor Mendel (Figure 27) reports the findings from his studies of the garden pea at two presentations to the Natural Science Society of Brünn, and the following year publishes "Experiments in Plant Hybrids" in the Society's *Proceedings*. The 42-year-old Mendel showed that "potentially formative elements" (later called *genes* by Wilhelm Johannsen) act like particles by maintaining their integrity across generations. Furthermore, Mendel showed that the combination of genes passed to offspring is analogous to the shuffling of a deck of cards, not the blending of ingredients to make a cake (as predicted by the then-popular "blending" model of heredity). Mendel's results did not immediately generate much interest. His 1865 presentation was attended by only 40 people and little discussion followed his talk; likewise, his 1866 paper attracted little attention. Most readers interpreted his conclusions as merely confirming that hybridization eventually leads to reversion to ancestral forms. However, after publication of *On the Origin of Species*, the issue of heredity became of greater interest, and in 1900, Mendel's work was "rediscovered" by Hugo de Vries and others. Eventually, Mendel's results were fused with Darwin's theory during the Modern Synthesis of the first half of the 20th century, which laid the foundation for the contemporary understanding of biological evolution. Mendel, who became abbot at his monastery, later abandoned his breeding experiments and devoted considerable effort to combating attempts by the government to tax monasteries.

1865 Robert FitzRoy, depressed and in poor health, awakes early on the morning of April 30, kisses his daughter goodbye, and commits suicide by slashing his throat.

1865 New York Governor Reuben Fenton (1819–1885) signs into law the bill establishing Cornell University. The university's first president was Andrew Dickson White, who—unlike many other university administrators of his day—refused to apply religious tests to staff and students. White also provocatively pledged that his university would be "an asylum for *Science*—where truth shall be sought for truth's sake, not stretched or cut exactly to fit Revealed Religion." In 1869, *The New York Daily Tribune* included an article titled "The Battle-Fields of Science" in which White argued that "[i]n all modern history, interference with Science in the supposed interest of religion—no matter how conscientious such interference may have been—has resulted in the direst evils to Religion and Science, and *invariably*; and, on the other hand, all

Figure 27 Gregor Mendel's studies of inheritance in the garden pea established the foundation of modern genetics. In 1910—the centennial of Mendel's death—this statue of Mendel was unveiled overlooking the courtyard of the monastery where Mendel worked. The statue's dedication was attended by many famed biologists, including William Bateson and Erich von Tschermak. In 1965, the Mendelanium Museum opened in the monastery.

untrammeled scientific investigation, no matter how dangerous to religion some of its stages may have seemed for the time to be, has invariably resulted in the highest good both of religion and science."

1865 French naturalist Pierre Trémaux (1818–1895) complains "there are as many [definitions of species] as there are naturalists."

1865 Sir John Lubbock (1834–1913), banker, entomologist, and Charles and Emma Darwin's neighbor at Downe, publishes the influential archaeology book *Pre-historic Times, as Illustrated by*

Ancient Remains, and the Matters and Customs of Modern Savages. Lubbock, who became the first president of the Royal Anthropological Institute and also served as president of both the BAAS and the Linnean Society, was elected to Parliament in 1869. Lubbock, who was Charles Darwin's only student, defended Darwinian evolution at the British Association's 1860 meeting, at which Samuel Wilberforce and Thomas Huxley (Figure 23) purportedly clashed in their famous "debate."

1866 Famous American evangelist Henry Ward Beecher (1813–1887) writes to Herbert Spencer that "[t]he peculiar condition of American society has made your writings far more fruitful and quickening here than in Europe." Beecher was right; Social Darwinism became popular throughout the United States in the first half of the 1900s.

1866 Famed biologist Louis Agassiz concludes comments to NAS by claiming "[s]o here is the end of the Darwinian theory."

1866 In the 4th edition of *On the Origin of Species*, Charles Darwin expands his discussion of beauty (Chapter VI) by invoking an argument that he made in *On the Various Contrivances by Which British and Foreign Orchids are Fertilised by Insects, and on the Good Effects of Intercrossing* (1862)—namely, that features we consider beautiful are adaptations for survival and reproduction that are not directed at humans. In the 4th edition of *Origin*, Darwin also suggested that some species may undergo rapid bursts of change followed by longer periods in which change is slow or absent. This argument was a response to Scottish paleontologist Hugh Falconer's (1808–1865) report in 1863 that some species of elephants changed very little over long periods of time (even through the glacial periods). Although Falconer concluded his paper by noting that Darwin "laid the foundation of a great edifice," Falconer added that Darwin "need not be surprised if, in the progress of the erection, the superstructure is altered by his successors."

1866 The Ku Klux Klan is formed in Pulaski, Tennessee, by Confederate hero Nathan Bedford Forrest (1821–1877) to protect widows and orphans of the Confederate dead. White Southerners, frustrated by federal Reconstruction policies, later used the Klan to lash out against occupying troops and blacks that were benefiting from Reconstruction. Like virtually all racist organizations, the Klan based its beliefs on special creation; its members were horrified by the thought that people of color, or simply of different ancestry, might be their relatives, however distant. A separate origin for whites and blacks—as could be gleaned from creationism but not evolution—enabled the Klan and similar groups to claim that blacks and other ethnic minorities were not entitled to the same rights as whites. The Klan provided the antievolution movement with powerful support.

1866 *The Scientific Aspects of the Supernatural* is the first of several publications by Alfred Russel Wallace about spirits, miracles, and the occult. Wallace presented what he believed was evidence for the supernatural, proclaimed the existence of an "unseen universe of spirit," and argued that natural selection had been used by higher powers "for a special end." Thomas Huxley (Figure 23) dismissed Wallace's claims as "disembodied gossip."

1866 In *General Morphology of the Organism*, Prussian biologist Ernst Haeckel (1834–1919) introduces his biogenic law—"ontogeny recapitulates phylogeny." Haeckel's law claimed that the embryological development of an individual (ontogeny) is a fast-forward version of the species' evolutionary history (phylogeny). Haeckel believed that recapitulation was a mechanism for evolutionary change, not just a passive record of such changes. Haeckel supported his idea by comparing drawings of embryos of various organisms to demonstrate, for example, that mammals pass through a "fish" stage, an "amphibian" stage, and so on. Critics such as embryologist Ludwig Rütimeyer (1825–1895) charged that Haeckel's drawings overemphasized the similarity among particular stages of the embryos and that some details were conveniently omitted; Haeckel responded that he was trying to capture the overall morphology of each embryo in a general way that may or may not be present in any particular specimen. Creationists have often cited Haeckel's drawings as evidence for a supposed conspiracy to undermine religion and as proof against evolution. However, within 30 years after Haeckel had produced his drawings, science had already cast aside his biogenic law. (Textbooks did not do the same as quickly.) There are identifiable developmental similarities among organisms, and these similarities do reflect some aspects of common ancestry. However, a full-fledged recapitulation of evolutionary history does not occur, as Karl von Baer had noted in the 1830s: "The embryo of a vertebrate is at the beginning already a vertebrate." Haeckel described his 1866 meeting with Charles Darwin "as if some exalted sage of Hellenic antiquity . . . stood in the flesh before me," but Haeckel's extreme political and social views offended Darwin. In 1866, Haeckel coined the term *ecology* (*oekologie*, meaning "the comprehensive science of the relationship of the organism to the environment"), thereby giving expression to what biologists had vaguely described as the "economy of nature." Haeckel also coined the terms *phylum* and *phylogeny* and proposed the kingdom *Protista*.

1866 Charles Darwin responds to a question about religion by noting that "it has always appeared to me more satisfactory to look at the immense amount of pain and suffering in this world, as the inevitable result of the natural sequence of events, i.e., general laws, rather than from the direct intervention of God. . . . I am not responsible if [the meeting of science and theology] should still be far off." Darwin later suggested that religious belief is little more than an inherited instinct, similar to a monkey's fear of snakes.

1867 Amateur naturalist George Campbell (1823–1900), the 8th Duke of Argyll, argues that natural selection cannot explain the origin of species because it selects variations that originate by some other "law." Campbell claimed that variations are not the product of natural processes, but are designed by God, and therefore evolution is providentially guided.

1867 Famed fossil-hunter Edward Drinker Cope, repeating Robert Chambers's arguments from *Vestiges*, claims that the path of evolution was "conceived by the Creator according to a plan of His own, according to His pleasure."

1867 In the 10th edition of *Principles of Geology*, Charles Lyell refutes creationism and endorses Charles Darwin's ideas. Darwin knew that it was no small matter for Lyell to abandon the proposal of divine guidance that he had supported for so long in the book upon which he had established his reputation, noting that "considering [Lyell's] age, his former views and position in society, I think that his action has been heroic."

1867 In the second volume of *The Principles of Biology*, Herbert Spencer tells readers that "with a higher moral nature will come a restriction on the multiplication of the inferior." Spencer's optimistic views of social evolution were popular; by the time he died in 1903, American sales of authorized editions of his work exceeded 350,000 copies.

1868 Alfred Russel Wallace, writing in *Quarterly Review*, depicts humanity as the purpose of a divinely guided evolutionary process. Darwin is appalled by Wallace's claims, but politely responds that he hopes Wallace has "not murdered too completely your own & my child. . . . I differ greatly from you, and I feel very sorry for it." Wallace responded by telling Darwin that his experiences at séances had convinced him of the reality of spirit forces and that he now believed that God or spirits must have had an important role in human evolution.

1868 At the BAAS meeting in Norwich, Thomas Henry Huxley (Figure 23) delivers his famous "On a Piece of Chalk" lecture, noting that the Norwich area is built on vast deposits of chalk formed by the bodies of organisms in seawater that once covered Britain. Huxley concluded "[it] must have taken some time for animalicules of a hundredth of an inch in diameter to heap up such a mass as that. . . . The earth, from the time of the chalk to the present day, has been the theater of a series of changes as vast in their amount as they were slow in their progress."

1868 When Alfred Lord Tennyson meets Charles Darwin on the Isle of Wight, Tennyson comments that "What I want is an assurance of immortality."

1868 Charles Darwin publishes his two-volume *Variation of Animals and Plants under Domestication*, in which he struggles with the origin of genetic variation. Darwin incorrectly proposed that heredity is controlled by circulating "gemmules" in animals' bodies because in "animals which may be bisected or chopped into pieces, and of which every fragment will reproduce the whole, the power of regrowth must be diffused throughout the whole body." Darwin used an analogy involving rocks that had fallen off a cliff to discredit the claim that variation is divinely created: "Can it be reasonably maintained that the Creator intentionally ordered . . . that certain [rocks] should assume certain shapes so that the builder might erect his edifice?"

1868 Benjamin Waterhouse Hawkins is commissioned to create a "Paleozoic Museum" in New York City. Hawkins set up a studio on what is now the site of AMNH, but the studio was destroyed by vandals hired by infamous politician William "Boss" Tweed (1823–1878), and the planned museum never opened. Hawkins, who illustrated the fish and reptiles for Charles Darwin's *Zoology* *of the Voyage of the HMS Beagle*, later told *The New York Times* that Darwin's theory of evolution "could not be believed by any comparative anatomist" and that the "unity of design in nature showed the hand of God himself."

1868 Joseph Leidy (1823–1891), Benjamin Waterhouse Hawkins, and Edward Drinker Cope display a mounted skeleton of an 80-million-year-old, duck-billed *Hadrosaurus* at The Academy of Natural Sciences in Philadelphia (Figure 28). This was the only dinosaur skeleton on display anywhere in the world for the next 15 years. The 26-foot-long, 14-foot-tall skeleton boosted annual visitation to the Academy from 30,000 visitors to more than 100,000 visitors only two years later (despite the fact that the museum was open to the public only two afternoons per week and closed throughout August). With this display, Hawkins brought dinosaur-mania to America and forever changed the way dinosaurs were displayed to the public.

1868 William Thomson (Figure 25) uses conductive-cooling calculations to conclude that the earth is less than 100 million years old. Although Thomson's immense stature in the scientific community gave his claim much credibility, some scientists—notably, Thomas Huxley (Figure 23)—rejected Thomson's claims, suspecting that the earth is much older (see also Appendix A).

1868 Othniel Marsh, the first professor of vertebrate paleontology in the United States, discovers bones of a small horse (*Equus parvulus*, now *Protohippus*) while excavating in Nebraska. Marsh later teamed with Thomas Huxley (Figure 23) to describe the evolution of modern horses from a four-toed ancestor and, in the process, provided evidence for how species evolve. Several months after Marsh and Huxley predicted that a more ancient, five-toed animal probably existed, fossils of *Eohippus* ("the dawn horse") were discovered. In 1871, Marsh found the first American pterosaur fossils, and in 1878 he named *Diplodocus*, casts of which were distributed throughout the world by Andrew Carnegie (including the one currently on display at NHM). Marsh's discoveries made him one of the most famous paleontologists of his era and convinced many people of the validity of Darwin's theory. In 1866, Marsh's uncle, George Peabody (1795–1869), had provided $150,000 for the establishment of the Peabody Museum of Natural History at Yale College; many of the fossils in that and many other museums can be traced to Marsh and fellow paleontologist Edward Cope, a

Figure 28 Benjamin Waterhouse Hawkins stands beneath the skeleton of *Hadrosaurus foulkii* in 1868 in the Academy of Natural Sciences in Philadelphia. This articulated mount, which was the only dinosaur skeleton displayed anywhere in the world for the next 15 years, set the standard for exhibiting fossil bones. (*The Academy of Natural Sciences, Ewell Sale Stewart Library and the Albert M. Greenfield Digital Imaging Center for Collections*)

neo-Lamarckian with whom Marsh was a fierce rival. Their highly publicized rivalry was dubbed "The Bone Wars." Marsh named more genera of dinosaurs (19) than anyone else.

1868 Railroad workers in Les Ezyies, France, unearth several human skeletons that are later called Cro-Magnon (French for "big hole"). The skeletons were estimated to be 30,000 years old (see also Appendix C).

1868 After examining *Archaeopteryx* (Figure 24) and the small, bipedal theropod *Compsognathus* (both of which were found in the same limestone sediments in Germany), Thomas Huxley (Figure 23) proposes in "On the Animals which are Most Nearly Intermediate between Birds and Reptiles" that birds are descendants of dinosaurs. Huxley's suggestion was not taken seriously for nearly a century.

1868 In the speculative *Natürliche Schöpfungsgeschichte* (titled *The History of Creation* when translated into English by biologist Edwin Lankester in 1876), Ernst Haeckel—a student of Rudolf Virchow—discusses the common evolutionary history of apes and humans. Haeckel's book contains the first evolutionary trees for groups of taxa. In his tree for humans, Haeckel proposed 12 separate species and claimed that human evolution consisted of 22 phases, the 21st being a "missing link" represented by *Pithecanthropus alalus* ("speechless ape-man"), a yet-undiscovered ancestor that Haeckel predicted would be found in the fossil record. Haeckel also identified two characteristics that would mark this "missing link": bipedality and changes in the larynx that in later forms would allow for complex language. Haeckel's work later inspired Eugène Dubois (1858–1940) to name his discovery of hominin fossils in Java *Pithecanthropus erectus* ("upright ape-man") in homage to Haeckel (see also Appendix C). Haeckel later described Virchow—who claimed that Neanderthal Man was a hapless, relatively recent human afflicted with rickets and arthritis—as someone who "has regarded it as his special duty as a scientist to oppose Darwinian theory" and who claims that "it is quite certain that man did not descend from the apes . . . not caring in the least that now almost all experts of good judgment hold the opposite conviction." Haeckel's ideas extended far beyond science (e.g., "politics is applied biology"), claiming, among other things, that helping the poor and disabled was a waste of time and a recipe for added suffering later because these "good-for-nothings" were

unfit and genetically predestined to fail. Haeckel advocated Aryan superiority and a mythical German history, which he claimed had been destroyed by Christianity. For Haeckel, the "lowest" humans were various "species" of Africans and New Guineans and at the summit were Europeans, which he designated *Homo mediterraneus* (within *H. mediterraneus*, Haeckel's fellow Germans were at the pinnacle). Haeckel believed that knowledge of evolution and human origins would help people understand nature without clerical interference. Eventually, Haeckel claimed, most other races would "succumb in the struggle for existence to the superiority of the Mediterranean races." *History of Creation* ultimately went through 12 editions.

1869 A 10-foot-high, 2,990-pound statue of a human—commissioned for $2,600 by cigar manufacturer George Hull—is "discovered" behind the barn of Hull's cousin William "Stub" Newell in Cardiff, New York (Figure 29). The gypsum statue became famous as the Cardiff Giant, which some biblical literalists claimed as confirmation of the claim in Genesis 6:4 that "there

Figure 29 Cardiff Giant was a purported "petrified man" whose "discovery" in 1869 generated worldwide publicity. Showman P.T. Barnum also lured throngs of customers to his display of a replica of the Cardiff Giant, which he claimed was the "real" giant (Barnum's display was the origin of the quotation, "There's a sucker born every minute"). Today, the original Cardiff Giant is displayed as "America's Greatest Hoax" at Farmer's Museum in Cooperstown, New York, not far from the Baseball Hall of Fame. (*Library of Congress*)

were giants in the Earth in those days." At one point, more than 3,500 people per day were paying to see the Cardiff Giant.

1869 In his review of Charles Lyell's *Principles*, Alfred Russel Wallace praises Lyell's endorsement of evolution, but again argues that some human traits are not explicable by natural selection and instead result from the intervention of "an Overruling Intelligence" that has "guided the action of those laws [of organic development] in definite directions and for specific ends." The following year in *The Limits of Natural Selection as Applied to Man*, Wallace claimed that the universe "is not merely dependent on, but actually is, the WILL of higher intelligences or of one Supreme Intelligence." Lyell endorsed Wallace's claims about human evolution, not Darwin's.

1869 Graduate student Johann Friedrich Miescher (1844–1895) examines bandages from hospital patients and determines that cell nuclei contain a substance largely composed of nitrogen, phosphorous, and chromatin. He named the substance *nuclein*, which was later shown to be DNA.

1869 Francis Galton (1822–1911), a cousin to Charles Darwin, publishes *Hereditary Genius*, an early attempt to distinguish "nature" from "nurture." Galton later argued that people "of really good breed" should be encouraged to reproduce and that inferior people should be discouraged from reproducing. Galton—who believed that heredity linked physical health to mental health—claimed that this would improve humanity the way selective breeding improves livestock. Galton's sophisticated analyses of inheritance also contributed significantly to the then-developing field of statistics.

1869 Thanks largely to lobbying by naturalist Albert Bickmore (1839–1914), AMNH is founded in Central Park in New York City (Figure 30). The cornerstone for the first building was laid five years later by President Ulysses S. Grant (1822–1885), and the first building opened in 1877. Throughout its history, AMNH hosted numerous scientists who made important discoveries about evolution. Today, AMNH hosts four million visitors annually.

1869 Thomas Huxley (Figure 23)—his era's most vocal publicist for evolution—coins the term *agnostic* when, upon joining London's Metaphysical Society, he is asked to categorize his metaphysical

Figure 30 New York City's American Museum of Natural History has hosted numerous scientists who have made important discoveries about evolution.

philosophy. Huxley, who believed that science would "prove destructive to the forms of supernaturalism which enter into the constitution of existing religions," defined an agnostic as someone who suspends judgment about the existence of God. He explained that the perspective he followed is "not a creed but a method" of understanding based on observation.

1869 *Nature*, the first magazine-format journal devoted to science, is founded by Thomas Huxley and his colleagues, and publishes its inaugural issue. *Nature* became a leading scientific journal that continues to publish many of the major discoveries about evolution.

1869 Alfred Russel Wallace's *The Malay Archipelago: The Land of the Orang-utan and the Bird of Paradise* documents two different types of Malaysian animals: animals in the western islands resemble those of India, while animals in the east resemble those of Australia. Wallace dedicated the book to Charles Darwin, "not only as a token of personal esteem and friendship but also to express my deep admiration for his genius and his works." *The Malay Archipelago*—one of the most popular journals of scientific exploration of the 19th century—remained in print for more than 50 years; it was praised by scientists and others, including novelist Joseph Conrad, who called it his "favorite bedside companion" (he used it as a source for several of his novels, especially *Lord Jim*). In 1869, Wallace further embraced spiritualism, séances, and miracles and recanted parts of his endorsement of evolution by natural selection. Darwin dismissed séances as "rubbish."

1869 In the 5th edition of *On the Origin of Species*, Charles Darwin includes Herbert Spencer's phrase survival of the fittest in the title of Chapter IV. Darwin expanded his discussions of the origin of variation and the age of the earth, noting that evolution might have been more rapid during the early history of earth. Darwin also changed the Introduction's final sentence from "I am convinced that Natural Selection has been the main but not exclusive means of modification" to "has been the most important, but not the exclusive, means of modification."

1869 When challenged by antievolutionist Francis Orpen Morris (1810–1893) to cite support for Darwin's theory of evolution, Thomas Huxley (Figure 23) responds that answers to Morris's objections could be found in "principles and practices of inductive logic" supplemented by "five or six years' serious and practical study of physical and biological science" and "a return to the 'Origin of Species' . . . with the same earnest desire to grasp their real meaning as, I doubt not, animates you when you read your Bible."

1870s Famed evangelist Dwight Moody (1837–1899) begins his attacks on evolution, arguing that it contradicts biblical truth.

1871 William Thomson (Figure 25), who entered the University of Glasgow at the age of 10, was knighted in 1866 and named Lord Kelvin in 1892 by Queen Victoria, uses thermodynamic calculations of tidal friction and global cooling to claim that geologic history "must be limited within some such period of past time as one hundred million years" (see also Appendix A). Thomson again attacked Huttonian uniformitarianism by claiming that an unending sun would violate the laws of thermodynamics. Thomson—one of the greatest scientists of the 19th century—opposed evolution, and his enormous scientific stature was a major hurdle for the acceptance of Charles Darwin's idea.

1871 Charles Darwin publishes the two-volume *The Descent of Man, and Selection in Relation to Sex* to address what he called "the highest and most interesting problem for the naturalist." Part I of the

book described evidence for human evolution (e.g., homologies and vestigial structures shared with apes; see also Appendix C). Part II described sexual selection, which was used in Part III to explain human diversity and the origin of unique human traits. Darwin knew that his conclusions would "be highly distasteful to many," but that "there can hardly be a doubt that we are descended from barbarians." In *The Descent of Man*, which Ernst Haeckel positively described as "anti-Genesis," Darwin made a brave prediction: "In each great region of the world the living mammals are closely related to the extinct species of the same region. It is, therefore, probable that Africa was formerly inhabited by extinct apes closely allied to the gorilla and chimpanzee; and as these two species are now man's nearest allies, it is somewhat more probable that our early progenitors lived on the African continent than elsewhere." Darwin was vindicated in 1924 when Raymond Dart announced the discovery of *Australopithecus africanus* in Africa. *The Descent of Man* also marked the first time Darwin used the word *evolution*. The most cited naturalist in *The Descent of Man* was Alfred Russel Wallace, with whom Darwin often disagreed. Darwin closed *The Descent of Man* with a summary of his ideas: "We are not here concerned with hopes or fears, only with the truth as far as our reason permits us to discover it; and I have given the evidence to the best of my ability. We must, however, acknowledge, as it seems to me, that man with all his noble qualities . . . still bears in his bodily frame the indelible stamp of his lowly origin." More than 4,500 copies of *The Descent of Man* were sold in its first two months on sale.

1871

In *On the Genesis of Species*, English biologist St. George Mivart (1827–1900) challenges several claims made by Charles Darwin in *On the Origin of Species*. Mivart's book maintained that the earth is too young for evolution to have produced such diversity of species; that intermediate steps toward new anatomical structures served no purpose; that natural selection alone could not account for evolutionary history; and that Darwin's theory endangered the morals of British society. Mivart advocated "specific genesis" in which an innate force generated new species as "harmonic self-consistent wholes." Darwin addressed many of Mivart's criticisms in the 6th edition of *Origin* (1872), but failed to appease Mivart. Thomas Huxley (Figure 23) claimed that Mivart simply could not be "both a true son of the Church and a loyal soldier of science." A Catholic convert, Mivart was later denied the sacraments because of his views concerning evolution.

1871 William Thomson (Figure 25) suggests "the germs of life might have been brought to the earth by some meteorite." Kelvin's idea was resurrected under the name *panspermia* a century later.

1871 In *Christianity and Positivism*, Scottish philosopher James McCosh (1811–1894) claims that theism is consistent with evolution when evolution is "properly limited and explained."

1872 In *The Scriptural Doctrine of Creation*, theologian Thomas Rawson Birks (1810–1883) claims that Charles Darwin violated the laws of induction by neglecting "direct evidence" in the Bible about the origin of species.

1872 In his three-volume *Systematic Theology*, Princeton theologian Charles Hodge (1797–1872) argues that evolution is an unproved "hypothesis" inconsistent with both the Bible and natural theology.

1872 Geologist Louis Agassiz visits the Galápagos Islands (Figure 17) and uses his observations to try to undermine Darwin's proposals. Agassiz described happy, playful porpoises and portrayed the Galápagos as a harmonious place far from the Malthusian world of Darwin. When Agassiz returned from his voyage, he wrote in *Nature* that "Darwin's hypothesis of gradual variation of species, and the natural selection for preservation of those whose variations were favorable to them in the struggle for life, seems to me to have few facts to sustain it, and very many to oppose it." Agassiz, who died a year after visiting the Galápagos, was the last great American biologist to proclaim the special creation of species.

1872 The final (i.e., 6th) edition of *On the Origin of Species*, which is one-third longer than the 1st edition, contains an additional chapter dealing with objections that have been raised about Darwin's theory ("Miscellaneous Objections to the Theory of Natural Selection"). Upset that his conclusions had "lately been much misrepresented," Darwin reminded readers that the 1st edition of *Origin* had noted that "I am convinced that natural selection has been the main but not the exclusive means of modification." He then lamented, "[t]his has been of no avail. Great is the power of steady misrepresentation." In the 6th edition of *Origin*, Darwin also dropped "On" from the title and altered one of his most famous metaphors, changing "an entangled bank" to "a tangled bank." Darwin decided that the 6th edition of *Origin* would be the last edition; as he wrote to American philosopher Chauncey

Wright (1830–1875), "I have resolved to waste no more time in reading reviews of my works or on evolution." During Darwin's lifetime, John Murray Publishers sold more than 25,000 copies of the English version of *Origin* and more than 56,000 copies by 1899. From 1859 to 1881, Darwin's books earned him an average of £465 per year; in 1871, these royalties constituted only about 6% of Darwin's total income. In all, Darwin published 17 books in 21 volumes consisting of more than 9,000 pages of text.

1872 An attempt to elect Charles Darwin a Corresponding Member of the Zoology Section of the French Institute fails when he receives only 15 of 48 votes. Darwin was rejected because members claimed that *On the Origin of Species* and *The Descent of Man* are "not science, but a mass of assertions and absolutely gratuitous hypotheses, often evidently fallacious." However, six years later, Darwin was elected a Corresponding Member in the Botanical Section of the French Institute, prompting the modest Darwin to write to friend and botanist Asa Gray, "[i]t is rather a joke that I should be elected in the Botanical Section, as the extent of my knowledge is little more than that a daisy is a Compositous plant and a pea is a Leguminous one."

1873 Botanist George Allman (1800–1884) uses the phrase *intelligent design* in an address to the 1873 annual meeting of the BAAS.

1873 Karl Marx (1818–1883)—who described humans as "apes of a cold God"—inscribes a copy of the 2nd edition of *Das Kapital* to Charles Darwin "on the part of his sincere admirer." Darwin did not read the book.

1874 In *History of the Conflict Between Religion and Science*, American (English-born) chemist and physician John William Draper (1811–1882) claims that "[w]e are now in the midst of a controversy respecting the mode of government of the world, whether it be by incessant divine intervention, or by the operation of primordial and unchangeable law" and "the history of Science is not a mere record of isolated discoveries; it is a narrative of the conflict of two contending powers, the expansive force of the human intellect on one side, and the compression arising from traditionary faith and human interests on the other." From the title of Draper's book arose the historical conflict thesis, whose proponents contend that religion is, and always has been, in opposition to scientific advances.

1874 In *Outlines of Cosmic Philosophy*, American philosopher John Fiske (1842–1901) claims "there has never been any conflict between religion and science." Fiske also argued that it is unreasonable to use evolution to affirm God's goodness because the course of evolution has been "attended by the misery of untold millions of sentient creatures for whose existence their Creator is responsible." In 1874, Charles Darwin wrote in a letter to Fiske "I never in my life read so lucid an exposition (and therefore thinker) as you are."

1874 In his presidential address to the BAAS, Irish physicist John Tyndall (1820–1893) defends Epicurus's materialism and rejects "the notion . . . that nature has been in any way determined by intelligent design."

1874 Joseph LeConte (1823–1901), the University of California's first professor of geology and a student of Louis Agassiz, publishes *Religion and Science*, followed by *Evolution: its History, its Evidence, and its Relation to Religious Thought* (1888). These books discussed how LeConte—who was initially a self-proclaimed "reluctant evolutionist"—came to enthusiastically endorse evolution as being "entirely consistent with a rational theism." LeConte questioned apologists' claims that evolution confirms God's wisdom and goodness, noting that "[w]hat we call evil is not a unique phenomenon confined to man, and the result of an accident, but must be a great fact pervading all nature, and a part of its very constitution." During this period, LeConte was one of the most famous reconcilers of evolution and religion, and his definition of evolution—continuous progressive change resulting from resident forces according to certain laws—was a standard for many years.

1874 Charles Hodge—who had earlier noted, "the Bible is to the theologian what nature is to the man of science"—claims in *What Is Darwinism?* that "the first objection to [Darwin's] theory is its *prima facie* incredibility," "the grand and fatal objection to Darwinism is [the] exclusion of design in the origin of species or the production of living organisms," and "the conclusion of the whole matter is that denial of design in nature is virtually the denial of God." Hodge famously concluded that "Darwinism is atheism" because it denies divine design in nature. Hodge—an old-earth creationist—was one of the first prominent theologians to publicly declare that Darwinism threatens religion.

1874 Louis Agassiz's final essay rejecting evolution ("Evolution and the Permanence of Type") is published in *The Atlantic Monthly*.

1875 Sir Charles Lyell dies in London and is buried in Westminster Abbey (Figure 7) beneath a marker of fossil marble. Lyell's death prompted Charles Darwin to acknowledge, "I never forget that almost everything which I have done in science I owe to the study of his great works." Thomas Huxley (Figure 23) also paid homage to Lyell by noting "I cannot but believe that Lyell was for others, as for me, the chief agent in smoothing the road for Darwin." The 12th and final edition of Lyell's *Principles of Geology* was published posthumously.

1876 Alfred Russel Wallace's *Geographical Distribution of Animals* documents extinct fauna and the geographical movements of species over time. This two-volume book established modern zoogeography and was the definitive text about zoogeography for the next 60 years.

1876 Asa Gray and theologian George Frederick Wright (1838–1921) publish *Darwiniana*, a collection of essays that delineate a middle position—a compromise between evolution and religion—in the Darwinian debate. Later, Wright claimed that philosophy poses a greater challenge for Christianity than does evolution, noting, "Hume is more dangerous than Darwin."

1876 Johns Hopkins University opens the first modern Department of Biology. After famed biologist and agnostic Thomas Huxley (Figure 23) delivered the university's inaugural address, a local preacher noted, "it was bad enough to invite Huxley. It were better to have asked God to be present. It would have been absurd to ask them both." Huxley was criticized for not beginning his lecture with a prayer.

1876 Charles Darwin begins writing a short autobiography for his family. Darwin did not intend the autobiography for publication, but a censored version of the manuscript was published in 1887 by Francis Darwin (1848–1925) as *The Autobiography of Charles Darwin*. Years later, Darwin's daughter Nora published the entire original manuscript. To the end of his life, Darwin never published anything about religion, noting that what he believed was "of no consequence to anyone but myself." However in this autobiography, Darwin—acknowledging that "life is nearly over with me"—commented that "[c]onsidering how fiercely I have been attacked by the orthodoxy, it seems ludicrous that I once intended to be a clergyman" and that "the mystery of the beginnings of all things is insoluble by us; and I for one must be content to remain an agnostic." As Darwin noted in a letter to a friend, "I was very unwilling to give up my belief . . . disbelief crept over me at a very slow rate, but was at last complete."

1877 Charles Darwin publishes the 2nd edition of *Fertilisation of Orchids*. That same year, Darwin received an honorary doctorate from the University of Cambridge.

1877 John Murray, who had published Charles Lyell's and Charles Darwin's greatest books, writes and publishes *Skepticism in Geology* to attack Lyell's uniformitarianism and uphold Flood geology. Murray published this book under the pseudonym "Verifier."

1877 Famed American paleontologist Edward Cope publishes *The Origin of the Fittest*, which argues that natural selection might cull poorly adapted organisms but that Lamarckian processes produce variation. In 1895, after losing more than $250,000 in a mining fraud, the deeply religious Cope sold most of his dinosaur collection to Henry Fairfield Osborn at AMNH for $32,149. Cope remained famous; a brand of cigars was later named after him.

1877 Rudolf Virchow claims that Ernst Haeckel's efforts to have evolution taught in Germany's lower schools will promote socialism and communism.

1877 Biologist Othniel Marsh famously proclaims that "[t]o doubt evolution is to doubt science, and science is only another name for truth." The following year, Marsh noted that Darwin's *On the Origin of Species* had in two decades "changed the whole course of scientific thought. . . . Darwin spoke the magic word—'Natural Selection,' and a new epoch in science began."

1877 Sociologist Richard Dugdale (1841–1883) publishes *The Jukes: A Study in Crime, Pauperism, Disease and Heredity*. While working as a volunteer jail inspector, Dugdale discovered several members of the same extended family incarcerated together, and by analyzing five generations of the Jukes's (a pseudonym) pedigree, proposed that criminal behavior is an inherited trait. Such conclusions, combined with Francis Galton's proposals on eugenics, propelled efforts in the United States to limit reproduction of the supposedly less fit, culminating in the establishment of the Eugenics Record Office (ERO) and, in several states, the passage of laws requiring sterilization of those deemed unfit.

1878 In *Adamites and Pre-Adamites*, geologist and theistic evolutionist Alexander Winchell (1824–1891)—the first president of the American Geological Society—claims that humans existed before Adam. Winchell argued that because Negroes were too biologically inferior to have descended from Adam (who was white), humans had therefore existed before Adam. Winchell was later fired from Vanderbilt University for "holding questionable views on Genesis." Winchell helped popularize geology in America.

1878 Thirty-nine skeletons of *Iguanodon* are found in a mine more than 1,000 feet below Bernissart, Belgium. The miners believed that they had discovered a vein of gold, for the *Iguanodon* suffered from pyrite disease, a malady which causes fossilized bones—when moistened—to crack and dissolve into pyrite (fool's gold). Today, 11 of these skeletons greet visitors to the Royal Belgium Institute of Natural Science.

1879 German biologist Walther Flemming (1843–1905) describes and names *mitosis* in stained and living cells of salamanders. The term *chromosome* was not coined until 1888.

1879 Astronomer George Darwin (1845–1912), Charles and Emma Darwin's son, uses measurements of tidal friction to estimate that the earth is at least 56 million years old (see also Appendix A).

1879 In *Evolution, Old and New; or, The Theories of Buffon, Dr. Erasmus Darwin, and Lamarck, as Compared with that of Mr. Charles Darwin*, iconoclastic writer Samuel Butler (1835–1902) notes that "I attacked the foundations of morality in *Erewhon*, and nobody cared two straws. I tore open the wounds of my Redeemer as he hung upon the Cross in *The Fair Haven*, and people rather liked it. But when I attacked Mr. Darwin they were up in arms in a moment."

1879 The eccentric Englishman Herbert Spencer publishes the first volume of *Principles of Ethics*, which popularizes Lamarckian evolution by insisting on a scientific basis for right and wrong actions. Spencer's followers developed his ideas into Social Darwinism, which argued that the disparity of rich and poor is not an injustice, but simply biology: society should eliminate unfit individuals because nature does the same thing. Industrial pioneer Andrew Carnegie referred to Spencer as "Master Teacher," and Alfred Russel Wallace named one of his children Herbert Spencer Wallace. Although Darwin rejected claims that evolution is progressive and goal-oriented, Spencer believed that Social Darwinism would eliminate the "excrement of society" and eventually produce a final, and perfect, society. Social Darwinism had virtually nothing in common with Darwin's theory of evolution; Darwin never advocated extending natural processes into human social structures.

1880s Conservative evangelical institutions begin opening Bible colleges, the most famous of which will become Chicago's Moody Bible Institute, Minneapolis's Northwestern Bible School, and the Bible Institute of Los Angeles. During subsequent decades, these schools produced thousands of conservative preachers, most of whom denounced evolution.

1880 Charles Darwin publishes *The Power of Movement in Plants*. That same year, Darwin told his cousin Hensleigh Wedgwood (1803–1891) "there have been too many attempts to reconcile Genesis and science."

1880 Alfred Russel Wallace publishes *Island Life: The Geographical Distribution of Animals*, which he dedicates to Joseph Hooker. *Island Life* became a standard authority in biogeography.

1880 Othniel Marsh's lavish book *Odontornithes: A Monograph on the Extinct Toothed Birds of North America* is hailed by Richard Owen as "the best contribution to Natural History since Cuvier," and by Charles Darwin as "the best support to the theory of evolution which has appeared within the last 20 years." However, in 1892, Alabama Congressman Hilary Herbert (1834–1919) branded the book "atheistic rubbish" and cut the paleontology budget of the United States Geological Survey.

1880 American philosopher William James (1842–1910) begins his famous *Great Men and Their Environment* lecture by noting "[a] remarkable parallel . . . between the facts of social evolution on the one hand, and of zoological evolution as expounded by Mr. Darwin on the other."

1880 *Science*, "a weekly record of scientific progress," publishes its first issue with a $10,000 endowment from Thomas Edison (1847–1931). In the second issue, Thomas Huxley's (Figure 23) article titled "The Coming of Age of the *Origin of Species*" noted that "evolution is no longer a speculation, but a statement of historical fact." *Science* is one of the most prestigious journals in science.

1881 Charles Darwin publishes *Formation of Vegetable Mould through the Action of Worms*. More than 7,000 copies of *Formation of Vegetable Mould* were sold in its first two years. Like his other books, this book affirmed Darwin's ideas about evolution. Darwin's argument resembled those of Charles Lyell by documenting that small events acting over long periods can produce major changes (e.g., the burrowing of worms is seemingly trivial, but can be dramatic if it persists for a long time). Similarly, slight modifications of a species, if continued long enough, can change the species dramatically. During his lifetime, Charles Darwin's books brought in ~£10,000.

1881 The Natural History Museum (NHM) opens in London. Today, the NHM receives over 3.5 million visitors annually.

1881 Charles Darwin declines Edward Aveling's (Karl Marx's son-in-law) request for permission to dedicate the secularist book *The Student's Darwin* to Darwin, adding, "direct arguments against Christianity & theism produce hardly any effect on the public. . . . It has, therefore, been always my object to avoid writing on religion, & I have confined myself to science." That same year, Darwin was invited to dinner with the Prince of Wales (1841–1910; the future King Edward VII) and published a letter in *The New York Times* supporting vivisection.

1881 Despite objections by Joseph Hooker and others about Alfred Russel Wallace's endorsement of the supernatural ("Wallace has lost caste considerably . . . by his adhesion to Spiritualism"), Charles Darwin and Thomas Huxley convince Queen Victoria (1819–1901) to give Wallace an annual civil pension of £200. That same year, Darwin made out a will leaving £1,000 to his friends Hooker and Huxley "as a slight memorial of my lifelong affection and respect."

1881 Jack Collier (1850–1934) is commissioned by the Linnean Society to paint a three-quarter length oil-based portrait of Charles

Darwin. Darwin liked the painting, which today hangs in the headquarters of the Linnean Society in London (the image was also copied for the National Portrait Gallery, the Royal Society, and Down House). Collier married Marian Huxley (1859–1887; a daughter of Thomas Huxley) in 1879 and Ethel Gladys Huxley (1866–1941; another daughter of Thomas Huxley) in 1889. Collier also painted portraits of his father-in-law Huxley and Rudyard Kipling (both of which are displayed in London's National Portrait Gallery). A portrait of Wallace, who never acquired Darwin's celebrity status, was not painted for the Linnean Society until 1998.

1882 Francis Galton calls for a national set of labs to measure humans. When people showed little interest in the idea, Galton began to collect data himself. Galton later offered a £500 prize for the best analysis of hereditary diseases, but got no takers.

1882 Clarence Dutton's (1841–1912) *Tertiary History of the Grand Cañon District,* and the accompanying atlas, is the first stratigraphic history of the Grand Canyon region. The canyon's iconic status as a site where deep time is on display has induced young-earth creationists to try to co-opt it for their own purposes, publishing research monographs and tourist guides, and even organizing river expeditions, all arguing that the canyon was formed by Noah's Flood.

1882 Canadian-born George Romanes's (1848–1894) *Animal Intelligence* is the first attempt to describe animal behavior in the context of evolution. Romanes was later known as "the father of comparative psychology."

1882 George Frederick Wright's *Studies in Science and Religion* argues that scientists accept evolution because it agrees with observed facts, just as does Christianity. The writings of Wright and Asa Gray helped calm many people's fears about atheism in evolution.

1882 In his final paper in *Nature* (published 13 days before his death), Charles Darwin discusses the dispersal of freshwater mollusks. That paper was coauthored by shoemaker and amateur naturalist William Drawbridge Crick (1857–1903), the grandfather of Francis Crick (who, with James Watson, would in 1953 describe the double-helical structure of DNA). Thomas Huxley (Figure 23) contributed Darwin's obituary to *Nature* the same month.

1882 On April 19, after telling his wife Emma to "remember what a good wife you have been" and that he "was not the least afraid to die," Charles Darwin dies in his upstairs bedroom at Down House. He had wanted to be buried in St. Mary's churchyard in Downe (Darwin referred to Downe as "the happiest [place] on earth") beside the bodies of his dead children. However, six days after his death—at a standing-room-only funeral service attended by Britain's leading politicians, clergy, and scientists (Joseph Hooker, Alfred Russel Wallace, Thomas Huxley, John Lubbock, and Abraham Lincoln biographer James Russell Lowell were among the pallbearers)—Darwin was buried in a white-oak coffin in the northeast corner of the nave of London's Westminster Abbey beside astronomer John Herschel and near his friend Sir Charles Lyell. As Darwin's body was lowered into the abbey's floor, the choir sang "[h]is body is buried in peace, but his name liveth evermore." Darwin's tombstone bears the simple inscription: "Charles Robert Darwin, Born 12 February 1809, Died 19 April 1882." The *Pall Mall Gazette* described Darwin as "the greatest Englishman since Newton," the Vienna *Allegemeine* declared "[o]ur century is Darwin's century, we can suffer no greater loss," and the London *Times* noted, "[t]he Abbey needed [Darwin] more than [Darwin] needed the Abbey." Darwin was the only naturalist to be buried in Westminster Abbey (Figure 7).

1883 Francis Galton coins the term *eugenics* ("best born") for the study of using selective breeding to improve society in his book *Inquiries into Human Faculty and Its Development*. Galton, who believed that "character, including the aptitude for work, is heritable like every other faculty," summarized his attempts to make a science out of human breeding this way: "What Nature does blindly, slowly, and ruthlessly, man may do providently, quickly, and kindly."

1883 After becoming skeptical of alleged reconciliations of science and religion, American theologian George Fisher (1827–1909) writes in *Princeton Review* that "the progress of natural science has taught in repeated instances, and taught impressively, that the traditional views taken of the Scriptures contain error."

1883 The death of Karl Marx prompts German philosopher Friedrich Engels (1820–1895) to note "[j]ust as Darwin discovered the law of evolution in organic nature, so Marx discovered the law of evolution in human history." Only nine people attended Marx's funeral.

1884 Arnold Guyot (1807–1884), a Swiss-born geographer at Princeton University, publishes *Creation; or, The Biblical Cosmogony in the Light of Modern Science*, which interprets the six days of creation as separate epochs of time, including three instances of special creation: of matter, of life, and of humans. Guyot, a friend of Louis Agassiz, integrated his understanding of the natural world with his devout Christian beliefs to advocate day-age creationism. Guyot claimed that the Bible "is in perfect contrast with the fanciful, allegorical, intricate cosmogonies of all heathen religions" because it "leaves room for all scientific discoveries."

1884 Future Archbishop of Canterbury Frederick Temple's *The Relations between Religion and Science* openly supports evolution as the mechanism for creation.

1884 American zoologist Alpheus Hyatt (1838–1902) coins the phrase *neo-Lamarckism* to describe ideas formulated by himself and Edward Drinker Cope. Today, this phrase is used to explain the inheritance of acquired characteristics in a modern context.

1884 German botanist Eduard Adolf Strasburger (1844–1912), zoologist Oskar Wilhelm August Hertwig (1849–1922), and zoologist August Friedrick Weismann (1834–1914) independently conclude that the nucleus plays a central role in passing genetic traits from parent to offspring.

1884 Austrian monk Gregor Mendel (Figure 27), whose work would later revolutionize genetics, dies unknown to science. His funeral, however, was notable: Czech composer Leoš Janáček (1854–1928) played the organ. Not long before his death, Mendel told a colleague, "Má Doba Prijde" ("My time will come"). Today, that sentence greets visitors to the Mendelianum Museum at the monastery in Brno where Mendel worked.

1884 

In *The Man Versus the State*, Herbert Spencer—who sold more than one million copies of his books during his lifetime—denounces government-sponsored aid programs as contrary to natural laws (i.e., contrary to survival of the fittest). "'They have no work,' you say. Say rather that they either refuse work or quickly turn themselves out of it. They are simply good-for-nothings, who in one way or other live on the good-for-somethings." Spencer then quoted the Bible—"if any would not work, neither should he eat"—to argue that it was natural that "a creature not energetic enough to maintain itself must die" (II Thessalonians 3:10). During the 1870s and 1880s, Spencer's reputation rivaled that of Charles Darwin.

1884 James Woodrow (1827–1907), a professor of Natural Science at the Columbia Theological Seminary in Georgia, claims in *Southern Presbyterian Review* that Adam was a product of natural evolutionary forces, but with a soul imparted by God. Woodrow was fired from the seminary in 1896. Four years later, the Presbyterian General Assembly voted 139 to 31 to uphold Woodrow's firing, declaring that Adam had been formed from dirt "without any natural animal parentage of any kind." Throughout his ordeal, Woodrow affirmed his belief in the "absolute inerrancy" of the Bible, but emphasized his claim that "the Bible does not teach science; and to take its language in a scientific sense is grossly to pervert its meaning." Woodrow's views of evolution were not accepted by the Presbyterian General Assembly until 1969.

1885 David Starr Jordan (1851–1931) becomes president of Indiana University. While there, Jordan wrote *Darwinism* (1888) and *Footnotes to Evolution* (1898) and taught what he claimed was the world's first course about evolution. In 1891, Jordan became the first president of Leland Stanford Junior University (later known as Stanford University) and while in that position, coauthored with entomologist Vernon Kellogg (1867–1937) *Evolution and Animal Life* (1907). In 1925, Jordan chaired the Tennessee Evolution Case Defense Fund Committee for the American Civil Liberties Union (ACLU), gathering more than $11,000; the fund paid John Scopes's legal costs and contributed to Scopes's graduate education at the University of Chicago. Jordan later argued that "[t]he line of separation between religion and science is that all things in religion are assumed true because some people at some time said they were true. That is not knowledge and not science."

1885 In *The Idea of God as Affected by Modern Knowledge*, John Fiske acknowledges that "Nature is full of cruelty and maladaptation," but claims that "god is in the deepest sense a moral Being."

1885 Famed preacher Henry Ward Beecher's *Evolution and Religion* fuses religion and evolution into a new form of spiritual evolution. Beecher, who taught that Genesis "is a poem, not a treatise on cosmogony," claimed that it was inefficient for God to design each species separately, so He designed laws that generated everything.

1885 Thomas Huxley (Figure 23) unveils Sir Joseph Edgar Boehm's (1834–1890) 2.5-ton marble statue of Charles Darwin at NHM. The ceremony was attended by the Prince of Wales Edward VII as well as by Admirals J. Sullivan and A. Mellersh, both of whom had sailed with Darwin aboard the *Beagle*. That same year, Huxley—who Charles Darwin considered "the best talker I have known"—deplored the "refined depravity among the upper classes" that were studying spiritualism instead of science. In 1900, a marble statue of Huxley joined the statue of Darwin at the museum.

1886 Hugo de Vries's (1848–1935) studies of evening primrose (*Oenothera lamarckiana*) lead him to the two main concepts with which he is associated: pangenesis as a model of heredity and mutation as a mechanism of evolution. Using Charles Darwin's theory of pangenesis (i.e., circulating "gemmules" within an organism determine the characters exhibited by that individual) as a starting point, de Vries proposed that organisms' traits are determined by cellular particles he called *pangenes* (the origin of the term *gene*, which was introduced in 1909 by Wilhelm Johannsen), with each specific pangene affecting a particular trait. Each cell in an organism contained the same pangenes; however, because pangenes exist in two forms—active (expressed) and latent (not expressed)—cells differed because they had unique combinations of active and latent pangenes. De Vries developed his theory of pangenesis in *Intracellulare Pangenesis* (1889). De Vries's mutational model of evolution—namely, that evolution is saltational and driven by mutations rather than selection—became popular during the early part of the 20th century, before the basis of heredity was determined. De Vries's model united August Weismann's idea of the germplasm and the model of discrete hereditary units proposed by Charles Darwin, Herbert Spencer, and—unbeknownst to him at the time—Gregor Mendel.

1886 Traveling-shoe-salesman-turned-evangelist Dwight Moody founds the Chicago Evangelization Society (later renamed the Moody Bible Institute). The Moody Bible Institute interpreted the Bible literally, claiming that "the Bible, in its original documents, is free from error in what it says about geography, history and science as well as in what it says about God." Moody, who avoided public controversy, was his era's greatest evangelist, and many preachers active in the evolution-creationism controversy were trained at the Moody Bible Institute.

1886 While touring England, American evangelist Henry Ward Beecher proclaims that "I regard evolution as being the discovery of the Divine method in creation. . . . Evolution, so far from being in antagonism with true religion, will develop it with more power than any other presentation of science that ever has occurred in this world. The day will come when men will render thanks for that which now they deprecate." At the time, Beecher was the highest paid preacher in the United States; in addition to his $20,000 annual salary (he had summers off), he also delivered about 50 public lectures each year for $1,000 each.

1887 George Thomas Bettany (1850–1892) publishes *Life of Charles Darwin*. Aside from obituaries and a published lecture delivered by biologist Louis Miall (1843–1921) to the Leeds Philosophical and Literary Society, this was the earliest biography of Darwin. Darwin's life later became the basis of a large industry.

1887 Two near-complete skeletons of Neanderthal Man are found in a cave in Spy, Belgium, along with tools and bones of extinct mammals. This discovery, coupled with subsequent discoveries of Neanderthal skeletons in La Chapelle-aux-Saints, France, in 1908, convinced scientists that Neanderthal Man was an earlier type of human that lived in Europe from 150,000 to 35,000 years ago (see also Appendix C).

1887 Edward Drinker Cope's *Theology of Evolution* argues that consciousness is not a product of evolution, but instead comes from the divine mind of the universe and governs evolution by directing animals to new goals.

1887 Francis Darwin (Charles' and Emma's son) edits *The Autobiography of Charles Darwin* and Thomas Huxley's *On the Responses of the Origin of Species*. Francis also produced *The Life and Letters of Charles Darwin* (1887) and *More Letters of Charles Darwin* (1905).

1887 Ignoring Charles Darwin's prediction that Africa represented the cradle of human evolution, Dutch anatomist Eugène Dubois (1858–1940) accepts Ernst Haeckel's claim that Asia was where humanity arose and embarks on a 43-day voyage to Sumatra to search for Haeckel's "ape-like man," *Pithecanthropus*. Although hominin fossils had been found and studied before, Dubois was the first anthropologist to purposefully search for them (see also Appendix C).

1888 Heinrich Wilhelm Gottfried Waldeyer-Hartz (1836–1921) names the thread-like structures visible during mitosis *chromosomes*.

1888 Writing in *The Presbyterian Review*, Princeton theologian Benjamin Warfield (1851–1921)—a famed advocate of biblical inerrancy—commends Charles Darwin's "unusual sweetness" while noting that "on the quiet stage of this amiable life [Darwin] played out before our eyes the tragedy of the death of religion out of a human soul. . . . No more painful spectacle can be found in all biographical literature." Warfield also claimed that as a result of Darwin's idea, "God became an increasingly unnecessary and therefore an increasingly incredible hypothecation." Warfield believed that there is no "general statement in the Bible or any part of the account of creation, either as given in Genesis 1 and 2 or elsewhere alluded to, that need be opposed to evolution."

1889 In *Darwinism*—Alfred Russel Wallace's most-cited book—Wallace explains and develops natural selection.

1889 In *Essays upon Heredity and Kindred Biological Problems*, August Weismann develops his idea (suggested in an 1885 lecture) that sexual reproduction increases variability and therefore increases the effectiveness of natural selection. This is the first plausible hypothesis for the evolution of sexual reproduction.

1890 Francis Galton announces that he is interested only in exceptional people, adding that "an average man is morally and intellectually an uninteresting being [who is] of no direct help towards evolution, which appears to our dim vision to be the goal of all living existence."

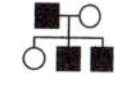

1890 In "The Deadlock in Darwinism," novelist Samuel Butler (1835–1902) claims that Darwinian evolution produces no progress, but only a "nightmare of waste and death."

1890 Seventh-Day Adventist Ellen White's *Patriarchs and Prophets* (part of her *Conflict of the Ages* series) describes White's visions of creation, the Fall, and the biblical Flood. Although many geologists believed that Noah's Flood was a regional event in the Middle East, White rejected these conclusions as being geologically motivated "compromises" and insisted that the Flood was worldwide and had created all of the geologic strata. As White noted: "[t]he entire surface of the earth was changed at the Flood. . . . Everywhere were strewn

the dead bodies of men and beasts. . . . At this time immense forests were buried. These have since been changed to coal, forming the extensive coal beds that now exist and yielding large quantities of oil." Much of the modern young-earth creationist movement can be traced to White's writings about Noah's Flood.

1890 Scientist and priest John Zahm (1851–1921) writes in *Evolution and Dogma* that "I am not unaware of the fact that Evolution has had suspicion directed against it, and odium cast upon it, because of materialistic implications and its long anti-Christian associations. . . . But this does not prove that Evolution is ill-founded or that it is destitute of all elements of truth, because it explains countless facts and phenomena which are explicable on no other theory." Although Zahm believed that evolution was compatible with church doctrine, the church put Zahm's book on its *Index of Forbidden Books*. Zahm avoided a public condemnation when he withdrew his book from publication and recanted his ideas. Zahm never again commented publicly about evolution.

1890 Episcopal priest Thomas MacQuery, influenced by Joseph LeConte's efforts to bridge science and religion, publishes *The Evolution of Man and Christianity*, another attempt to reconcile evolution and religion. The book was praised by *Popular Science Monthly*, but outraged the Episcopal Church and led to his being tried for heresy. The following year, MacQuery was removed from his post.

1890 The bitter feud between Othniel Marsh and Edward Cope is reported in *The New York Herald*, further stimulating the public's growing interest in dinosaurs. In their celebrated careers, Cope produced 1,400 publications and named 1,200 species (Figure 22); Marsh published 270 papers and named 500 species. By the end of the 1890s, Marsh and Cope had discovered 136 species of dinosaurs, although some of these species were later disallowed. During their feud, Marsh and Cope each spent over $200,000 of their own money to finance their expeditions.

1891 After moving his search for *Pithecanthropus* to Java, Dutch anatomist Eugène Dubois finds a molar of an apelike hominid along the Solo River outside the tiny village of Trinil. Within a month, Dubois and his colleagues also found a 5-inches by 7.5-inches piece of a fossil skullcap, and in 1892, a near-complete left femur. Dubois determined that this organism had been bipedal with a cranial capacity of about 1,000 cm^3, indicating a brain size intermediate between chimpanzees and humans. Dubois's discovery was the first of many fossils proposed to be a "missing link" with humans. In a subsequent report summarizing his findings (published as a pamphlet in Batavia [now Jakarta] instead of a recognized journal), Dubois concluded that the fossil "is the first known transitional form linking Man more closely with his next of kin among the mammals" and suggested that it supported the contention that "the first step on the road to becoming human taken by our ancestors was acquiring upright posture." Dubois named his discovery *Pithecanthropus erectus*, combining Ernst Haeckel's proposed genus with a specific epithet that emphasized the distinctive bipedal nature of the organism (see also Appendix C). (When Haeckel learned of Dubois's discovery, he wired Dubois "congratulations to the discoverer of *Pithecanthropus* from its inventor.") Dubois referred to the fossil informally as "P.e.", but the rest of the world knew it as "Java Man." Dubois was confident of the importance of his discovery: "P.e. is the transitional form which, according to the theory of evolution, must have existed between Man and the anthropoids; he is Man's ancestor." Famed pathologist Rudolf Virchow was not convinced; he claimed that Dubois had found the skullcap of a gibbon and dismissed Dubois's claim that he had found "Man's ancestor" as "fantasy . . . beyond all experience." It would be several decades before anyone found more evidence of ancient hominids in Asia. Ultimately, *Pithecanthropus erectus* was reclassified as an example of *Homo erectus*, a widespread species that existed in Asia, Africa, and Europe 1–2 million years ago. Dubois's famous fossilized skullcap—now known as *Trinil 2*—is the type specimen for *H. erectus*.

1892 In *The Race Problem in the South*, Joseph LeConte—a respected geologist and president of AAAS—argues that Negroes have traits that make them appropriate for enslavement, and that for more-specialized "redskin" races, "extermination is unavoidable."

1892 George McCready Price's (1870–1963) first publication attacking evolution claims that evolution and socialism are destroying morality. In subsequent books such as *The Geological-Ages Hoax*, Price tried single-handedly to overturn two centuries of geologic research; he rejected faunal succession and the stratigraphic column and instead advocated the biblical Flood as the cause of geologic formations. Price urged a return to "primitive" Christianity and insisted that

deviations from James Ussher's chronology (Figure 8) are "the devil's counterfeit" and "theories of Satanic origin."

1892 Theologian John Davis (1854–1926) advocates what will become known as *framework creationism*, in which the seven days are a literary (i.e., a topical or symbolic framework rather than chronologic) device for presenting creation.

1892 The University of Pisa awards an honorary degree to Galileo, 250 years after his death.

1892 German physician and academic August Weismann publishes *Das Keimplasma: Eine Theorie der Vererbung* (*The Germ-Plasm, A Theory of Heredity*). Weismann proposed a machinery of inheritance, including "idiants" (chromosomes), "determinants" (genes), and "biophores" (nucleic acid bases). Weismann is also known for his strong opposition to Lamarck's theory of heredity and for a series of experiments in which he removed tails from 1,592 mice (over 22 generations) to demonstrate Lamarck's faulty premise about the inheritance of acquired characteristics. (All of the offspring of the de-tailed mice were born with tails.) However, Lamarck never claimed that the effects of mutilations or injuries are inherited. Instead, he proposed that acquired traits resulted from an individual's internal needs.

1892 Joseph Hooker oversees preparation of *Index Kewensis*, a two-volume description of 400,000 genera of plants. *Index,* paid for by the estate of Charles Darwin, registered the scientific names of plants, an effort Darwin supported because of the difficulties he encountered in applying Linnaeus's original taxonomy.

1892 The official name of the Galápagos Islands is changed from "Archipelago de Ecuador" to "Archipiélago de Colón" (Figure 17).

1893 In his encyclical *Providentissimus Deus* ("On the Study of Holy Scriptures"), Pope Leo XIII (1810–1903) defends the inerrancy of the Bible while claiming "there can never, indeed, be any real discrepancy between the theologian and the physicist, as long as each confines himself within his own lines." Pope Leo's encyclical cited St. Augustine (taking the same position used in 1615 by Galileo in his letter to Grand Duchess Christina) to argue that the Bible does not aim to teach science.

1893 In his *Evolution and Ethics* lecture, Thomas Huxley (Figure 23)—after noting nature's "moral indifference"—rejects Social Darwinism by remarking that ethical progress in society is based not on imitating or avoiding natural processes such as natural selection, but in combating them.

1893 In *The Secret Doctrine*, Elena Gan (1831–1891, better known as Helena Blavatsky) proposes a form of evolution that, like many such theories of the 19th century, is progressive and leads to a higher form of spiritual consciousness. Gan disliked special creation, miracles, and Darwinian evolution, but used the ideas of Charles Darwin, Charles Lyell, and Ernst Haeckel to support her claims. Gan claimed that humans were the source of all mammalian life, and that modern humans originated on the lost continent of Lemuria in the Indian Ocean.

1893 Clarence King (1842–1901), the first director of the United States Geological Survey, uses tidal stability to estimate that the earth is 24 million years old (see also Appendix A). William Thomson (Lord Kelvin; Figure 25) later reviewed King's calculations and concluded "I am not led to differ much from [King's] estimate."

1894 George Romanes's review of Alfred Russel Wallace's *Darwinism* notes that "we encounter the Wallace of spiritualism and astrology, the Wallace of vaccination . . . the Wallace of incapacity and absurdity." The following year, Romanes coined the term *neo-Darwinism* to describe the idea that evolution occurs through natural selection (i.e., shorn of Lamarckism). Today, *neo-Darwinism* is used by many biologists to refer to the product of the Modern Synthesis.

1894 With *Social Evolution*, philosopher Benjamin Kidd (1858–1916) creates a firestorm by claiming that natural selection favors the preservation of "nonrational" institutions such as religion.

1894 Adam Sedgwick (1854–1913), Lecturer in Animal Morphology at Trinity College of Cambridge, publishes one of the first written criticisms of Ernst Haeckel's biogenic law in *Quarterly Journal of Microscopical Science*, noting "a blind man could distinguish" the embryos of different classes of animals.

1896 After his electrifying "Cross of Gold" speech at the Democratic National Convention in Chicago, 36-year-old William Jennings

Bryan becomes the Democratic nominee for president. In the audience was delegate Clarence Darrow, who would confront Bryan at the Scopes Trial in 1925. At the time, "Boy Bryan" was the youngest person ever nominated for the presidency. During the campaign, Bryan—who was outspent 20-to-1 by his Republican opponent William McKinley (1843–1901)—traveled almost 20,000 miles and was the first presidential candidate to take his message directly to voters, often from the back of railroad cars (this was a new tactic since presidential candidates had traditionally stayed home and let others speak on their behalf). McKinley defeated Bryan by an electoral vote of 271 to 176 and by a popular vote of 51% to 47%. Many other Democratic candidates for national office lost during that election, including Darrow, who ran for Congress. Bryan later failed in two more attempts to become president.

1896 Antoine Becquerel (1852–1908), a professor of applied physics at the Muséum d'Histoire Naturelle in Paris, discovers strange radiations emitted by uranium salts. This observation ultimately led to the discovery of radioactivity and radiometric dating, which helped establish that the earth is 4.55 billion years old—enough time for evolution by natural selection, as proposed by Charles Darwin, to produce life's diversity.

1896 Former British Prime Minister William Gladstone (1809–1898) publishes *The Impregnable Rock of Holy Scripture*, which rebuts Thomas Huxley's attacks on the scientific truth of the Bible and promotes "literary" creationism—namely, that the six days of creation are a literary (rather than a chronologic) device for presenting creation.

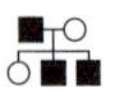

1896 Connecticut becomes the first state to enact marriage laws with eugenic criteria. These laws banned "epileptic, imbecile, or feebleminded" people from marrying.

1896 Emma Darwin dies peacefully at age 88 in Cambridge and is buried near her children at St. Mary the Virgin Church in Downe. She rests in the same tomb as Charles's brother Erasmus.

1897 Former baseball star William "Billy" Sunday (1862–1935) holds his first revival in Garner, Iowa. Before his death 38 years later, the theatrical "all-American" Sunday became the richest and most famous evangelist in America and a harsh critic of the teaching of evolution.

1897 Artist Charles Knight (1874–1953), despite being legally blind for most of his life, becomes the first artist to work with paleontologists to produce paintings of dinosaurs for the AMNH (Figure 30). Knight's work defined dinosaurs for the world; virtually every dinosaur book published in the first six decades of the 20th century included Knight's illustrations. Knight's work also influenced numerous filmmakers, including Willis O'Brien (1886–1962), who first realized prehistoric animals on film (e.g., *The Lost World*, *King Kong*). *Godzilla* was also based on Knight's paintings. Evolutionary biologist Stephen Jay Gould (1941–2002), who featured one of Knight's paintings on the cover of his 1991 book *Bully for Brontosaurus*, memorialized Knight this way: "I cannot think of a stronger influence ever wielded by a single man in such a broad domain of paleontology. . . .[Knight] painted all the canonical figures of dinosaurs that fire our fear and imagination to this day."

1897 American clergyman Lyman Abbott (1835–1922) promotes theistic evolution by claiming that evolution is "God's way of doing things."

1898 Chemist Marie Sklodowska Curie (1867–1934) coins the term *radioactivity*, and her husband Pierre discovers that radium releases heat as it decays. This discovery led to the development of radiometric dating, which showed that the earth is much older than many people believed.

1898 Edwin Lankester (1846–1929), a protégé of Thomas Huxley, replaces Richard Owen as director of NHM and shifts the emphasis of the museum from classification to evolution. This change in emphasis was manifest in Lankester's moving the life-size statue of Charles Darwin to the museum's most prominent position, atop a stairwell overlooking the Great Hall. Lankester, who was openly hostile to Alfred Russel Wallace's defense of the supernatural, wrote *Degeneration, A Chapter in Darwinism* (1880) and *Extinct Animals* (1905), the latter of which inspired the creatures in Sir Arthur Conan Doyle's *The Lost World*.

1899 Irish geologist John Joly (1857–1933) speculates that when the oceans originally formed, they contained freshwater and that the ocean's present salinity is due to salts leaching from rocks. Joly assumed that the rate of leaching had been constant and that therefore the age of the earth's first ocean equaled the mass of sodium in the oceans divided by the annual rate of sodium input;

this calculation—using a method originally proposed by English astronomer Edmond Halley—produced an age of 89 million years. A year later, Joly raised his estimate to 100 million years (see also Appendix A). (Ironically, Halley had believed that his method would confirm "the evidence of the Sacred Writ [that] Mankind has dwelt about 6,000 years.") Joly later argued that radioactive decay was an unreliable clock because decay rates were not constant throughout geologic history. Many young-earth creationists continue to make this claim.

1899

Social Darwinist and steel magnate Andrew Carnegie (1835–1919) funds an expedition to the western United States in which Jacob Wortman (1856–1926) unearths an 84-foot-long dinosaur skeleton of what was later named *Diplodocus carnegii* at the Morrison Formation in Wyoming (now known as Dinosaur National Monument; Figure 31). The Carnegie Institute funded other expeditions to search for dinosaur remains, and today the museum houses some of the world's most important dinosaur fossils, including one of the first *T. rex* remains ever found. A cast of the 150-million-year-old *Diplodocus* skeleton is a centerpiece of the Carnegie Museum of Natural History (CMNH) in Pittsburgh, Pennsylvania.

Figure 31 Steel magnate Andrew Carnegie funded several geologic expeditions, including the one that uncovered this 150-million-year-old *Diplodocus carnegii* in the Morrison Formation (Late Jurassic) in Albany County, Wyoming. Carnegie gave casts of the giant skeleton to European royalty, and one of these casts dominates The Natural History Museum. (*Carnegie Museum of Natural History*)

1899 In an essay in *Science* titled "The Age of the Earth as an Abode Fitted for Life," Lord Kelvin (Figure 25) continues his attack on uniformitarianism while claiming that the earth formed "more than 20 and less than 40 million years ago; and probably much more nearer 20 than 40." By the time Kelvin died in 1907, his calculations of the earth's age were undermined by the discovery that radioactivity could supply heat within the earth and sun (see also Appendix A).

1899 The death of famed preacher Dwight Moody ends a prominent era of Christian evangelism. Moody profoundly influenced many of the fundamentalists who later propelled the evolution-creationism controversy, including Amzi C. Dixon, Billy Sunday, William Bell Riley, and Cyrus Scofield, who officiated at Moody's funeral.

1900 John Scopes (1900–1970) is born to Mildred (Mary) and Thomas Scopes in Paducah, Kentucky. In 1925—one year after he graduated from the University of Kentucky—Scopes's trial in Dayton, Tennessee, became the most famous event in the history of the evolution-creationism controversy.

1900 American embryologist Thomas Hunt Morgan (1866–1945) visits geneticist Hugo de Vries in Holland and decides to start studying heredity. Morgan's work later revolutionized genetics.

1900 English mathematician Karl Pearson (1858–1936) pioneers the application of statistics to biology with the development of the chi-squared test for comparing theoretical results of an experiment to actual data. Pearson, along with W.F.R. Weldon (1860–1906), helped found the biometric school of evolution by seeking mathematical descriptions of continuous traits instead of experimental verifications of discontinuous traits. Both Pearson and Weldon dismissed the importance of Mendelian genetics in heredity and evolution.

1900 Hugo de Vries claims that he independently discovered the principles of heredity described earlier by Mendel (Figure 27). At the time, this elevated de Vries to the level of "rediscoverer" of Mendelian inheritance (along with Carl Correns [1864–1933] and Erich Tschermak von Seysenegg [1874–1962], who each in 1900 also claimed independent discovery of Mendel's principles). However, comparison of de Vries's presentation of his data before and after he is known to have read Mendel's work suggested that he may have modified his data and his interpretations in response to having Mendel's conclusions in hand (e.g., a 2:1 phenotypic ratio reported by de Vries in 1897 became a 3:1 ratio, as found by Mendel, in 1900). In later publications, de Vries credited Mendel's work, although he never completely harmonized his concepts of heredity with those of Mendel. The work of de Vries, Tschermak, and Correns (who included Mendel's name in his paper's title) rescued Mendel's ideas from near oblivion, began the widespread acceptance of Mendelian inheritance, and ultimately helped bring about the full fruition of Charles Darwin's theory.

1900 William Jennings Bryan—nicknamed "The Great Commoner" because of his faith in the goodness of common people—again loses the presidential election to William McKinley, and in 1908 he loses to William Taft (1857–1930). Nevertheless, Bryan's belief in the majority remained strong; as he often asked, "[b]y what logic can the minority demand privileges that are denied to the majority?" Bryan did not separate politics and religion, and his policies were often described as "applied Christianity." At his Sunday school classes in Miami, which attracted thousands of worshipers, Bryan often attacked evolution, claiming "more of those who take evolution die spiritually than die physically from smallpox."

1902 William Bateson (1861–1926) subsidizes the publication of *Mendel's Principles of Heredity: A Defense*, which he sends to leading geneticists. This was the first genetics textbook.

1902 The Bible League of North America is founded to combat "current destructive teachings" and restore faith by "common sense and rational, or truly scientific, method." Its *Bible Student and Teacher*, whose contributors included future antievolutionists such as William Bell Riley, harshly criticized evolution.

1902 Barnum "Mr. Bones" Brown (1873–1963), the greatest fossil collector of the 20th century, discovers in Hell Creek, Montana, the first partial skeleton of *Tyrannosaurus rex*. Brown described humanity's first encounter with *T. rex* this way in a letter to Henry Fairfield Osborn: "Quarry No. 1 contains [several bones] of a large carnivorous dinosaur not described by Marsh. . . . I have never seen any thing like it from the Cretaceous." Today, that specimen—which has a 4-foot skull and foot-wide neck vertebrae—glares menacingly at visitors to the CMNH in Pittsburgh, Pennsylvania (Brown's largest *T. rex* is displayed in the AMNH in New York; Figure 30). Many of Brown's subsequent expeditions were funded by Harry Sinclair (1876–1956), who incorporated an image of *Diplodocus* into his company's logo to emphasize corporate might. In return, Brown—the "Father of the Dinosaurs"—wrote the dinosaur booklets that were distributed at Sinclair gas stations in the 1930s and 1940s. Brown, who was also the scientific consultant for the Disney movie *Fantasia* (1940), was one of the greatest scientific celebrities of his day.

1902 Rudyard Kipling (1865–1936) writes the children's book *Just-So Stories*, which includes fanciful stories about how animals get their traits (e.g., "How the Giraffe Got Its Long Neck" and "How the Camel Got Its Hump"). For example, Kipling told readers that an elephant has a long nose because a crocodile grabbed it and pulled. The concept of a "just-so story" has been used disparagingly by both evolutionary biologists and creationists to describe the lack of rigor sometimes evident in evolutionary explanations. For example, in an article published in 1979, Stephen Jay Gould and Richard Lewontin used Voltaire's fictional Dr. Pangloss—who in *Candide* noted "things cannot be other than as they are, for since everything was made for a purpose, it follows that everything is made for the best purpose. Observe: our noses were made to carry spectacles, so we have spectacles"—to ridicule the claim that all biological traits can be explained via an adaptationist "just-so" story.

1902 American geneticist Walter Sutton (1877–1916) lays the foundation for the chromosomal theory of inheritance when he deduces that hereditary factors in grasshoppers are located on chromosomes and that chromosomes separate for reproduction.

1903 In *Principia Ethica*, English philosopher G. E. Moore (1873–1958) attacks Social Darwinist Herbert Spencer for committing what Moore calls the *naturalistic fallacy*—namely, basing ethical conclusions (e.g., what is "good") on facts of nature (e.g., natural selection).

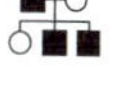

1903 John Joly and George Darwin confirm that radioactive decay generates heat.

1903 Alexander Patterson's *The Other Side of Evolution* argues that evolution, which "violently" opposes the Bible, "originated in heathenism and ends in atheism."

1904 Johns Hopkins zoologist Maynard Metcalf (1868–1940) publishes *An Outline of the Theory of Organic Evolution.* Metcalf later served as the only expert witness to testify at the Scopes Trial and one of eight scientists to provide written statements attempting to reconcile evolutionary theory with the Bible. At Scopes's trial, Metcalf's explanations of evolution—including his conclusion that "no normal man can hold any doubts of the facts of evolution"—were described by reporters as being so clear that "they caused the courtroom audience to lean forward."

1904 William Bateson, after being awarded the Royal Society's Darwin Award, notes "the campaign against the teaching of evolution is a terrible example of the way in which truth can be perverted by the ignorant." Bateson is remembered for coining the terms *genetics, homozygous, heterozygous,* F_1, F_2, and *allelomorph*, which was shortened to *allele*.

1904 William Jennings Bryan first publishes his views of evolution in *The Prince of Peace*. Bryan warns scientists "you shall not connect me with your family tree [of monkeys] without more evidence."

1905 Ernest Rutherford (1874–1937), who would become known as the father of nuclear physics, proposes radiometric dating.

1905 Alfred Russel Wallace's autobiography, *My Life*, includes a chapter describing his friendship with Charles Darwin. Wallace—a towering figure in the transition from old-fashioned natural history to modern biology—was gracious and kind and showed no envy, regret, or resentment.

1905 Russian biologist Constantine Mereschkowsky (1855–1921) proposes that plastids are the evolutionary descendants of endosymbiotic bacteria-like organisms. Mereschkowsky's idea was revived and extended in 1967 when Lynn Sagan proposed the endosymbiotic theory for the evolution of eukaryotic cells.

1906 While working as a handyman at a Seventh-Day Adventist sanitarium in southern California, self-taught geologist George McCready Price writes a small book titled *Illogical Geology: The Weakest Point in the Evolutionary Theory*. Price dismissed Charles Darwin's theory as "a most gigantic hoax" and offered $1,000 for proof that fossils have different ages. In 1961, Henry Morris and John Whitcomb, Jr., developed Price's ideas into "scientific creationism."

1906 The Hereditary Commission—an advisory group of the U.S. government—is founded by American breeder Willet Hays (1859–1927) "to investigate all proper means" of using heredity to "better the race" and to determine if "a new species of human being could be consciously evolved."

1906 African pygmy tribesman Ota Benga (1881–1916) becomes the only human to ever be exhibited at an American zoo. Benga lived in the Monkey House in the New York Zoological Park (better known as the Bronx Zoo) with monkeys, chimps, and America's first gorilla. The exhibit was highly controversial for many reasons, including the implication that it promoted a Darwinian view of human origins. Sixteen years after his release from zoo custody, Benga, unable to return to his native Congo, committed suicide.

1906 In his first scientific paper, Sergei Chetverikov (1880–1959) proposes that periodic "waves" of population decline could enable chance events to overcome selection. This anticipated the work of American biologist Sewall Wright (1889–1998) on genetic drift as a force of nonadaptive evolutionary change.

1906 American engineer Daniel Barringer (1860–1929) suggests that a giant crater (now known as Barringer Crater, Meteor Crater, and Canyon Diablo Meteor Crater) in Arizona was caused by a meteor impact (Figure 32). Barringer's claim was not fully confirmed until 1960 by astronomer Eugene Shoemaker (1928–1997). Barringer Crater, which was made a National Natural Landmark in 1967, was a source of material that Claire Patterson used to determine the earth's age in 1956.

1907 American physicist Bertram Boltwood's (1870–1927) studies of the radiometric decay of uranium to lead establish that the earth is at least 400 million years old (see also Appendix A). This was the first successful application of radiometric dating.

1907 In *Father and Son*, Edmund Gosse (1849–1928) claims that his father's (Philip Gosse) *Omphalos* (1857) represents "a record of educational and religious conditions which, having passed away, will never return." The younger Gosse was wrong, for in 1961 Henry Morris and John Whitcomb, Jr., resurrected Philip Gosse's ideas while promulgating young-earth creationism in their monumental *The Genesis Flood*.

Figure 32 Barringer Crater (also known as Meteor Crater) in northern Arizona is a popular tourist-site; it is more than 4,000-feet across and 550-feet deep. In the 1950s, Claire Patterson analyzed meteor fragments from Barringer Crater to determine that Earth is approximately 4.5 billion years old. (*Charles O'Dale*)

1907 Moscow's Darwin Museum (now the State Darwin Museum of Natural History) is founded by zoologist Alexander Kohts (1880–1964) and primatologist Nadezdha Kohts (1890–1963) as the world's first museum explicitly devoted to evolution.

1907 French philosopher Henri Bergson's (1859–1941) *Creative Evolution* introduces his *élan vital*, an undetectable force that powers all life and shapes organisms through time. Bergson's alternative to unguided evolution was attractive to some (e.g., Catholic priest Pierre Teilhard de Chardin) who were uncomfortable with Charles Darwin's ideas.

1907 Drawing on the work of Francis Galton, Indiana enacts the first law allowing forced sterilization. That same year, Galton established the Laboratory of National Eugenics at University College, London; statistician and Galton disciple Karl Pearson was the laboratory's first director. Laws permitting forced sterilizations did not gain widespread popular approval in the United States until the 1920s, but between 1900 and 1935, 32 states enacted similar laws, and by 1941 more than 35,000 people in the United States had been forcibly sterilized. These procedures were common into the 1940s and were sanctioned by law until 1983. Virtually all non-Catholic western nations—including Sweden, Canada, Australia, Norway, Finland, Denmark, and Switzerland—applied similar policies to "hereditary and incurable drunkards, sexual criminals, and lunatics."

1907 Lord Kelvin (Figure 25) dies and is buried in Westminster Abbey (Figure 7) adjacent to Isaac Newton and eventually near Ernest Rutherford. By this time, most geologists accepted that the earth is much older than Kelvin had claimed.

1908 English mathematician Godfrey Harold Hardy (1877–1947) and German obstetrician Wilhelm Weinberg (1862–1937) independently develop a mathematical model showing how genes behave in populations and that populations have a vast reservoir of genetic material that can be expressed in future generations. With their null-model assumptions, Hardy and Weinberg established the theoretical conditions under which genetic frequencies remain constant and that evolution does not occur (i.e., when there are no mutations, no gene flow, no genetic drift, no natural selection, and random mating). The Hardy-Weinberg Principle is now a fundamental principle of population genetics.

1908 In *The Panorama of Creation,* David Holbrook claims that Chapter 1 of Genesis is literature, not science, and that descriptions in Genesis of God's work during creation "week" are "rhetorical" devices that add flourish to the writing.

1908 Alfred Russel Wallace attends the Linnean Society's 50th anniversary celebration of the presentation of the Wallace-Darwin papers, where he notes that his and Charles Darwin's success can be traced to their having been "ardent beetle-hunters." Wallace—one of the greatest field biologists of the 19th century—told attendees that Darwin's contribution to evolution by natural selection far exceeded his, "as twenty years is as to one week." That year, the Linnean Society began awarding the Darwin-Wallace Medal. In the first year, silver medals were awarded to Joseph Hooker, Ernst Haeckel, August Weismann, and Francis Galton; the only gold medal was given to the 85-year-old Wallace.

1908 French geologist Marcellin Boule's (1861–1942) study of the first complete skeleton of a Neanderthal (from Chapelle-aux-Saints, France) erroneously gives Neanderthals a stooped, primitive posture (see also Appendix C). The skeleton's spine was deformed by

osteoporosis, but Boule could not conceive that Neanderthals might have walked upright.

1908 William Jennings Bryan asks Clarence Darrow to help in his third campaign for the presidency, but Darrow declines.

1909 The *Times* of London notes the centennial of Charles Darwin's birth by noting that he produced "a revolution in human thought so large, so pervading, so sudden, and yet so enduring . . . [Darwin's] achievement has, in a sense, become so familiar . . . that we can hardly see it plain or measure its proportions. It is not a matter for the learned only, but for all of us."

1909 Self-taught Dutch botanist Wilhelm Johannsen (1857–1927) publishes *Elements of Heredity*, which integrates his studies on heredity (i.e., morphological variation within a pure line must be due to the effects of the environment) with Mendel's results. *Elements* introduced the terms *genotype* and *phenotype*, and Johannsen coined the term *gene* to describe the unit of heredity that influences a phenotypic character (although Johannsen did not consider genes to be material entities). Johannsen's work helped scientists understand heredity and set the stage for the integration of genetics with evolution—the Modern Synthesis—during the first half of the 20th century.

1909 Geologist Charles Doolittle Walcott (1850–1927) and his family discover an area in the Burgess Shale Outcrop in the Canadian Rockies of British Columbia (near the Burgess Pass) that contains thousands of exquisitely preserved fossils. The fossils were about 505 million years old and included thousands of bizarre, soft-bodied organisms that were trapped in mud deposited by under-water landslides. During nine seasons of work, Walcott brought more than 65,000 of these fossils to the Smithsonian Institution, and they were later popularized by American biologist Stephen Jay Gould's best-selling book *Wonderful Life: The Burgess Shale and the Nature of History.* Gould's book, which attacked the notion of "progress" in evolution, was named after Frank Capra's classic 1946 movie *It's A Wonderful Life*, in which Jimmy Stewart's character sees what the world would have been like if he had never been born. The Walcotts's discovery was one of the greatest fossil-collecting achievements in the history of invertebrate pale-ontology. Although Walcott never finished high school and had no degree, he became Director of the U.S. Geological Survey,

Secretary of the Smithsonian Institution, and President of NAS. "Walcott's Quarry," as the tennis-court-size rock exposure is now known, is still under excavation.

1909 In *The Finality of Higher Criticism*, Minneapolis preacher William Bell Riley (1861–1947) claims that evolution gives humans "a slime sink for origin and an animal ancestry," while Christianity makes humans "the creature and child of the most high." Riley, a tireless critic of the teaching of evolution, later organized the fundamentalist movement in the United States.

1909 While working in what is now Dinosaur National Monument in Utah, paleontologist Earl Douglass of the CMNH in Pittsburgh unearths a near-complete skeleton of *Apatosaurus*. During the next 15 years, paleontologists working near this area uncovered the largest known concentration of Jurassic dinosaurs, including *Allosaurus*, *Stegosaurus*, and *Diplodocus* (Figure 33).

Figure 33 Excavations of dinosaurs in the American West—especially at Carnegie Quarry (later Dinosaur National Monument)—fueled the public's interest in prehistoric life, and prompted many people to wonder about dinosaurs' fates. Although workers often braved baking sun, blizzards, and the threat of attacks from Native Americans while collecting fossils, their work had a tremendous impact; in some instances, entire museums were erected to display the discoveries. Shown here are Jacob Kay and Joe Ainge with Billy and Joe, the Carnegie Museum's team of mules. (*Carnegie Museum of Natural History*)

1909 American philosopher and educational reformer John Dewey's (1859–1952) *The Influence of Darwin on Philosophy and Other Essays* argues that Darwinian evolution can restructure Western thinking. As Dewey noted, "[o]ld ideas give way slowly; for they are more than abstract logical forms and categories. They are habits, predispositions, deeply ingrained attitudes of aversion and preference. . . . Old questions are solved by disappearing, evaporating, while new questions corresponding to the changed attitude of endeavor and preference take their place. Doubtless the greatest dissolvent in contemporary thought of old questions, the greatest precipitant of new methods, new inventions, new problems, is the one effected by the scientific revolution that found its climax in the 'Origin of Species.'"

1909 The dispensationalist *The Scofield Reference Bible* popularizes gap creationism and becomes the standard Bible for fundamentalists. Congregationalist pastor Cyrus Ingerson Scofield (1843–1921), who wrote the footnotes in the first edition of this Bible, was strongly influenced by gap creationist G.H. Pember's (1837–1910) *Earth's Earliest Age*. Scofield's copious notes, which included a justification of slavery (regarding the curse on Ham's descendants after the Flood), claimed that passages in Genesis "clearly indicate that the earth had undergone a cataclysmic change as the result of divine judgment" and that "the face of the earth bears everywhere the marks of such a catastrophe" that resulted from "a previous testing and fall of angels." Scofield claimed that if readers "relegate fossils to the primitive creation . . . no conflict of science with Genesis cosmogony remains." *The Scofield Reference Bible* reproduced Bishop James Ussher's claim that the creation week described in Genesis occurred in 4004 BCE (Figure 8), but after the gap that was inserted between Genesis 1:1 and 1:2. Scofield claimed that Genesis 1:1 "refers to the dateless past, and gives scope for all geological ages." Scofield's Bible became a bestseller; more than two million copies have been sold.

1910 Organized by Baptist fundamentalist Amzi Clarence Dixon (1854–1925), oil magnates Lyman and Milton Stewart and other wealthy patrons of a Bible institute in Los Angeles fund publication of *The Fundamentals: A Testimony to the Truth*. Lyman Stewart hired Dixon—the pastor at Chicago's Moody Church—to edit the first five volumes. Dixon denounced Catholicism, liquor, Henry Beecher's liberalism, and evolution, but conceded that gap creationists and theistic evolutionists could be Christians.

Another contributor, Calvinist Presbyterian James Orr (1844–1913), claimed that "the world is immensely older" than Ussher's claim and that "there is no violence done to the narrative in substituting in thought 'aeonic' days—vast cosmic periods—for 'days' on our narrower, sun-measured scale. Then the last trace of apparent 'conflict' disappears." *The Fundamentals*—which were sent to every pastor, professor, and theology student in the United States—generated much interest; after publication of the first two volumes, the editors and publisher received 10,000 letters, and after volume three, 25,000 letters. Subsequent editors of *The Fundamentals* included Dwight Moody's colleague Reuben Torrey (1856–1928), who claimed "evolution is a guess pure and simple, without one scientifically observed fact to build upon." Contrary to popular belief, only about one-fifth of the articles in *The Fundamentals* discussed evolution, and a six-day creation was virtually absent. Although there was no consensus about how Christians should view evolution, most of the leading fundamentalist thinkers at this point accepted an ancient earth, historical links between species, and progressive creationism. The fundamentalist movement that followed became synonymous with its condemnation of evolution, but most writers of *The Fundamentals* were relatively unconcerned about the topic. The appearance of *The Fundamentals* became a convenient marker for the origin of modern fundamentalism.

1910 German geographer Alfred Kirchhoff's (1838–1907) posthumously published *Darwinism Applied to Peoples and States* argues that morality will prevail if the morally advanced European races eliminate the morally deficient races (i.e., "the crude, immoral hordes").

1910 The National Museum of Natural History, a part of the Smithsonian Institution, opens in Washington, DC. In 1978, an evolution-based exhibit at the Smithsonian prompted fundamentalist Dale Crowley, Jr., to file a lawsuit (*Crowley v. Smithsonian Institution*) demanding that the Smithsonian either close the exhibit or give equal time and space to an exhibit promoting the biblical story of creation.

1910 The Eugenics Record Office (ERO) opens as its basic claim—namely, that a person's character and abilities are determined at birth by the power of inheritance—becomes increasingly popular. The ERO was established at Cold Spring Harbor, New York, with

Charles Davenport (1866–1944) as director. The eugenics movement subsequently advocated forced sterilization of individuals deemed unfit due to "feeblemindedness," "shiftlessness," promiscuity, and epilepsy.

1910 In *The World of Life: A Manifestation of Creative Power, Deductive Mind, and Ultimate Purpose*, Alfred Russel Wallace invokes God (an "Infinite and Eternal Being") as the director and instigator of life who requires "continuous coordinated agency of myriads of [spiritual] intelligences."

1910 British politician Winston Churchill (1874–1965) advocates sterilizing "the feebleminded and insane classes."

1911 Four years after beginning his studies of inheritance in fruit flies, American geneticist Thomas Hunt Morgan proposes that genes are arranged linearly on chromosomes. Morgan got his original population of flies from Fernandus Payne (1881–1977), who had collected them from bananas left in a windowsill in what is now Schermerhorn Hall at Columbia University.

1911 Geologist Arthur Holmes (1890–1965) uses Bertram Boltwood's findings about radioactive decay to determine that rocks from the Devonian are 370 million years old (at that time, the oldest claimed age of a rock) and that there is a strong correlation between dates obtained by uranium-lead dating and the relative stratigraphic age of the rock (see also Appendices A and B). Holmes continued this research throughout his career, gradually extending his estimates of the earth's age from 1,600 million years in 1944 to 3.45 billion years in 1956. By 1960, Holmes's geologic timescale was similar to that used today. As Holmes noted in 1964, "Earth has grown older much more rapidly than I have—from about 6,000 years when I was 10, to 4 or 5 billion years by the time I reached sixty."

1911 Dying too early to see the horrors that his ideas would inspire, Francis Galton—the father of eugenics—leaves £46,000 to University College, London, to establish a national eugenics laboratory (The Francis Galton Laboratory of National Eugenics) and endow a professorship of eugenics. Karl Pearson was Galton's choice for the position. Despite his accomplishments and eminence, and therefore high probability of passing on desirable traits, Galton died childless.

1911 French biologist Lucien Cuènot (1866–1951), a member of the first generation of geneticists inspired by Gregor Mendel (Figure 27), champions the term *preadaptation* to describe the changed functions of already existing structures and claims that these changes play critical roles in enabling organisms to invade unoccupied niches (*des places vides*). In 1955, Theodosius Dobzhansky dismissed preadaption as "a meaningless notion if it is made different from 'adaptation.'"

1912 British lawyer and amateur paleontologist Charles Dawson (1864–1916) and paleontologist Arthur Woodward (1864–1944) announce their discovery of *Eoanthropus dawsonii* ("Dawson's Dawn-Man") at a meeting of the British Geological Society. The fossil, consisting of an ape-like mandible and part of a human skull gathered from a gravel pit near Piltdown in southeastern England, became known as Piltdown Man. The discovery was hailed as a "missing link" and produced global press coverage, including an article in *The New York Times* titled "Darwin Theory is Proved True." NHM accepted the Piltdown skull from Dawson, who described it as the most important fossil ever found. In 1953, Piltdown Man was shown to be a hoax and fueled creationists' arguments about the tenuous nature of evolutionary claims.

1912 At the annual meeting of the Geological Association, geologist Alfred Wegener (1880–1930)—one of the many scientists indebted to the work of Alfred Russel Wallace—announces the theory of continental drift. This theory, which proposed that 200 million years ago the earth had one giant continent that broke and drifted apart, helped explain the distribution of organisms among present-day land masses (e.g., the biogeographic discontinuity marked by "Wallace's Line").

1912 Cornell University botanist George Atkinson's (1880–1918) *Botany for High School* notes that evolution "has been accepted because it appeals to the mind of man as being more reasonable that species should be created according to natural laws rather than by arbitrary and special creation." Three years later, Vernon Kellogg and Rennie Doane's (1871–1942) *Elementary Textbook of Economic Zoology and Entomology* claimed that "although there is much discussion of the causes of evolution there is practically none any longer of evolution itself. Organic evolution is a fact, demonstrated and accepted."

1912 Julian Huxley's (1887–1975) first book, *The Individual in the Animal Kingdom*, introduces the concept of a biological "arms race" between competing lines of organisms.

1912 President Woodrow Wilson (1856–1924) appoints William Jennings Bryan as Secretary of State. Bryan resigned three years later in response to the "war preparedness" that led the United States into World War I.

1912 *The Catholic Encyclopedia* reviews various claims about the earth's age (including evidence based on radiometric dating), but does not endorse any theory or claim.

1912 The First International Congress of Eugenics honors Francis Galton, who had died a year earlier. Charles and Emma Darwin's son Leonard Darwin (1850–1943) presided over the Congress, and vice presidents of the Congress included David Starr Jordan, Winston Churchill, August Weismann, and Alexander Graham Bell (1847–1922). At the Congress, Cambridge biologist Reginald Punnett (1875–1967)—who lent his name to the square of Mendelian segregation patterns familiar to biology students—claimed that feeblemindedness "is a case of simple Mendelian inheritance." Most geneticists agreed, with Thomas Morgan a notable exception.

1913 American geneticist Alfred Sturtevant (1891–1970) uses Thomas Morgan's data regarding crossing-over frequencies to produce the first linear map of genes (the first map involved six traits). Two years later, Morgan used Sturtevant's method to conclude that genes are too small to be seen with an optical microscope and that fruit flies have more than 1,000 genes.

1913 When Alfred Russel Wallace dies at age 90, *The New York Times* notes his "scientific follies," but adds that he was "the last of the giants. He belongs to that wonderful group of intellectuals composed of Darwin, Huxley, Herbert Spencer, Lyell, Owen, and other scientists, whose daring investigations revolutionized and evolutionized the thought of the century."

1913 After debating the "Feebleminded Persons (Control) Bill" the previous year, the British government passes the "Mental Deficiency Bill," which isolates the mentally deficient in lunatic asylums.

Unlike in the United States, the British government never instituted compulsory sterilization.

1913 In proposing his geologic timescale, British geologist Arthur Holmes, notes that "[i]t is perhaps a little indelicate to ask of our Mother Earth her age, but Science acknowledges no shame and from time to time has boldly attempted to wrest from her a secret which is proverbially well guarded" (see also Appendix A).

1913 George McCready Price's *Fundamentals of Geology* announces his Law of Conformable Stratigraphical Sequence, which becomes a foundation of his young-earth arguments: "Any kind of fossiliferous rock may occur conformably on any other kind of fossiliferous rock, old or young." Price again promoted Flood geology, claiming that the fossil record and the earth's geologic features are best explained by a Noachian flood. Price referred to Flood geology as the "new catastrophism" to distinguish it from the catastrophism advocated by Georges Cuvier and Louis Agassiz, both of whom had accepted an old earth.

1914 John Joly uses sediment accumulation data to argue that the earth is 47–188 million years old (see also Appendix A).

1914 The ERO develops a plan to sterilize 10% of the members of every generation, until 15 million people have been sterilized. ERO manager Henry Laughlin's (1880–1943) work with state legislatures as an "Expert Eugenical Agent" in crafting sterilization laws led him to propose in 1914 a "Model Eugenical Sterilization Law," which became a template to expedite the enactment of such legislation. By the mid-1930s, more than 30 states had such laws in place.

1914 George Hunter (1873–1948), a teacher at DeWitt Clinton High School in New York City, publishes *A Civic Biology: Presented in Problems*, the textbook used by John Scopes when he taught biology in Dayton, Tennessee. Hunter discussed how the notorious Jukes family (as first described in 1877 by Richard Dugdale) produced "24 confirmed drunkards, 3 epileptics, and 143 feebleminded," as well as 33 "sexually immoral" people. Hunter described these families as "true parasites" that were spreading "disease, immorality, and crime to all parts of the country. . . . If such people were lower animals, we would probably kill them off to prevent them from spreading." William Jennings Bryan said that Hunter's book "could not be more objectionable," and Bryan was especially upset by a diagram in Hunter's book grouping humans with mammals instead of being in a group of their own. After the Scopes Trial, Tennessee abandoned Hunter's book, which was later renamed *New Civic Biology* to distinguish it from the book associated with Scopes's sensational trial.

1915 A medallion commemorating Alfred Russel Wallace is placed in Westminster Abbey (Figure 7), not far from Charles Darwin's tomb.

1915 American geologist Charles Walcott identifies fossilized bacteria in stromatolites from the Precambrian era.

1915 Crowds flock to AMNH (Figure 30) to see the first public exhibition of *T. rex*. This specimen—the second *T. rex* found by Barnum Brown in Montana—produced headlines such as "The Prize Fighter of Antiquity" as people began imagining prehistoric life on earth. Even Brown was impressed, noting that "I have seen nothing like it before."

1915 British evangelist Elizabeth Reid "Lady" Hope (1842–1922) claims in the Baptist newspaper *Watchman Examiner* that when she visited Charles Darwin just before he died, he denounced evolution and became a Christian. Although Darwin's children denied Lady Hope's story (she was never present during any of Darwin's illnesses and he never repudiated evolution), the story became a favorite of antievolutionists.

1915 Calvin Bridges (1889–1938), who invented virtually all of the standard procedures for working with fruit flies (e.g., the recipe for fly food, anesthetization with ether), identifies strains of mutant fruit flies with extra wings. Decades later, strains of these mutants helped biologists understand how *Hox* genes control the development of basic anatomical structures. Bridges's scandalous love life and questionable morals made him dependent on mentor Thomas Morgan for financial support.

1915 English writer Virginia Woolf (1882–1941) notes "imbeciles . . . should certainly be killed."

1915 *Men of the Old Stone Age* presents a foundation for Henry Fairfield Osborn's theory of human evolution, but contains little original work. Osborn claimed that Asia was "the chief theater of evolution of both animal and human life." *Men of the Old Stone Age* was Osborn's best-selling work; H. G. Wells used it when writing his *The Outline of History*.

1915 Thomas Morgan and colleagues Arthur Sturtevant, Hermann Muller (1890–1967), and Calvin Bridges publish *The Mechanism of Mendelian Heredity*, which argues that genes are arranged linearly on chromosomes. Morgan's studies of fruit flies elucidated the evolutionary effects of mutations, autosomal linkage, and sex linkage and established a molecular basis for the inheritance patterns observed by Mendel and others. Morgan, who was originally skeptical about evolution by natural selection, eventually became a strong advocate of Charles Darwin's theory. Morgan's new theory of heredity helped revitalize Darwinism and later became a basis for the Modern Synthesis in biology. Morgan won the Nobel Prize in Physiology or Medicine in 1933 "for his discoveries concerning the role played by the chromosome in heredity."

1915 In *The Origin of Continents and Oceans*, geophysicist Alfred Wegener explains his ideas on continental drift and outlines evidence for the existence of a supercontinent, Pangaea (Greek for "all land"), present about 300 million years ago. Although it was later accepted, Wegener's work gained a tepid reception during his lifetime.

1916 In *The Belief in God and Immortality: A Psychological, Anthropological and Statistical Study*, psychologist James Leuba (1868–1946) reports that over half of the scientists in *American Men of Science* doubt or reject a personal God and personal immortality, and that belief in a personal god is substantially lower among "greater scientists" than among less-renowned scientists. Leuba also noted that many students had lost their religious faith during college after being exposed to modern ideas, including the theory of evolution. William Jennings Bryan viewed Leuba's results as proof that evolution was destroying moral standards. Bryan was especially troubled by the teaching of human evolution, dismissing change in other species because it "does not affect the philosophy upon which one's life is built." All subsequent laws banning the teaching of evolution in the United States banned only the teaching of *human* evolution.

1916 American lawyer and anthropologist Madison Grant's (1865–1937) *The Passing of the Great Race* invokes the evolutionary ideas of his friend Henry Fairfield Osborn (who wrote a praise-filled introduction) to retell Western history along racial lines, with Nordic supremacy as its foundation. Adolf Hitler later described Grant's book as "his Bible."

1916 Harry Sinclair forms the Sinclair Oil and Refining Company, which funds several dinosaur-hunting expeditions in the American West. Sinclair's models of dinosaurs in the "Dinoland" exhibit during the 1964–65 World's Fair in New York—viewed by 10 million visitors—fueled America's growing interest in ancient life. Some of Sinclair's models, which today are exhibited in Dinosaur Valley State Park in Glen Rose, Texas, were based on paintings by Charles Knight.

1917 In the influential and richly illustrated *On Growth and Form*, Scottish biologist D'Arcy Thompson (1860–1948) suggests that the shapes of different animals result from changes to a common underlying body plan.

1917 George McCready Price publishes *Q.E.D.; or, New Light on the Doctrine of Creation*, which again promotes a young earth, special creation in six literal days, and a fossil-forming Flood. Although believers considered the self-educated Price to be a scientist, most scientists dismissed his ideas as having no scientific basis. Price responded that his work was prophetic and quoted the Bible's claim (2 Peter 3:3) that "there shall come in the last days scoffers, walking after their own lusts."

1917 Harold Cook finds a tooth on his ranch in Nebraska, which paleontologist Henry Fairfield Osborn (1857–1935) will claim in 1922 belongs to *Hesperopithecus haroldcookii*, or "western ape." Osborn claimed that the discovery was "irrefutable evidence that the man-apes wandered over from Asia into North America," and in 1922 the *Illustrated London News* depicted *Hesperopithecus* as an archaic humanoid hunter. However, Osborn was wrong; in 1925, scientists showed that the tooth was from an extinct genus of javelina. Although the claim that the tooth was from an ape was retracted in 1927 (in an article in *Science*), creationists continued to claim that the discovery is evidence that evolutionary data cannot be trusted.

1917 Vernon Kellogg's *Headquarters Nights* links Darwinian evolution with German war ideology.

1917 William Bell Riley's *The Menace of Modernism* attacks intellectuals and educators who question fundamentalists' claims. Riley argued that liberal preachers, university officials, and teachers show their loyalty to modernism by "their insistence on Darwinism."

1918 Organizers of the New York Prophetic Bible Conference at Carnegie Hall are surprised when the meeting attracts overflow crowds. William Bell Riley used this conference as a blueprint for his World Conference on the Fundamentals of the Faith held in Philadelphia the following year, which led to the formation of WCFA.

1918 President Theodore "Teddy" Roosevelt (1858–1919), who established the U.S. National Park Service, writes that his interest in evolution and natural history results from studying "at the feet of Darwin and Huxley."

1918 Ronald Fisher (1890–1962) shows that the conflict between the statistical and Mendelian geneticists about the inheritance of continuously variable (quantitative) traits (e.g., height) could be overcome by a model that invokes several loci acting on the trait simultaneously. Fisher's work was critical in unifying the field of genetics. Fisher also introduced the concept *variance* to describe the statistical dispersion of a range of scores as the squared deviations of scores about a mean. Fisher later showed the mathematical compatibility of Mendelian genetics with population genetics.

1918 Militant fundamentalist John Roach Straton (1875–1929) becomes pastor of Calvary Baptist Church in New York City and begins a series of highly publicized campaigns against "utterly disgusting" evolution, dancing, low-cut dresses, and AMNH (Figure 30). During the next decade, Straton was featured in more than 900 stories in *The New York Times* and more than 70 articles in *Time*.

1918 Benjamin Kidd's *The Science of Power* convinces William Jennings Bryan that German militarists had used evolutionary theory to justify their actions and lead people away from Christianity. *The Science of Power* also examined Charles Darwin's influence on German philosopher Friedrich Nietzsche (1844–1900), who in 1882 famously proclaimed "God is dead."

1919 Enigmatic Viennese biologist Paul Kammerer (1880–1926; Figure 34) claims that he has documented the Lamarckian inheritance of nuptial pads in midwife toads (*Alytes obstetricans*). Most toads mate in water and have nuptial pads on their hind limbs that help them cling to each other while they mate. However, midwife toads mate on land and lack these pads. When Kammerer forced midwife toads to mate in water, he reported that they laid fewer eggs and developed the black nuptial pads. Kammerer's announcement made front-page news throughout the world and he was hailed as a successor to Charles Darwin. Because Kammerer's claims supported socialist ideals and were consistent

Figure 34 Paul Kammerer's alleged proof of Lamarckian inheritance—that is, the inheritance of acquired traits—involved nuptial pads on midwife toads. Soon after his evidence was shown to be a forgery, Kammerer committed suicide. (*Library of Congress*)

with agronomist Trofim Lysenko's (1898–1976) Lamarckian version of genetics, Kammerer was offered and accepted a job in Moscow. In 1926, when Kammerer's results were shown to be fraudulent, Kammerer committed suicide.

1919 In a speech titled "Brother or Brute?" William Jennings Bryan tells the World Brotherhood Congress that Nietzsche had carried Darwinism to its ultimate conclusion and that Darwinism is "the most paralyzing influence with which civilization has had to contend." Bryan argued that if Darwinism is true, humans cannot overcome their animal nature and therefore all attempts at reform will be pointless. In the same year, Bryan spoke at the high school graduation ceremony in Salem, Illinois; among the graduates meeting Bryan at the ceremony was John Scopes. In 1925, Bryan helped prosecute Scopes for allegedly violating the Tennessee law banning the teaching of human evolution.

1919 The National Civil Liberties Bureau is founded in New York. This organization later became the American Civil Liberties Union (ACLU), which participated in several court cases associated with the evolution-creationism controversy.

1919 Frederick Aston (1877–1945) develops modern mass spectrometry, which yields accurate and precise analyses and measurements of isotopes. These techniques, which were refined in the 1930s and 1940s by Alfred Nier (1911–1994), paved the way for more reliable radiometric estimates of the earth's age.

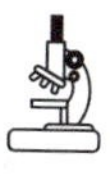

1919 William Bell Riley's World Conference on the Fundamentals of the Faith draws thousands to Philadelphia and ignites the fundamentalist movement in the United States. As Riley told the attendees, "[t]he importance of this occasion exceeds the understanding of its organizers." The meeting, which mobilized religious conservatives in unprecedented numbers, soon produced WCFA, the first and most formidable of the early fundamentalist organizations. Although the nine-point doctrinal statement written by Riley for the organization spanned denominational divisions, one goal of the WCFA was to eradicate the teaching of evolution "not by regulation, but by strangulation." The published proceedings of the Philadelphia meeting were titled *God Hath Spoken*.

1919 The Grand Canyon becomes a National Park. For scientists, the canyon documents the long spans of time that have been available for the earth's geology to be modified, in this case, by the slow action of water. However, many young-earth creationists cite the canyon as evidence for the Noachian Flood (Figures 35 and 36).

1920 In *The Watchman Examiner*, editor Curtis Laws (1868–1946) coins the term *fundamentalist* to describe someone willing to "cling to the great fundamentals" and "do battle royal" for the faith. Laws—a Baptist preacher—hoped to promote theological orthodoxy and biblical Christianity, and his broad definition of *fundamentalist* required neither inerrancy nor dispensationalism. Laws emphasized verification of the Bible's truth, not its value as a scientific document, and believed that these truths would be known by common sense.

1920 Speaking at the annual meeting of the AAAS in Toronto, British biologist William Bateson describes the origin and nature of species as "utterly mysterious." Bateson's comments were reported throughout the world as the "collapse of Darwinism" and were cited as support for restrictions on the teaching of evolution.

1920 In the two-volume *The Outline of History*, H. G. Wells (1866–1946) illustrates Neanderthal Man as "hairy, ugly, dimwitted . . . low browed and brutish," which was the prevailing opinion of the time.

Figure 35 Young-earth creationists often cite the Grand Canyon as evidence of a worldwide flood being responsible for Earth's geological formations. Geologists have determined that the canyon's rocks range in age from the 2-billion-year-old Vishnu schist (at the bottom of the canyon's Inner Gorge) to the 230-million-year-old Kaibab limestone on which these tourists are walking on the canyon's rim. (*U.S. Geological Survey*)

Figure 36 Landscapes in the American southwest often display spectacular geologic formations (see also Figure 35). The sediments shown here in the Grand Canyon support Nicolaus Steno's Principle of Superposition—that is, that undisturbed sediments lower in the geological column are older than sediments closer to Earth's surface. (*Randy Moore*)

1920 The ACLU is chartered in New York as a national nonprofit organization that works through litigation, legislation, and community education "to defend and preserve the individual rights and liberties guaranteed to every person in this country by the Constitution and the laws of the United States." The ACLU figured prominently in several evolution-related lawsuits, including *State of Tennessee v. John Thomas Scopes, Edwards v. Aguillard, McLean v. Arkansas Board of Education*, and *Kitzmiller et al. v. Dover Area School District*.

1920 William Bell Riley's "The Scientific Accuracy of the Scriptures" argues that scientific discoveries are predicted in the Bible (e.g., explosives are predicted in the Book of Job 38:22). Similar claims subsequently became a foundation for "creation science."

1920 The Kentucky Baptist Board of Missions demands that the state ban the teaching of the "false and degrading theory of evolution." The Baptists' cause was supported by William Jennings Bryan, who was invited to address a joint session of the Kentucky General Assembly. Instead of defending the literal "truth" of the Bible, Bryan focused more on its power to heal societal ills.

1920 The Research Science Bureau is founded by Presbyterian minister and self-proclaimed scientist Harry Rimmer (1890–1952) to abolish evolution with documentable evidence instead of authoritarian proclamation. The Bureau was never successful, but Rimmer—an outspoken advocate of day-age creationism—remained popular throughout the 1930s and 1940s with his books and "Bible and Science" lectures.

1920 At the annual meeting of WCFA in Chicago, organizers report growing concerns about the teaching of evolution and resolve to take official action against teaching evolution at its next meeting (Figure 37).

1920 Baptist preacher Thomas Theodore "T.T." Martin (1862–1939) strikes the first blow in the antievolution movement in North

Figure 37 The World's Christian Fundamentals Association (WCFA) was the most influential organization of fundamentalists in the United States. WCFA, which was headquartered in First Baptist Church of Minneapolis, Minnesota, organized resistance to the teaching of evolution, and in 1925 was responsible for recruiting William Jennings Bryan to help prosecute John Scopes at the famous Scopes Trial.

Carolina when he uses a series of articles in the Baptist publication *Western Recorder* to attack William Poteat, president of Wake Forest College, and demand Poteat's resignation. Martin denounced Poteat's reconciliation of Christianity with Darwin's theory of evolution, but failed to convince the state legislature to ban the teaching of evolution. Poteat, with the support of the college's Board of Trustees and the North Carolina Academy of Science, did not resign. At the Scopes Trial in 1925, Martin sold his *Hell and the High Schools: Christ or Evolution—Which?* (1923), a book that claimed that the acceptance of evolution dooms high school students to hell (in Chapter 1, Martin portrays the conflict between science and religion in stark terms by claiming "[e]volution says that there are ten lies in the first chapter of Genesis"). Martin demanded that schools not hire "any teacher who believes in evolution" because doing so would mean that the savior "was not Deity . . . only the bastard, illegitimate son of a fallen woman." Martin defended antievolution laws by claiming that they protected students' religious liberties and that "German evolution" sends people "to hell by the thousands." After the Scopes Trial, Martin was active in the Antievolution League of America and the Bible Crusaders of America, and helped secure passage of an antievolution law in Mississippi by likening evolution teachers to German soldiers who poisoned French children during World War I.

1920 Swedish zoologist Erik Nordenskiöld's (1872–1933) *The History of Biology*, published earlier in Swedish, judges Ernst Haeckel's books as "the chief source of the world's knowledge of Darwinism." Nordenskiöld also wrote that "Darwin's theory of the origin of species was long ago abandoned. Other facts established by Darwin are all of second-rate value."

1920 William Jennings Bryan delivers for the first time his famous "The Menace of Evolution" speech that decries the dangers of evolution, which he claims is "not science at all" but "guesses strung together." Bryan argued that his goal was to protect people "from the demoralization involved in accepting a brute ancestry." The subsequent printing of Bryan's lecture in pamphlets and newspapers reached millions of readers and produced one of the antievolution movement's most famous claims—namely, that "it is better to trust in the Rock of Ages than to know the age of the rocks." Bryan's lectures and pamphlets evolved into a book titled *In His Image* that was published the following year, and claimed that religion is the "only basis of morality," that evolution is the greatest threat to morality, and that Darwinism "leads logically to war" and "to a denial of God." Bryan implored taxpayers to refuse to support the teaching of evolution and demanded that atheists and evolutionists build their own schools; as Bryan noted, "if it is contended that an instructor has a right to teach anything he likes . . . [then] the parents who pay the salary have a right to decide what shall be taught. . . . A man can believe anything he pleases but he has no right to teach it against the protest of his employers." *In His Image*, which became a defining document in the antievolution movement, established Bryan as a national leader of the antievolution movement.

1921 In *Les Hommes Fossiles*, Marcellin Boule becomes one of the first to allege that Piltdown Man is a composite of a chimp jaw and a human skull.

1921 The Second International Congress of Eugenics opens in New York at the AMNH (Figure 30). Invitations to the Congress had been issued by the U.S. State Department, and Henry Fairfield Osborn was the presiding officer (inventor Alexander Graham Bell was the honorary president). At the Congress, Leonard Darwin pleaded for the "elimination of the unfit."

1921 The Hall of the Age of Man at AMNH becomes the first major exhibit in the United States to seriously investigate human evolution. The hall, which included murals by Charles Knight, was later renamed The Hall of Human Biology and Evolution, and today is known as The Bernard and Anne Spitzer Hall of Human Origins.

1921 The preface of Truman Moon's (1879–1942) best-selling *Biology for Beginners* states that biology is "based on the fundamental idea of evolution" and avers that "both man and ape are descended from a common ancestor." However, after the Scopes Trial in 1925, evolution disappeared from Moon's and most other biology textbooks.

1921 Playwright George Bernard Shaw (1856–1950)—an outspoken advocate of Lamarckism—claims in the preface to *Back to Methuselah* that "[i]f it could be proved that the whole universe had been produced by [natural] selection, only fools and rascals could bear to live."

1921 William Jennings Bryan introduces social science into the antievolution movement by citing James Leuba's *The Belief in God* *and Immorality* (1916). Leuba's findings—from surveys at nine representative colleges—suggested that America's colleges were undermining the faith of their students and that most faculty members in the sciences and social sciences believed neither in God nor immortality. Soon thereafter, fundamentalist leaders such as John Roach Straton claimed that colleges "are the places where Satan's seat is." By the end of 1921, Bryan was urging Christian taxpayers to assert their "legal rights" by banning evolution from all tax-supported classrooms.

1922 After forcing several professors to resign, famed militant fundamentalist J. Frank Norris (1877–1952) tours New England to energize the antievolution crusade in the North. Norris urged leaders to fight against evolution "without regard to State lines." In response, subscriptions to Norris's *Searchlight* (Figure 38) began "coming in avalanches." When William Jennings Bryan died in

THE SEARCHLIGHT

Entered as Second-Class Mail Matter at the Postoffice at Fort Worth, Texas, March 15, 1917, Under the Act of March 3, 1897
PUBLISHED BY THE SEARCHLIGHT PUBLISHING CO.

VOL. VII. FORT WORTH, TEXAS, FRIDAY, MAY 23, 1924. NO. 26.

WAR DECLARED AGAINST EVOLUTION AT SAN ANTONIO

RILEY ASKS BAPTISTS IN NORTH TO OUST FEDERAL COUNCIL

By W. B. RILEY.

The Milwaukee Convention is immediately at hand. It is approached with mingled feelings. For the true progress marked in the past year, every Northern Baptist believer ought to be and is grateful. In the problems to be faced, these same men and women ought to be and are interested. It is not the purpose of this writer to review either the victories won or to call attention to the battle lost. The present situation in the Northern Baptist Convention offers many opportunities for praise and it equally affords special occasions for criticism. We shall engage in neither. Not that the first would not be a pleasant exercise, nor yet that the second might not produce profit, but that the past is fixed and represents the practically unalterable. Yesterday can never be the subject of vital interest on the part of progressive men. Tomorrow beckons too loudly to them, and to it they rightfully turn.

There may be attendants at the Northern Baptist Convention who have no unselfish interest in the progress of our great and common cause. Such delegates can never contribute ought to the organization or bring any blessing to the denomination. We are inclined to think they are few in number, and we can afford to forget them as inconsequential.

This writer, at least, has nothing against the denomination. He came into it more than forty years ago, from conviction. From that

so! If we do not, let us quit business!

But a second subject of great present concern is the

NORRIS SENDS GENERAL BARTON MESSAGE

General Thomas D. Barton, Austin, Texas.

I have just read your address in today's Dallas News. Texas is facing the most serious crises since the days of the Alamo and the one issue above every other issue is law enforcement. I am in San Antonio holding big tabernacle evangelistic campaign and return tonight. You are eminently correct in all that you say, therefore regardless of other issues and personal friends who are in the governor's race, please accept my support. The time has come when all law abiding citizens, ministers, fathers and mothers should be united on a great and heroic law enforcement program.

(Signed)
J. FRANK NORRIS.

THOMAS D. BARTON SOUNDS TOCSIN OF WAR

In his address beginning his campaign for governor, General Thomas D. Barton said among other fine things the following:

THE OFFICE OF TEACHING

(Stenographically Reported by L. H. Evridge)

DR. NORRIS:

I want to give this morning a special message on the office of teaching. My text this morning is found in the 13th chapter and 1st verse of the Acts of the Apostles. the 13th chapter and 1st verse of the Acts of the Apostles.

"Now there were in the church that was at Antioch certain prophets and teachers; as Barnabas, and Simeon that was called Niger, and Lucius of Cyrene, and Manaen, which had been brought up with Herod the tetrarch, and Saul."

"Now there were in the church that was at Antioch certain prophets and teachers."

The first business of the New Testament prophets was to forthtell rather than to foretell. Now there was at Antioch certain prophets and teachers. Without doubt the church at Antioch influenced the whole world more than the church at Jerusalem. Barnabas was there, Paul was there and the Holy Spirit was the administrator of that church. Mark that—they didn't get orders from Rome or from Jerusalem or from New York or any where else. The Holy Spirit was their sole administrator. He called men into the ministry and called missionaries and sent missionaries forth. And that was the main reason that they had such a marvelous success in that day. Now the New Testament church had but one organization and that was the teaching, soul-winning organization and I will challenge any man to find where you have a thousand different side issues, wheels and machines in the church of the New Testament time. They didn't have it. They were too busy doing their main busi-

FROM THE PASTOR.

Rev. M. C. Eldson of San Antonio read the following letter from the Pastor Sunday morning and night:

To the Membership of the First Baptist Church:

As you well know I am in the most terrific fight of my life. The victory is already overwhelming and tremendous. Hundreds are being saved. Old-timers say never were such crowds, such interest and such results in San Antonio.

The opposition has been fierce. All Rome and Hell have conspired against the meeting. But "If God be for us who can be against us?"

If ever in your whole life you prayed for a man, do it now. There is a great army of the Lord's heroic elect in San Antonio and they are on the firing line in the front line trenches.

I am holding up most splendidly, though have never worked and fought so hard in all my life. I am going to stay until the victory is complete and decisive. I am preaching every day and night on sin, hell, judgment and redemption.

In great love,
J. FRANK NORRIS.

P. O. Box 266,
Ozona, Texas, 5-12-24.
The Searchlight Pub. Co.
Fort Worth, Texas.
Dear Sirs:

NORRIS IN GREATEST REVIVAL IN HISTORY OF ALAMO CITY

(From San Antonio Daily Light, Tuesday, May 20.)

NORRIS DENOUNCES EVOLUTIONISTS; DECLARES WHOLE BIBLE IS INSPIRE[D]

Does Not Mince Words in Attack on "Clique" Which P[re]vented Adoption of Resolution at Baptist Convention.

A short time ago the Fort Worth pastor received an [in]vitation to come to the famous Metropolitan Taberna[cle,] Spurgeon's Church, in August. Since he has been in [San] Antonio, he has received the invitation from Dr. F. B. Me[yer,] minister of Christ's Church, London, to preach in the fa[mous] Kingsway Hall, which has a capacity of over ten thousan[d.] The invitation asks him to speak on "the Advent Testimony," or "the Near Return of Our Lord."

The big tabernacle campaign has been characterized b[y] one shock or explosion following in quick succession afte[r] another. Monday night, before a capacity audience, the evan[ge]gelist, in clear, ringing voice, brought to his hearers a mos[t] terrific indictment against evolution on the theme "Th[e] Bible versus Evolution."

Boldly, and without mincing words, he launched an at[]tack on the forces, the inside clique which prevented th[e] adoption of the resolution that was offered at the Southern Baptist Convention, at Atlanta, against evolution, Dr. C. P. Stealey, of Oklahoma.

Figure 38 Famed Baptist preacher Frank Norris, who brought fundamentalism to the South, was a violent opponent of evolution. Norris's *Searchlight,* which had a circulation exceeding 60,000, featured Norris in the upper-left corner shining a searchlight on a cowering Satan on the opposite side of the page. This front-page from 1924 condemned the teaching of evolution while promoting the work of other antievolution crusaders (note the article about William Bell Riley in the left column). Norris—who was repeatedly arrested and acquitted for felonies including perjury, arson, and murder—was the most controversial person in the history of the evolution-creationism controversy. (*Courtesy of Arlington Baptist College*)

1925, *Searchlight* lamented that Bryan "is in heaven with his Lord, but—THE FIGHT MUST GO ON."

1922 WCFA, meeting in Los Angeles, announces its first official support of the antievolution movement by passing a resolution stating that "as taxpayers we have a perfect right to demand of public schools that they cease from giving to our children pure speculation in the name of science, and we have an equal right to demand the removal of any teacher who attempts to undermine . . . the Christian faith of pupils." Fundamentalist leader William Bell Riley proclaimed that "[w]e no longer need to advertise ourselves or our 'Association'; it is the best known movement of the twentieth century." Riley later noted that the fundamentalist movement "is stirring the nation from sea to sea."

1922 In a famous essay titled "The Root of Modern Evils," famed fundamentalist preacher Amzi Dixon claims that "[t]he beast jungle theory of evolution robs a man of his dignity, marriage of its sanctity, government of its authority, and the church of her power and Christ of his glory."

1922 Columbia University philosopher John Dewey notes "the campaign of William Jennings Bryan against science and in favor of obscurantism and intolerance is worthy of serious study. It demands more than the amusement and irritation which it directly evokes."

1922 Four days after hearing William Jennings Bryan denounce evolution in a speech to a joint meeting of the Kentucky legislature, Representative George W. Ellis introduces the nation's first antievolution bill. Frank McVey (1869–1953), president of the University of Kentucky, publicly opposed the proposed bill that would ban the teaching of "atheism, agnosticism, or the theory of evolution" in Kentucky's public schools. Thanks to McVey's opposition, the legislation—the nation's first vote on legislation to ban the teaching of evolution—was rejected by the Kentucky legislature by a vote of 42 to 41. Despite the defeat, Ellis encouraged "open war against Infidel Evolution." During the next 10 years, more than 40 similar bills were introduced in 20 different states. McVey's courage impressed University of Kentucky undergraduate John Scopes, who in 1925 agreed to be arrested to test the validity of an antievolution law in Tennessee. Seventy-two years after the vote on Ellis's legislation, Kentucky deleted the word *evolution* from its state educational guidelines, and in 1976 passed a law ensuring that (1) teachers who cover evolution in their classes can also teach biblical creationism, and (2) students who adhere to biblical creationism get credit for creationism-based answers on exams. As of late 2009, a reenacted version of that law remained in effect.

1922 A Bible conference in Raleigh, North Carolina—described by William Bell Riley as "one of the biggest religious events of the year"—famously features fundamentalist Jasper Massee's (1871–1965) condemnations of evolution as a cause of all societal ills because it produces "morality without Christ" and a "multitude of modern infidelities." Massee concluded his talk by noting that the teaching of evolution would cause Christianity and democracy to "topple in ruin" and that "it is impossible for a man to be a Christian and believe in a theory that denies the supernatural and makes Jesus Christ the bastard son of illegitimate intercourse between Mary and Joseph."

1922 Only weeks after the defeat of the Kentucky antievolution bill, William Jennings Bryan claims in an editorial in *The New York Times* that "[t]he real question is, did God use evolution as his plan? If it could be shown that man, instead of being made in the image of God, is a development of beasts, we would have to accept it, regardless of its effect, for truth is truth and must prevail. But when there is no proof we have a right to consider the effect of the acceptance of an unsupported hypothesis." Bryan then concluded that the teaching of evolution is heretical and "harmful, as well as groundless." In a rebuttal to Bryan's editorial, Henry Fairfield Osborn—who believed that "man has a long, independent, superior line of ascent . . . [and] has not descended from any known kind of monkey or ape"—cited St. Augustine while claiming that God uses evolution as part of his divine plan. Bryan then denounced Osborn as someone who believed that the discovery of fossils was more important than the birth of Jesus Christ.

1922 AAAS adopts the first of several resolutions supporting the teaching of evolution and opposing the teaching of creationism. That resolution, titled "Present Scientific Status of the Theory of Evolution," noted that evolution is not "a mere guess," that legislative restrictions on the teaching of evolution "could not fail to injure and retard the advancement of knowledge and of human welfare," and that "no scientific generalization is more strongly supported by thoroughly tested evidences than is that of organic evolution."

1922 In *Evolution—A Menace* published by the Southern Baptist Convention, John W. Porter claims that "[i]f evolution is true, the Bible, or at least portions of it, are absolutely false." Porter also argued that evolution produces immorality, that evolution transforms humans into "a developed beast," and that "evolution logically and inevitably leads to war."

1922 The Southern Baptists, meeting in Jacksonville, Florida, announce "no man can rightly understand evolution's claim as set forth in the textbooks of today, and at the same time understand the Bible."

1922 American explorer Roy Chapman Andrews (1884–1960) begins expeditions (funded by the Dodge automobile company) in the Gobi Desert, where he discovers numerous dinosaurs and dinosaur eggs. Andrews, who later became Director of the AMNH, is thought to be the inspiration for the movie character Indiana Jones. Until they were renovated in the 1990s, the fossil halls of AMNH were tributes to Andrews and Henry Fairfield Osborn; every dinosaur on display was either collected by Andrews or by an expedition sponsored by Osborn.

1922 Woodrow Wilson, who was strongly influenced by the proevolution views of James Woodrow (Wilson's uncle), notes "of course, like every other man of intelligence and education, I do believe in organic evolution. It surprises me that at this late date such questions should be raised." Three years later, Wilson's letter was included in Winterton Curtis's (1875–1966) testimony at the Scopes Trial.

1923 After declaring her to be mentally retarded, "feebleminded," and an "incorrigible" genetic threat to society, the state of Virginia tries to forcibly sterilize 17-year old Carrie Buck (1906–1983) of Charlottesville, Virginia. Buck resisted the sterilization and sued to stop the process. Her lawsuit produced *Buck v. Bell*, which was heard by the U.S. Supreme Court four years later. The Court supported Virginia's right to forcibly sterilize Buck. Although in 1942 the Supreme Court struck down a law allowing the forced sterilization of criminals, it never reversed the general concept of eugenic sterilization established by *Buck v. Bell*.

1923 An editorial in *The Los Angeles Examiner* notes "the evolution idea has poisoned and is poisoning, the minds of more people who are fools enough to listen to it. . . . Take the evolutionists, infidels and no-hell teachers out somewhere and crucify them, head downward, and we will have a better country to live in."

1923 Henry Fairfield Osborn's article "The Dawn Man" uses prehistoric nature to promote Social Darwinism and the racial superiority of the white race. Osborn showed Cro-Magnons at war with Neanderthals, noting that "[i]t was a case always of the complete extermination of the weak by the strong. The law of the survival of the fittest is not a theory, but a fact." The article also included an illustration by Charles Knight showing Neanderthal man attacking a mammoth. That illustration later appeared in Knight's children's book *Before the Dawn of History* (1935).

1923 Raymond Dart (1893–1988; Figure 39) becomes head of the anatomy department at the University of Witwatersrand in Johannesburg, South Africa. When Dart discovered that the university had no reference collection of bones and fossils, he offered his students a prize for the most interesting bones they could bring to him. The following summer, Josephine Salmons—Dart's only female student—brought Dart a baboon cranium that she found at the home of the director of the Northern Lime Company near Taung, South Africa. After Dart asked the director to notify him of any other fossils unearthed by miners, he received two boxes of rocks, one of which encased the hominin fossil that came to be known as Taung Child (see also Appendix C).

1923 At a WCFA meeting in Ft. Worth, Texas, hosted by Frank Norris's First Baptist Church, fundamentalists stage a two-hour mock trial of evolutionists that is attended by 3,000 supporters. The evolutionists were convicted and hanged in effigy.

1923 In response to William Jennings Bryan's campaign against the teaching of evolution, defense lawyer Clarence Darrow publishes a letter on the front page of the *Chicago Tribune* questioning Bryan about his beliefs. For example, did Bryan believe in the literal truth of the Bible? What about the Flood? The origin of man? Did Jonah live inside a whale for three days? Bryan did not respond. At the Scopes Trial in 1925, Darrow asked Bryan many of these same questions.

Figure 39 Raymond Dart unearthed the earliest known skull of an australopithecine in South Africa. This photo, taken in 1963, shows Dart at the site of some of his discoveries. (*Associated Press*)

1923 Oklahoma becomes the first state to pass an antievolution law. The legislation offered free textbooks to public schools whose teachers did not mention evolution and banned the use of books promoting Darwinism. The law was repealed in 1925.

1923 The Florida legislature passes a resolution declaring it "improper and subversive" for any teacher at a public school "to teach Atheism or Agnosticism, or to teach as true Darwinism, or any other hypothesis that links man in blood relationship to any other form of life." The resolution had little impact.

1923 In his 700-page *The New Geology*, George McCready Price again claims that the Noachian Flood (that included "great tidal waves sweeping daily around the earth . . . traveling 1000 miles per hour") explains most of the observed geologic phenomena, including fossil-containing rocks. Price attacked the validity of the geologic column, claiming that "[t]he alleged historical order of the fossils is clearly a scientific blunder." *The New Geology* inspired generations of young-earth creationists while making Price, as noted in the journal *Science* (1926), "the principal scientific authority of the Fundamentalists." Some fundamentalists were initially leery of Price's claims because of his link with Adventism and prophetess Ellen White's visions, but Price was soon cited by some fundamentalists as a credible scientist. Although Price's claims contradicted William Jennings Bryan's nonliteral view of biblical chronology, Bryan nevertheless invited Price to Dayton in 1925 as an expert witness at the Scopes Trial. Price declined because he was lecturing in England that summer. *The New Geology* established Price as the leading scientific authority among creationists.

1923

William Bell Riley (Figure 40), realizing that evolution unites fundamentalists across denominational lines, asks ministers from eight denominations to help form the Antievolution League to oppose the teaching of evolution in public schools. An initial meeting at the Swiss Tabernacle (the largest church in Minneapolis, Minnesota) drew hundreds of supporters and produced the prototype for numerous similar organizations. The league's first president was Kentucky theologian John W. Porter, who helped get an antievolution bill before the Kentucky legislature. T. T. Martin was the league's field secretary and editor of *The Conflict*, the organization's official publication. The league was endorsed and supported by William Jennings Bryan and soon after its formation began its national "Bible-and-Christ-and-Constitution Campaign

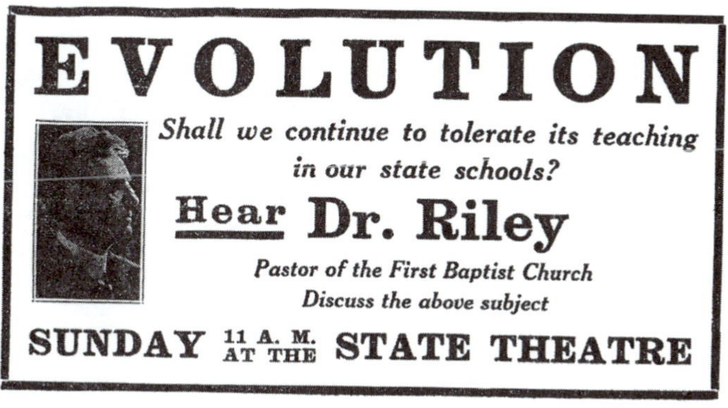

Figure 40 William Bell Riley, who had organized the WCFA, worked tirelessly to block the teaching of evolution in public schools. Shown here is an advertisement of one of Riley's many antievolution speeches, which often attracted thousands of people.

Against Evolution in Tax-Supported Schools." The league claimed to represent not just religious conservatives, but also "parents," "taxpayers," and "American citizens." In *The Conflict*, the league claimed that "[e]veryone is nerved for the battle that will never end until every evolutionist is driven from the tax-supported schools of America."

1923 Cambridge University biologist J. B. S. "Jack" Haldane's (1892–1964) *Daedalus; or, Science and the Future* describes his ideas about the future implications of eugenics. This and other books inspired Aldous Huxley (Thomas Huxley's grandson) to write the futuristic *Brave New World* (1932), which described a society based on the scientific control of behavior and reproduction. In Huxley's "World State," women used a contraceptive device called a "Malthusian belt." The French translation of *Brave New World* was titled *Le Meilleur des Mondes* (*The Best of All Worlds*), a reference to an expression used by Voltaire's fictional character Pangloss in *Candide*. In 1963, Haldane coined the term *clone*.

1923 William Jennings Bryan blames evolution for World War I, describes evolution as "the only menace to religion that has appeared in the last 1900 years," and claims that the teaching of evolution is "poison." The following year, Bryan expanded his indictment of evolution by claiming that the teaching of evolution is responsible for "all the ills from which America suffers."

1924 Aleksandr Oparin's (1894–1980) *Proiskhozhedenie Zhizni* develops his ideas about the early earth's atmosphere and first organisms. Oparin's theory centered on the idea that the early earth had a reducing atmosphere, in which complex organic molecules could form from simpler, inorganic compounds. Oparin's idea, combined with similar work by J. B. S. Haldane in Britain, was often referred to as the Oparin-Haldane "primordial soup" theory. Oparin's work was published in English in 1967. By the 1990s, however, many scientists questioned Oparin's claims about the early atmosphere.

1924 The Bible League of North America notes that "Fundamentalists draw the weapon of their warfare from the arsenal of God's Word; modernists draw theirs from the evolutionary philosophy."

1924 Famed American taxidermist Carl Akeley's sculpture *The Chrysalis* depicts a modern human (resembling a youthful Akeley himself) emerging from a gorilla. Akeley noted that humans and modern apes "undoubtedly had a common ancestor," and that "science is on the trail of this ancestor and will locate it." The sculpture, which outraged creationists, was eventually obtained by Reverend Charles Potter of New York's West Side Unitarian Church. Potter, an advocate of evolution, noted that "I know of no concrete symbol which so well expresses the religious message which I am trying to preach every Sunday." When asked about his religious faith, Akeley told a reporter that "most of my worshipping has been done in the cathedral forests of the African jungles with the voices of the birds and animals as music."

1924 George McCready Price's *The Phantom of Organic Evolution* devalues Charles Darwin's contribution to evolutionary thought while rejecting geologic evidence for an ancient earth (Figure 41).

1924 AAAS Vice President Edward Rice uses his address at the society's annual meeting to confront statements about evolution made by William Jennings Bryan. Rice's statements gained a wide professional readership in *Science* the following year.

1924 Raymond Dart (Figure 39) uses his wife's knitting needles to chip the first known skull of an australopithecine from a piece of limestone from Harts Valley at Taung, South Africa. The brain case

Figure 41 George McCready Price often used Montana's Chief Mountain to argue that geology is a false science. Price claimed "All of Glacier National Park, of which [Chief Mountain] is a part, shows a similar antievolutionary sequence." Today, geologists know that Chief Mountain is part of the Lewis Overthrust, a fault in which softer Cretaceous rocks are atop rocks that are 1,400 million years younger.

was larger than that of a chimpanzee but smaller than that of known human ancestors. The following year, Dart's paper in *Nature* described his discovery as "an extinct link between man and his simian ancestor." Dart named his discovery *Australopithecus africanus* ("australis" meaning "south" and "pithecus" meaning "ape"), and it became known as Taung Child. Dart's controversial claim was strengthened when it was authenticated by renowned Scottish paleontologist Robert Broom (1866–1951), who noted that "[i]n *Australopithecus* we have a connecting link between the higher apes and one of the lowest human types." Broom, who considered himself the "scientific son" of Richard Owen, agreed with Dart that the discovery vindicated "the Darwinian claim that Africa would prove to be the cradle of mankind" (see also Appendix C).

1924 United States Representative John William Summers (1870–1937) of Washington introduces an amendment to a District of Columbia appropriations bill that would ban funding for all school officials who allowed the teaching "of partisan politics, disrespect of the Holy Bible, or that ours is an inferior form of government." The so-called Summers Amendment passed the House and Senate without debate and generated little interest until the following year when Treasury Department employee Loren Wittner used it as a test case against District of Columbia school administrators. Wittner's challenge failed.

1924 North Carolina Governor Cameron Morrison (1869–1953) convinces the North Carolina Board of Education to reject biology textbooks that include evolution because they are "unsafe," adding that he does not want his "daughter or anybody's daughter to have to study a book that prints pictures of a monkey and a man on the same page. . . . I don't believe in any missing links. If there were any such things as missing links, why don't they keep on making them?"

1924 Shailer Mathews (1863–1941), dean of the University of Chicago Divinity School, publishes *The Faith of Modernism,* a book promoting modernism. Mathews (who was prepared to testify for the defense at the Scopes Trial) was often attacked by fundamentalists for claiming "Genesis and evolution are complementary to each other." After the sensational Scopes Trial, John D. Rockefeller, Jr., gave $1,000,000 to Mathews's Divinity School. Antievolution crusader William Bell Riley (Figure 40) claimed in *Evolution: A False*

Philosophy that Mathews's school taught "that there is no God" and "destroy[ed] Faith in God."

1924 The California State Board of Education instructs public school teachers to present evolution "as a theory only." Throughout the country, similar demands continue to the present day.

1924 Unitarian preacher and modernist Charles Potter faces militant antievolutionist and flamboyant fundamentalist John Roach Straton in four highly publicized debates in New York's Carnegie Hall. Descriptions of the debates in newspapers fueled the escalating evolution-creationism controversy. Potter, who spoke extensively at the Scopes Trial the following year, regarded all of the Bible as obsolete except Jesus' moral teachings.

1924 In *Christianity and Liberalism*, fundamentalist J. Gresham Machen (1881–1837) presents a classic defense of orthodox Christianity that indicts modernistic theology as unscientific and un-Christian. Machen was especially upset that many modernists were emphasizing "human goodness" instead of humans being "sinners under just condemnation of God."

1924 William Bell Riley (Figure 40) collapses in the pulpit following an automobile accident. Riley required months to recover, after which he never regained his position as fundamentalism's preeminent figure.

1924 Worried that his children will be corrupted by the teaching of evolution in the public schools, Tennessee state legislator John Butler (1875–1952) drafts what will be known as the Butler Law (House Bill No. 185). This law made the teaching of human evolution (i.e., "Any theory that denies the story of the Divine Creation of man as taught in the Bible, and to teach instead that man has descended from a lower order of animals") in any of Tennessee's public schools a misdemeanor punishable by a fine of $100–$500. Although editors of the *Chattanooga Times* urged that Butler's legislation be ignored, his proposed ban on teaching human evolution was otherwise unopposed. Butler's legislation, which became law the following year, was the basis for the Scopes Trial, the most famous event in the history of the evolution-creationism controversy.

1924 Zoologist H. H. Newman (1875–1957) writes in *Outlines of General Zoology* that evolution has triumphed over creationism and that "there is no rival hypothesis to evolution except the out-worn and completely refuted one of special creation, now retained only by the ignorant, dogmatic, and prejudiced."

1924 James Gray (1851–1935), the president of the Moody Bible Institute, claims that evolution threatens democracy and that its acceptance promotes communism and the overthrow of the U.S. government. The previous year, the institute described the evolution controversy as "the greatest battle, or rather war, known to ecclesiastical history."

1925 Texas governor Miriam "Ma" Ferguson (1875–1961)—the first woman to be elected governor in the United States—bans public schools' use of biology textbooks that include evolution. Ferguson threatened to fire and prosecute any teacher who used an unapproved book and justified her edict by reminding Texans that she was "a Christian mother." For the next several decades, the ban imposed by Ferguson forced publishers to produce special editions of their biology textbooks for Texas classrooms.

1925 While John Straton, William Bell Riley, and Frank Norris tour the country denouncing evolution and praising George McCready Price's *The New Geology*, Tennessee Governor Austin Peay (1876–1927) signs John Butler's legislation into law. The Butler Law made it a crime to teach human evolution in Tennessee and led to the prosecution of John Scopes four months later. William Jennings Bryan telegrammed Governor Austin Peay that "[t]he Christian parents of the State owe you a debt of gratitude for saving their children from the poisonous influence of an unproven hypothesis." Bryan described the upcoming trial of John Scopes as "the contest between evolution and Christianity" and a "duel to the death . . . the two cannot stand together." *The New York Times* later complained that the so-called "'duel to the death' has become a battle of statements."

1925 In response to the newly passed Butler Law, the ACLU places an ad in the *Chattanooga Daily Times* and other Tennessee newspapers: "Looking for a Tennessee teacher who is willing to accept our services on testing the law in the Courts." Dr. George Rappleyea (1894–1966), manager of the Cumberland Coal and Iron Company in Dayton, Tennessee, sees the ad and meets with some of Dayton's leaders; he hoped to bring Dayton some publicity and to boost the area's struggling economy. The meeting ultimately produced the Scopes Trial (i.e., *State of Tennessee v. John Thomas Scopes*).

1925 In *The Earth Speaks to Bryan*, Henry Fairfield Osborn of AMNH uses "Nebraska Man" to ridicule the antievolutionary views of William Jennings Bryan (who had served as a congressman from Nebraska). As Osborn noted, "[t]he earth speaks to Bryan from his own state" but "he fails to hear a single sound."

1925 John Scopes's contract with Rhea County High School expires on May 1, but four days later he agrees to be arrested for the teaching of human evolution and is charged the following day. George Rappleyea wired the ACLU a plan for a "four-round fight" that would culminate with a hearing at the United States Supreme Court. Two days later, the *Washington Post* announced the story on its front page: "J.T. Scopes, of the science department of the Rhea County High School, was arrested by a deputy sheriff, charged with violating the Tennessee law prohibiting the teaching of evolution in the state public schools." Fundamentalist leader William Bell Riley (Figure 40) asked William Jennings Bryan (Figures 42) to represent WCFA at Scopes's trial. On May 12, Bryan—who

Figure 42 William Jennings Bryan arrives in Dayton, Tennessee, to help prosecute coach and substitute-teacher John Scopes for allegedly teaching human evolution in the local high school. Bryan's death in Dayton five days after the trial, as well as his subsequent portrayal in *Inherit the Wind,* made Bryan an icon of the evolution-creationism controversy. (*Associated Press*)

hoped that the upcoming trial would return him to the front pages of the nation's newspapers—responded that he would "be pleased to act for your great religious organizations and without compensation." Bryan, who thanked Riley for "the opportunity the Fundamentalists have given me to defend the faith," believed the Scopes Trial would "end all controversy." In Dayton, Bryan and his entourage eventually took over four rooms of the home of local druggist F. R. Rogers.

1925 *Baltimore Sun* journalist H. L. Mencken (1880–1956; Figure 43) meets with famed attorney Clarence Darrow on May 14 to urge him to defend John Scopes. Mencken—who coined the phrases

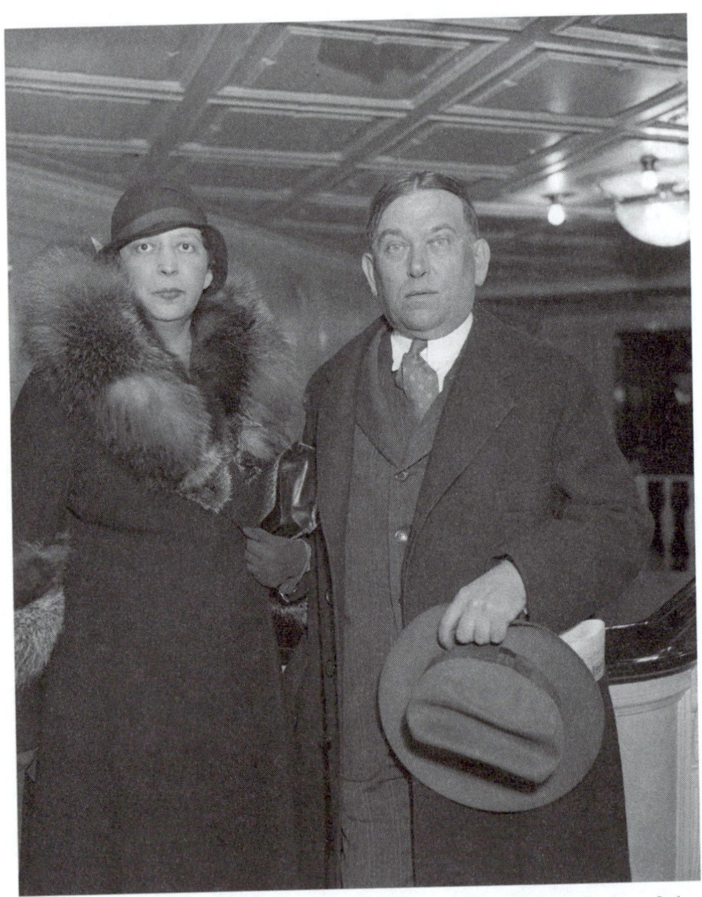

Figure 43 H. L. Mencken was a famous writer whose coverage of the Scopes Trial is regarded as some of the greatest journalism in American history. Mencken, whose columns enthralled readers and shaped the trial, coined the terms *Bible Belt* and *Monkey Trial*. This photo of Mencken and his wife was taken in 1932. (*Associated Press*)

Bible Belt and *Monkey Trial*—covered the trial in Dayton (where he was described as "the most respected, hated, reviled, feared, and loved person" in Tennessee) and wrote 13 articles for the *Sun*, work that is regarded as some of the greatest journalism in American history. In the most famous of his articles, Mencken used the front page of the *Sun* to describe Scopes's trial as a "religious orgy." Mencken later described fundamentalists as being "everywhere where learning is too heavy a burden for mortal minds to carry." Mencken shaped, as well as reported, the trial, and even Scopes admitted that the trial "was Mencken's show" and that "a mention of the Dayton trial more likely invokes Mencken than it does me." Although William Jennings Bryan (Figure 42) once described Mencken as "the best newspaperman in the country," Mencken despised Bryan, and his first comment upon hearing of Bryan's death was "[w]e killed the son-of-a bitch." Bryan's death did not slow Mencken's attack; indeed, Mencken told readers that if Bryan was sincere, "then so was P. T. Barnum." Mencken's "In Memoriam: W.J.B.," a masterpiece of invective, was taken at face value by Jerome Lawrence and Robert Lee when creating the Bryan-esque character Matthew Brady for their influential play *Inherit the Wind*. Just before the start of the Scopes Trial, Mencken marked the 100th anniversary of Thomas Huxley's (Figure 23) birth by praising him as "the greatest Englishman of the Nineteenth Century—perhaps the greatest Englishman of all time."

1925 David Scott Poole (1858–1955) introduces a resolution to the North Carolina state legislature declaring that Darwinism is "injurious to the public welfare." Although Poole claimed that "the religion of the Lord Jesus is on trial," the resolution was defeated by a vote of 67 to 46. Two years later, a similar bill introduced by Poole was defeated in committee.

1925 In *The Predicament of Evolution*, George McCready Price claims "Marxism, Socialism, and the radical criticism of the Bible . . . are now proceeding hand in hand with the doctrine of organic evolution to break down all those ideas of morality [upon which] civilization has been built."

1925 Fiery fundamentalist Billy Sunday (Figure 44) conducts an 18-day crusade in Memphis, Tennessee, at which he denounces evolution, declares Darwin an "infidel," and claims that education is "chained to the devil's throne." Sunday's crusade was attended by

Figure 44 Billy Sunday, a former baseball-player who became the most famous preacher of his era, used theatrical sermons to condemn the evils of evolution. At the height of his popularity, Sunday's revivals were attended by tens of thousands of people. (*Associated Press*)

more than 10% of Tennessee's residents. In North Carolina, just before the start of the Scopes Trial, Sunday denounced evolutionary biologists as "theological bootleggers" while noting that "[i]f you believe that you came from a monkey, then take your ancestors and go to the Devil."

1925 John Godsey (1874–1932), one of John Scopes's original attorneys, files a motion seeking to have the charge against Scopes dropped because the Butler Act is unconstitutional. The motion was rejected, and Godsey bowed out of the case before the trial began.

1925 Alfred McCann (1879–1931), the author of *God—or gorilla*, refuses William Jennings Bryan's invitation to participate in the Scopes Trial by writing to Bryan that "[e]ven though we have succeeded in bludgeoning the world with Volsteadism, we can't hope to bottle up the tendencies of men to think for themselves."

1925 Clarence Darrow, at the height of his powers and fame, volunteers to defend John Scopes at the Scopes Trial. Darrow was America's greatest criminal lawyer and its most famous champion of anticlericalism. Darrow denounced Tennessee's antievolution law "as brazen and bold an attempt to destroy learning as was made in the Middle Ages." When Darrow heard that Bryan would be at Dayton, Darrow responded "at once I wanted to go. . . . I realized that there was no limit to the mischief that might be accomplished unless the country was aroused to the evil at hand." Darrow's objective in Dayton was to prevent "bigots and ignoramuses from controlling the education of the United States, and that is all. . . . My object, and my only object, was to focus the attention of the country on [William Jennings Bryan] and the other fundamentalists."

1925 Bainbridge Colby (1869–1950), who succeeded William Jennings Bryan as Secretary of State under Woodrow Wilson, is asked by George Rappleyea and the ACLU to help defend John Scopes. Colby accepted the invitation and urged Scopes's attorney John Neal (1876–1959) to try to transfer the trial to a federal court, where he could argue the constitutionality of the newly passed Butler Law. A federal judge denied the request, and Colby resigned from the defense team two days before the trial started.

1925 As the Scopes Trial approaches, the American Telephone & Telegraph Company installs 10.5 miles of temporary lines to speed the transmission of stories (*The New York Times* reporters alone telegraphed more than 100,000 words about the trial). On the first day of the trial, Vernon Dalhart (born Marion Try Slaughter; 1883–1948)—a popular country singer—popularized the trial by recording Carson Robison's (1890–1957) *The John T. Scopes Trial* for the Columbia Phonograph Company (#15037-D) and Edison Records (#51609-R; Figure 45). Weeks later, Dalhart also recorded *Bryan's Last Fight* (Columbia Records #15039), which proclaimed that Bryan "stood for his own convictions, and for them he'd always fight."

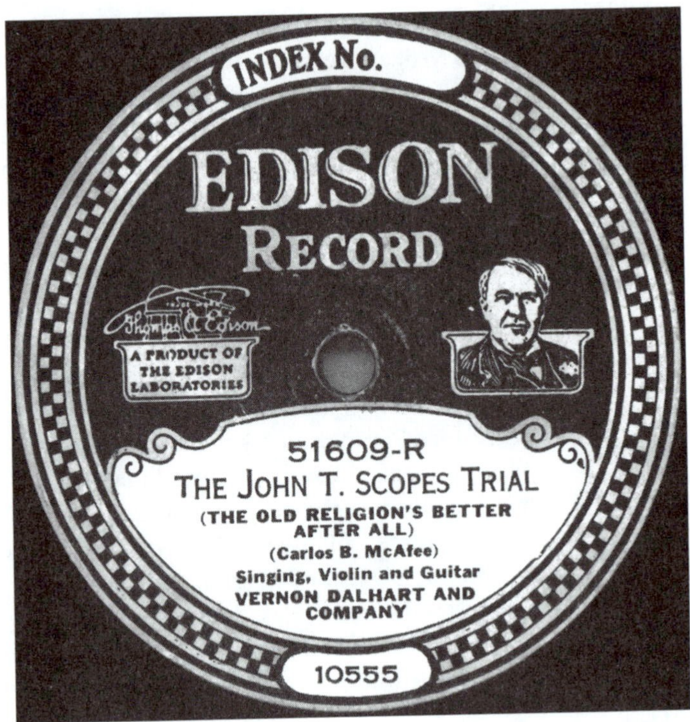

Figure 45 The Scopes Trial made headlines throughout the world and generated a variety of associated products, including recordings of songs such as "The John T. Scopes Trial." (*Randy Moore*)

1925 American evangelist Mordecai Ham (1877–1961) announces that opponents of Tennessee's antievolution law are "anti-Christ Communists."

1925 Famed evangelist Aimee Semple McPherson (1890–1944; Figure 46) promises William Jennings Bryan that 10,000 members of her church will be praying for his success. From her huge Angelus Temple (now a National Historic Landmark) in Los Angeles, California, the flamboyant McPherson—who often preached in an iconic long white gown—proudly proclaimed her willingness "to sacrifice science rather than religion." McPherson—who was often referred to as "a female Billy Sunday"—wanted to abolish barriers between church and state and urged Christians to seize control of government by boycotting schools that taught evolution. In 1927, "Sister Aimee" denounced evolution as "the greatest triumph of Satanic intelligence in 5,931 years of devilish warfare against the Hosts of Heaven. It is poisoning the minds of the children of the nation. It is responsible for jazz, bootleg booze, the crime wave,

Figure 46 Aimee Semple McPherson was a famous preacher who used modern marketing and sensational theater to attract huge crowds to her Angelus Temple in Los Angeles, California. Although her reputation was later tarnished by an alleged "kidnapping" (that was used to hide a romance with one of her employees), McPherson's sermons demonizing evolution were among her most famous. McPherson continues to be regarded by her followers as a prophetess, but has often been portrayed as a religious hypocrite and sexual vixen. This photo shows McPherson leading a worship service at Angelus Temple in 1943. (*Associated Press*)

student suicides, Loeb and Leopold, and the peculiar behavior of the younger generation." McPherson participated in several highly publicized debates with atheist Charles Smith (1887–1964), who also debated fundamentalists William Bell Riley and John Straton. Despite her fame as an evangelist, McPherson became best known for her alleged "kidnapping" in 1926, a claim that had numerous inconsistencies and that suggested McPherson had in fact "disappeared" to Mexico with her lover. After reappearing a few weeks later, a grand jury investigating McPherson's alleged kidnapping adjourned without delivering an indictment. In 1944, McPherson died of a drug overdose.

⚖ **1925** The Scopes Trial (Case No. 5232; see also Appendix D), which produced a carnival-like atmosphere in Dayton, Tennessee (Figure 47), begins with a fundamentalist prayer that John Scopes describes as "interminable." On the fifth day of the trial, Dudley Field Malone (1882–1950)—an international divorce attorney and law partner to Arthur Hays (1881–1954) aiding in the defense of John Scopes—delivered a 25-minute speech that, according to John Scopes and others, was the turning point of the trial. When Malone finished, Scopes said he could see the "tragedy on [Bryan's] beaten face," and H. L. Mencken (Figure 43) reported that Malone's words "roared out of the open windows like the sound of artillery." Other members of the press, breaking their customary silence of neutrality, gave Malone a standing ovation. Even William Jennings Bryan (Figure 42) acknowledged that Malone had given "the greatest speech I've ever heard," to which Malone responded, "I am sorry it was I who had to make it." Years after the trial, Malone admitted that his famous oration in Dayton was the only extemporaneous speech he ever made.

⚖ **1925** Almost 2,000 spectators watch the Scopes Trial reach its climax on the lawn of the courthouse. After earlier baiting Bryan by

Figure 47 The carnival-like atmosphere surrounding the Scopes Trial included preachers, trained apes, vendors, and countless other sideshows. Shown here is a booth selling evangelist T. T. Martin's book *Hell and the High School*. (*Associated Press*)

saying that "Bryan has not dared test his views in open court under oath," Hays announced to the court that "[t]he defense desires to call Mr. Bryan as a witness." Bryan did not have to testify (Judge Raulston left the decision to Bryan), but Bryan—falling for Darrow's trap—took the witness stand. In the 90-minute examination, Darrow referred to Bryan's "fool religion" and questioned Bryan about his "fool ideas" (e.g., Jonah being swallowed by a whale, Joshua's commanding the sun to stand still to lengthen the day). Bryan, who was less concerned about the earth's age than the influence of evolution on societal morals, eventually admitted that he did not believe in a literal interpretation of the Bible and instead endorsed day-age creationism. *The New York Times* described the Darrow-Bryan encounter as "an absurdly pathetic performance" and noted that "Darrow succeeded in showing that Bryan knows little about the science of the world." Throughout Scopes's trial, John Straton provided a daily editorial for the Hearst newspapers. Straton warned readers that "[w]ithin a year there will be a struggle in nearly every state; the moral decline of the present day started two generations ago when the dark and sinister shadow of Darwinism fell across the fair field of human life. America's educational system will ultimately be wrecked if the teaching of evolution is allowed to continue."

1925 John Scopes, who did not testify at his trial, is convicted and fined $100, which is paid by the *Baltimore Sun* (see also Appendix D). After being convicted, Scopes told the court: "Your honor I feel that I have been convicted of violating an unjust statute. I will continue in the future, as I have in the past, to oppose this law in any way I can. Any other action would be in violation of my ideal of academic freedom—that is, to teach the truth as guaranteed in our constitution, of personal and religious freedom." Scopes was later offered a new contract to continue teaching at Rhea County High School (at a salary of $150 per month), provided he adhered "to the spirit of the evolution law." Scopes declined the offer and enrolled in graduate school at the University of Chicago.

1925 Five days after the Scopes Trial, William "Jimmy" McCartney (the chauffeur for William Jennings Bryan and his wife) discovers Bryan's corpse in an upstairs bedroom of the home of F. R. Rogers. Eulogizing fundamentalists compared Bryan to Jesus Christ, and Scopes's defenders to Pontius Pilate and other biblical villains. Reporter H. L. Mencken (Figure 43) began Bryan's obituary in the *Baltimore Sun* by asking, "[h]as it been duly marked by historians that William Jennings Bryan's last secular act on this globe of sin was to catch flies?" Bryan was buried atop a tree-covered hill in the south end of Arlington National Cemetery, where he rests with his wife Mary (1861–1930) beneath the tiny inscription, "He Kept the Faith." Bryan's death triggered discussions throughout the country over who would succeed him as leader of the fundamentalist movement. This story, which appeared in the *Detroit Free Press* was typical: "With the passing of William Jennings Bryan, recognized leader of the Fundamentalists in their campaign to prevent the teaching of the theory of evolution in every tax-supported school in the United States, as well as to quash Modernism within the Christian Church itself, the question arises, 'On whose shoulders will the mantle of leadership fall? Who in the movement is capable of bringing the ambitious program to fruition?' Fundamentalists who have been closely connected with the fight look only to one person for leadership. He is the one who recognized the movement, or largely so, and has been its directing head, as Mr. Bryan has been its contact with the lay or outside world. This man is William B. Riley, executive secretary of WCFA and pastor of the First Baptist Church, Minneapolis."

1925 In an article in the *American Fundamentalist*, New York's antievolution firebrand John Straton praises fundamentalists in the South, "where women are still honored, where men are still chivalric, where laws are still respected, where home life is still sweet, where the marriage vow is still sacred, and where man is still regarded, not as a descendant of the slime and beasts of the jungle, but as a child of God." Straton asked if the AMNH (Figure 30) was "poisoning the minds of school children by false and bestial theories of evolution? Ought not the Bible to be exhibited at the museum as well as lot of musty old bones? Isn't Genesis right?. . . . It is treason to God Almighty and a libel against the human race. Better wipe out all the schools than undermine belief in the Bible by permitting the teaching of evolution." The museum dismissed Straton's attack as "rhetoric and rubbish."

1925 In *Concerning Evolution*, biologist J. A. Thomson (1861–1933) claims that creationism is a thing of the past, adding "we do not know of any competent naturalist who has any hesitation in accepting evolution."

1925 Adolf Hitler's (1889–1945) *Mein Kampf*, written during his imprisonment following a failed coup attempt, invokes a crude and distorted version of biological evolution to justify Nazi aggression. Hitler declared that destroying the Jews is the "Lord's work" and that nature showed that race-mixing is "original sin." Hitler's "evolutionism" invoked Lamarckian progress while stressing "racial purity." *Mein Kampf* does not mention Charles Darwin, natural selection, or biological evolution. Hitler suggested that he believed in a young earth, noting that the earth "will, as it did thousands of years ago, move through the ether devoid of men."

1925 The Bible Crusaders of America is established by Florida real estate tycoon George Washburn to continue William Jennings Bryan's crusade against the teaching of evolution in public schools. The Crusaders enlisted virtually all of the leading antievolutionists for their cause, including John Straton and William Bell Riley. Washburn claimed that "[t]he Bible Crusaders are destined to spread over the entire universe before many months have passed." The Bible Crusaders of America's *Crusaders' Champion*, which absorbed John Roach Straton's *Fundamentalist* and T. T. Martin's *Conflict*, promoted Washburn as "the successor of William Jennings Bryan."

1925 Despite his vocal opposition to the teaching of evolution, T. T. Martin adopts the pseudonym "J. J. Boyer" and enters an eloquent essay titled "Why Evolution Should Be Taught in Our Schools Instead of the Book of Genesis" in a contest sponsored by Science League of America. Martin did not win the $50 prize.

1925 The Bryan Bible League is founded by evangelist Paul W. Rood to continue William Jennings Bryan's campaign against the teaching of evolution while defending "the historic position of evangelical Christianity." The league, which was endorsed by Bryan's widow, was one of the few antievolution organizations to form on the West Coast.

1925 Thirty thousand members of the KKK meet in Washington, DC and pay tribute to William Jennings Bryan (Figure 42). At this meeting, the Klan became the first national organization to call for "equal time" for evolution and creationism in public schools. This demand was renewed in the 1970s and 1980s and culminated with *McLean v. Arkansas Board of Education* (1982) and *Edwards v. Aguillard* (1987), which stipulated that laws requiring "equal time" and "balanced treatment" for creationism were unconstitutional. The Klan supported, and was supported by, several leaders of the antievolution movement in the early 1900s, including Bob Jones, Frank Norris, Billy Sunday, and William Jennings Bryan.

1925 Famed plant-breeder Luther Burbank (1849–1926) labels the Scopes Trial "a great joke, but one which will educate the public and thus reduce the number of bigots."

1926 After months of preaching about the evils of evolution and other aspects of modernism, Robert "Bob" Jones (1883–1968) founds Bob Jones College in Panama City, Florida. The school later moved to Greenville, South Carolina, and in 1947 was renamed Bob Jones University. The university—whose creed includes belief in "the creation of man by the direct act of God"—banned the admission of blacks until 1970 and continues to be a citadel of creationism. The biology department promotes young-earth creationism by training "Christian biologists who see the living world indelibly marked with the fingerprints of a God of limitless wisdom and power." Today, instructors at Bob Jones University produce pro-creationism textbooks and other materials that denounce evolution. They claim evolution produces immorality and that conflicts between science and religion are due to mistakes and biases of scientists. Bob Jones University Press is the largest textbook supplier to home school families in America.

1926 Edward Young Clarke, a membership director of the KKK, founds the Supreme Kingdom in Atlanta to continue William Jennings Bryan's crusade against the teaching of evolution. At its first meeting of the Supreme Kingdom, Clarke claimed that "it is the theory of evolution which has swept the country that is causing the very foundations of liberty, morals and Christianity to totter." Clarke organized the Supreme Kingdom like the Klan, hired Fred Rapp—Billy Sunday's former business manager—to promote the organization, and offered to pay militant antievolution evangelist John Straton $30,000 to give 60 antievolution lectures. Clarke had ambitious plans for the organization, including the construction of a home in Florida "for those who grow old in the war against evolution." The Supreme Kingdom's official publication, *Dynamite*, announced a goal of four million members, and Clarke proclaimed in the *Birmingham Post* that "[i]n another two years, from Maine to California and from the Great lakes to the Gulf, there will be lighted in this country countless bonfires, devouring those damnable and detestable books of evolution." The Supreme Kingdom dissolved soon after the *Macon Telegraph* reported that Clarke was pocketing two-thirds of every $12.50 membership fee.

1926 Geologist Giorgio Bartoli's *The Biblical Story of Creation* defends Genesis against "infidel science" and claims that if man evolved, then God is a liar.

1926 Harvard geologist Kirtley Mather (1888–1978) publishes "The Psychology of the Antievolutionist" in *The Harvard Graduates' Magazine*, noting inconsistencies in antievolutionists' claims, such as their opposition to human evolution but not the evolution of plants and other animals. Mather believed that evolution did not contradict Genesis, but rather "affirms that story and gives it larger and more profound meaning." He attended the Scopes Trial as an expert witness for the defense and provided testimony that concluded with "comparing the body structure of monkeys, apes, and man, it is apparent that they are all constructed upon the same general plan." Later editions of Mather's popular book, *The Earth Beneath Us*, incorporated the new idea of plate tectonics.

1926 G. Kingsley Noble (1894–1940), a biologist at the AMNH who had examined Paul Kammerer's (Figure 34) midwife toads, claims in an article in *Nature* that Kammerer's results regarding the acquired inheritance of nuptial pads were faked. William Bateson agreed, claiming that the alleged pad "was no nuptial pad at all, but just a spot of black pigment." Subsequent examinations of one of Kammerer's pickled toads showed that the black pads—the trait allegedly acquired via Lamarckian inheritance—were actually black ink that had been injected into the toad's foot. Kammerer claimed to be astonished by Noble's accusation and denied any wrongdoing. Six weeks later, while on a walk in the Theresien Hills of Austria, Kammerer committed suicide by shooting himself in the head.

1926 Henry Fairfield Osborn's article titled "Why Central Asia?" launches Osborn's "Dawn Man" crusade by claiming that humans originated in Central Asia. Osborn believed that the name "Dawn Man"—which he took from Arthur Woodward's name for Piltdown Man, *Eoanthropus*—gave humans a history "quite distinct from that of the anthropoid apes." Newspapers later published cartoons showing an ape thanking Osborn for claiming that humans did not descend from apes.

1926 In *Inspiration or Evolution?*, William Bell Riley (Figure 40) claims that evolution promotes anarchy, that "science is now the subtle word of Satanic employment . . . [i]n its conception, development and application, evolution is utterly false . . . and so Scripture and this unproven and unprovable hypothesis can never speak together. . . . If it were in my power, I would take every false [non-Fundamentalist] teacher out of every pulpit and professorship in the land." Later, Riley also equated the teaching of evolution with "Hitlerism" (Figure 48).

HITLERISM

OR THE PHILOSOPHY OF

Evolution in Action

BY DR. W. B. RILEY

Figure 48 Although famed antievolutionist William Bell Riley had earlier praised Adolf Hitler's efforts "to foil the Jew's nefarious plot," Riley later turned against Hitler, arguing in *Hitlerism* (1941) "that Hitlerism is nothing other than the philosophy of Evolution in action" that would produce "World-wide . . . Devastation." (*Courtesy of the First Baptist Church, Minneapolis*)

1926 Mississippi Governor Henry Whitfield (1868–1927) signs a law banning the teaching of human evolution in the state's public schools. This law, which became the last surviving antievolution law, was not declared unconstitutional until 1970 (see also Appendix D).

1926 Russian geneticist Sergei Chetverikov's most influential paper, "On Several Aspects of the Evolutionary Process from the Viewpoint of Modern Genetics," predicts that because mutations are random events that tend to produce recessive alleles, mutations can persist in populations (populations should "soak up mutations like a sponge"), thereby providing significant genetic variation upon which natural selection can act. Chetverikov's research verified that natural populations house substantial genetic variability, thereby linking genetics with Darwin's ideas about adaptive evolutionary change. However, because his relatively few publications (26 during his career) appeared only in Russian journals, Chetverikov's work was relatively unknown by the founders of the Modern Synthesis.

1926 Speaking in Berlin at the Fifth International Congress of Genetics, Hermann Muller claims that mutations can be induced by X-rays, that genes are the basis of life, and that the evolution of life is traceable to the first gene. Muller won the Nobel Prize in Physiology or Medicine in 1946.

1926 Tennessee Governor Austin Peay breaks ground in Dayton for what becomes Bryan College (named for William Jennings Bryan). The ceremony was attended by more than 10,000 people.

1926 The ACLU uses a mass mailing to AAAS members to raise money to pay debts generated by the Scopes Trial. For several years after Scopes's conviction, the ACLU searched for another challenger to the Butler Law, but found no volunteers. Laws banning the teaching of human evolution in Tennessee, Arkansas, and Mississippi remained in place for more than 40 years.

1926 The Southern Baptist Convention repudiates "as unscriptural and scientifically false every claim of evolution that declares or implies that man evolved to his present state from some lower order of life. . . . [We] accept Genesis as teaching that man was the special creation of God, and reject every theory, evolution or other, which teaches that man originated in, or came by way of, a lower animal ancestry."

1926 Young-earth creationist Gerald Winrod (1900–1957)—an anti-Semitic Nazi-sympathizer—founds the Defenders of the Christian Faith to continue the antievolution work of William Jennings Bryan. The Defenders soon shifted their focus to the "Negro menace." Winrod was the prototype for the Buzz Windrip character in Sinclair Lewis's *Elmer Gantry* (1927).

1926 United States Representative Bill Lowrey (1862–1947) of Mississippi, a friend of T. T. Martin, tries to attach the Summers Amendment to an appropriations bill for the District of Columbia. That same year, United States Senator Cole Blease (1868–1942) of South Carolina—proclaiming that he was "on the side of Jesus Christ"—attached an amendment to the Dill Radio Control Bill that would have banned radio broadcasts about "the subject of evolution." Blease's amendment was defeated.

1926 Southern Baptist firebrand Frank Norris declares "we have heroically and triumphantly delivered the knockout blow against evolution and evolutionists."

1927 Arthur Holmes (1890–1965) claims in *The Age of Earth: An Introduction of Geological Ideas* that the earth "is just over 3,000 million years" old (see also Appendix A).

1927 Scottish anatomist Arthur Keith (1866–1955) tells BAAS that the human mind is merely a reflection of nervous activity in the brain.

1927 Despite the defeat of antievolution legislation in his home state of Minnesota, William Bell Riley (Figure 40) urges his followers to join the "fight to the finish, a fight that asks no quarter from the world, the flesh, or the devil." The same year, Riley lamented that the work of the WCFA no longer made headlines in newspapers.

1927 At its annual meeting in Atlanta, WCFA begins to formulate an antievolution bill to be presented in legislatures in every state as well as in Europe, China, and South America. This meeting was the organization's high-water mark; soon thereafter, membership began to decline, and by 1930 the association included no scheduled talks about evolution.

1927 English geologist and mountaineer Noel Odell (1890–1987) describes fossilized seashells atop Mt. Everest.

1927 British urologist George Buckston Browne (1850–1946) buys Down House (Figure 18) from the Darwin heirs for £4,250. After spending £10,000 on repairs and providing an endowment of £20,000, Browne gave Down House to the BAAS in 1929.

1927 In *Buck v. Bell*, the U.S. Supreme Court rules (by a vote of 8 to 1) that the state of Virginia can forcibly sterilize Carrie Buck. Invoking the "public welfare," Justice Oliver Wendell Holmes, Jr., (1841–1935) wrote the famous decision upholding the statute instituting compulsory sterilization "for the protection and health of the state," noting that "it is better for all the world if, instead of waiting to execute degenerate offspring for crime, or to let them starve for their imbecility, society can prevent those who are manifestly unfit from continuing their kind. . . . Three generations of imbeciles are enough." Buck was sterilized five months later at Virginia's State Colony for Epileptics and Feebleminded, which was headed by Dr. James H. Bell. Virginia's law was not repealed until 1974. By the end of the 1970s, more than 60,000 people had been forcibly sterilized in the United States (more than half in California alone). *Buck v. Bell* is often cited as the U.S. Supreme Court's worst decision.

1927 Maynard Shipley's *The War on Modern Science: A Short History of the Fundamentalist Attacks on Evolution and Modernism* notes that "the armies of ignorance are being organized, literally by the millions, for a combined political assault upon modern science. . . . For the first time in our history, organized knowledge has come into open conflict with organized ignorance."

1927 Sinclair Lewis (1885–1951) writes *Elmer Gantry*, a best-selling novel. The main character in the satirical story—a hypocritical evangelist—was based partly on antievolution crusader John Roach Straton. Lewis dedicated the book to H. L. Mencken (Figure 43) "with profound admiration."

1927 While appealing John Scopes's conviction, Clarence Darrow—who defended Scopes without pay—chides fundamentalists with his now-famous line: "With flying banners and beating drums, we march back to the glorious ages of medievalism." Darrow's villainy at the Scopes Trial became fodder for countless fundamentalist sermons across the United States. The Tennessee Supreme Court set aside Scopes's conviction and recommended that he not be retried

("We see nothing to be gained by prolonging the life of this bizarre case"), thereby ending one of the most famous court cases in American history. Scopes had no role in the appeal and did not return to Tennessee for the hearing or decision. When motions for a new hearing were rejected, Darrow admitted, "[i]t will probably take another case to clear up the matter." That case did not come along until 1965, when Arkansas biology teacher Susan Epperson (b. 1941) challenged the Arkansas law banning the teaching of human evolution (see also Appendix D). Although John Scopes's legal proceedings were now over, his case continued to be celebrated in popular culture and dissected in legal casebooks.

1928 Arkansans vote 108,991 to 63,406 to enact Initiative Act No. 1, banning the teaching of human evolution in public schools. This law, the only antievolution law ever passed by a popular vote, was not overturned for 40 years, by *Epperson v. Arkansas*.

1929 American astronomer Edwin Hubble (1889–1953) provides evidence for Belgian priest and astronomer Georges-Henri Lemaître's (1894–1966) suggestion that the universe is expanding. Hubble's observation—that light from other galaxies is red-shifted—suggested that the universe began as a "primeval atom," a concept now known as the Big Bang Theory.

1929 Charles and Emma Darwin's Down House opens as a public museum (Figure 18). Today, a restored Down House is maintained by English Heritage and is open for tours most of the year.

1929 Harold Clark's (1891–1986) *Back to Creationism* describes George McCready Price's ideas as "creationism." Clark, a former student of Price, dedicated his book to Price, whom he described as a "Teacher, Friend, Fellow-warrior, and Prophet of the New Catastrophism." Clark also claimed "the world has had enough of evolution . . . in the future, evolution will be remembered only as the crowning deception which the arch-enemy of human souls foisted upon the race in his attempt to lead man away from the Savior. The science of the future will be creationism. . . . The time is ripe for a rebellion against the dominion of evolution."

1929 In response to Presbyterian preacher Harry Rimmer's offer of $1,000 for anyone who could prove evolution, New York atheist William Floyd cites mathematical problems associated with feeding dead quail to all the children of Israel (*a la* Numbers 11:31). Despite litigation, Rimmer never paid Floyd.

1929 In a revolt against modernism at Princeton Seminary, J. Gresham Machen—a fundamentalist who often proclaimed, "I never called myself a fundamentalist"—establishes Westminster Theological Seminary in Philadelphia. Unlike most other fundamentalists, Machen had a scholarly approach to science and theology, and he refused to denounce evolution. Machen, a day-age creationist, believed that science and religion were concerned with the same thing—"facts"—and that "the church is perishing through a lack of thinking, not through an excess of it." Some followers of William Jennings Bryan wanted Machen to be the first president of Bryan Memorial University.

1929 Swedish paleontologist Gunnar Säve-Söderbergh (1910–1948) discovers an animal (a tetrapod from the Devonian; see also Appendix B) having traits of fish and amphibians. The transitional animal—the earliest tetrapod—was later named *Ichthyostega soderberghi*.

1930 Bryan College (Figure 49) opens in the old high school building where John Scopes allegedly taught evolution. Scopes Trial instigator George Rappleyea hoped to create a liberal college in Dayton to offset Bryan College, but that college was never built.

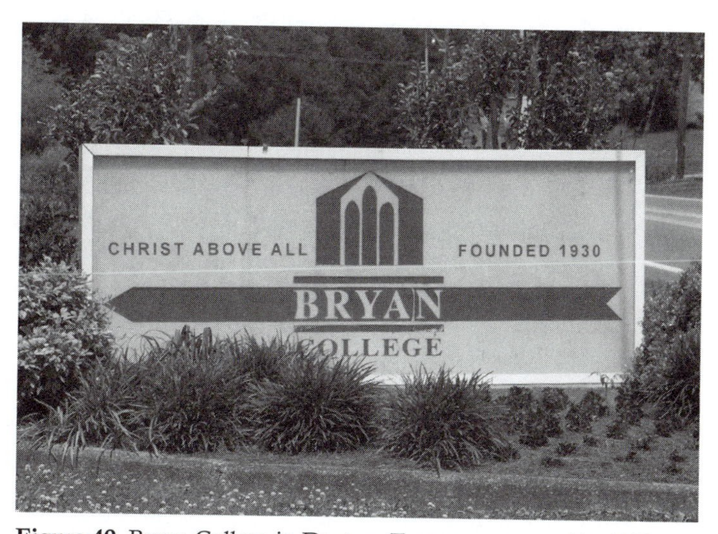

Figure 49 Bryan College in Dayton, Tennessee, opened in 1930 as a monument to the ideals of William Jennings Bryan. Today, Bryan College is a highly regarded Christian college. (*Randy Moore*)

1930 Historian William Sweet's (1881–1959) popular book *The Story of Religion in America* depicts Clarence Darrow's questioning of William Jennings Bryan at the Scopes Trial as "Fundamentalism's last stand."

1930 Walter Lammerts (1904–1996), a disciple of George McCready Price who later helped found the Creation Research Society (CRS), becomes the first prominent creationist to earn a Ph.D. in a scientific discipline. His degree was from the University of California at Berkeley. Lammerts claimed that there can be "no discrepancies" between nature and the Bible and insisted on "the absolute fixity of species." His approach to evolution was simple: "If a man is such a stupid fool he can't see that evolution is wrong, I'm not going to try to convince him."

1930 In his influential *The Genetical Theory of Natural Selection*, Ronald Fisher (Figure 50) becomes the first biologist to merge Darwinian natural selection with Mendelian genetics. Fisher argued that natural selection can accumulate the effects of otherwise random mutations; if a gene confers an advantage and increases the rate of reproduction, its frequency in the population increases. The first two chapters of the 12-chapter book stated Fisher's main points, and the final chapters promoted eugenics. Harvard evolutionary biologist Stephen Jay Gould later praised *The Genetical Theory of Natural Selection*—which was dedicated to Charles Darwin's son Leonard Darwin, an ardent supporter of eugenics—as "the keystone for the architecture of modern Darwinism."

1931 Frederick Allen's (1890–1954) influential *Only Yesterday* portrays the Scopes Trial as blind fundamentalism versus enlightened skepticism and becomes the standard interpretation of events in Dayton. This interpretation was later reinforced by *Inherit the Wind*.

1931 Father Ernest Messenger's *Evolution and Theology* confronts the disagreement between the dogma of the Catholic Church and the implications of continuing scientific discoveries, especially in relation to evolution: "From the theological point of view . . . Scripture neither teaches nor disproves the doctrine of the evolution of the human body. . . . Scripture really teaches spontaneous generation." Messenger advocated theistic evolution, in which organic change—including human evolution—was allowed but guided by God. In particular, the human soul was strictly the product of divine intervention, an accommodation between

Figure 50 Ronald Fisher was an architect of the Modern Synthesis and an outspoken eugenicist. Fisher later claimed that famed geneticist Gregor Mendel's data were faked. (*American Philosophical Society*)

theology and science that would become the stated position of the Catholic Church.

1931 In a paper titled "Evolution in Mendelian Populations" published in *Genetics*, Sewall Wright proposes a "three-phase shifting-balance theory of evolution" that depicts evolution as a shifting balance between natural selection, genetic drift (random changes in gene frequencies in small populations), and migration. Wright's ideas, along with those of J. B. S. Haldane, Ronald Fisher, and Theodosius Dobzhansky, produced the Modern Synthesis, a

unification of evolution by natural selection with Mendelian genetics and population genetics. Wright was a student of William Castle (1867–1962), the first biologist to use fruit flies (*Drosophila melanogaster*) to study genetics.

1932 After a seven-year effort, the ACLU abandons its search for another volunteer to challenge laws in Arkansas, Tennessee, and Mississippi banning the teaching of human evolution. The various antievolution laws remained unchallenged until 1965 (see also Appendix D).

1932 J. B. S. Haldane publishes *The Causes of Evolution,* which unifies classical genetics, cell biology, and biochemistry with population genetics and evolutionary biology. The eccentric Haldane simultaneously studied a variety of topics in biology, from enzyme kinetics to inheritance and mutation, and this broad perspective allowed him to appreciate how evolution operates. Haldane was an ardent communist until after World War II, and in 1961, he left England and became an Indian citizen. Haldane, like Ronald Fisher (Figure 50), did little experimental work, instead focusing on theoretical approaches to evolution and population genetics. *The Causes of Evolution* became a classic textbook in theoretical population genetics.

1932 John Scopes campaigns to be congressman-at-large from Kentucky on the Socialist ticket. Scopes lost, but made a respectable showing in the election, leading all the other minority-party candidates in the race. After his loss, Scopes returned to the oil business, working for companies in Houston and Louisiana while living quietly with his wife Mildred and their two children (and occasionally testifying before Congress as an oil expert). While George Rappleyea and others continued to exploit their association with Scopes's famous trial, Scopes avoided the spotlight for the next 30 years.

1932 At the Third (and last) International Congress of Eugenics, held at the AMNH (Figure 30), American geneticist Hermann Muller denounces negative eugenics (e.g., state-enforced sterilization of the "unfit"). Charles Davenport, a member of NAS and a strong advocate of negative eugenics, tried to bar Muller's presentation. The Congress included a presentation by Ronald Fisher (Figure 50; on behalf of Leonard Darwin) that predicted the doom of civilization if more eugenics measures were not implemented. In the United States, exhibits and discussions about eugenics were often very popular (Figure 51).

Figure 51 In the 1920s and 1930s, the American Eugenics Society and other groups began sponsoring "Fitter Family Contests" that included awards for families judged to be superior "human stock." Many churches proposed requiring certificates of "eugenic fitness" before approval of church weddings. Although the involuntary sterilization of hundreds of thousands of people and the killing of millions more by the Nazis forced a reevaluation of eugenics, notables such as Julian Huxley and James Watson subsequently advocated the benefit of eugenics. (*American Philosophical Society*)

1932 Ronald Fisher's (Figure 50) ideas become the basis for Sewall Wright's influential "adaptive landscape" visualization of fitness peaks and valleys. Wright, who studied the fates of genes in small, isolated populations, noted that not all genetic changes in species are adaptive, while showing that natural selection drives populations toward increased fitness. This apparent, and often inaccurately portrayed, downplaying of the role of natural selection was rejected by many biologists and cited by antievolutionists as a refutation of mainstream evolutionary biology.

1932 The Evolution Protest Movement is founded in England, with Sir John Ambrose Fleming (1849–1945; developer of the first workable electronic vacuum tube) as president. The organization is now known as the Creation Science Movement and exists to combat "[t]he hard-nosed humanism of evolutionism."

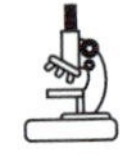

1933 German physicist Ernst Ruska (1906–1988) builds the first electron microscope, which provides greater resolution, and at higher magnifications, than the light microscope. This invention helped biologists discover cellular aspects of evolution.

1933 The Chicago World's Fair opens, with Sinclair Oil's seven life-size dinosaurs as the main attraction (Figure 52). Millions of people toured the Sinclair exhibit.

CHICAGO WORLD'S FAIR EDITION

CIRCULATION 2,000,000 | BIG NEWS | SECOND EDITION

Published by Sinclair Refining Company (Inc.), 45 Nassau Street, New York, N. Y.

MILLIONS SEE WEIRD SINCLAIR DINOSAURS

Story on Page 2.

Figure 52 In 1933–1934, the Chicago World's Fair (also referred to as The Century of Progress International Exposition) celebrated Chicago's centennial. Among the Fair's top attractions was an exhibit of life-size dinosaurs sponsored by Sinclair Oil. The popularity of the exhibit fueled dinosaur-mania throughout the United States. (© *Sinclair Oil Corporation. All rights reserved, used by permission*)

1933 Herbert W. Armstrong (1892–1986) founds the Worldwide Church of God, an influential church that adamantly opposes evolution. Armstrong, who decided in 1927 that the rejection of evolution was one of seven "conclusions" that would guide his life, claimed that evolution is "a false theory" because it contradicts the Bible.

1933 In H. G. Wells's *The Shape of Things to Come*, the world is governed by a benevolent dictatorship, which creates a global utopia by promoting science and destroying religion. Wells was a former student of Thomas Huxley.

1933 Richard Goldschmidt (1878–1958) coins the phrase *hopeful monster* to describe the products of sudden evolutionary jumps he felt were required to account for speciation. *Monster* referred to macromutations that might arise in a single generation and that provided a selective advantage in a changing environment; *hopeful* referred to Goldschmidt's belief that a new macromutation might be so adaptive that it would be selected as the new norm. The concept of evolution by way of hopeful monsters was soon discredited, but creationists continued to portray it as science's accepted model for evolutionary change, thereby suggesting the improbability of "random" evolution producing complex adaptations in a single step.

1934 Georgyi Frantsevich Gause's (1910–1986) *The Struggle for Existence* introduces the *competitive exclusion principle* (i.e., that two species utilizing the same resource in the same way cannot permanently coexist). Competition between species became viewed as a dominant selective force structuring natural communities. Gause's proposal later influenced David Lack's conclusions about the Galápagos finches.

1934 In *Modern Discoveries Which Help Us to Believe*, George McCready Price dismisses Charles Darwin as being "of the slow, unimaginative type . . . singularly incapable of dealing with the broader aspects of any scientific or philosophic problem."

1934 Barnum Brown's discovery of a giant "dinosaur graveyard" in the Bighorn Mountains leads him to speculate that the dinosaurs had died while seeking water in a drying lakebed. Six years later, Brown's imagined scene was popularized in Walt Disney's *Fantasia*, in which the dinosaurs turned into oil (Harry Sinclair advertised his higher-priced "premium" gasoline as if its quality depended on a paleontological "aging" process). Among paleontologists, Brown is almost universally recognized as the greatest fossil hunter of all time; more than 30 of his discoveries remain on display at AMNH today (Figure 30). Dozens of crates of Brown's fossils remain unopened at the AMNH.

1934 Laura FitzRoy—the captain's daughter—tells Nora Darwin (Charles' and Emma's granddaughter) that "Charles Darwin was a great man—a genius—raised up for a special purpose. But he overstepped the mark."

1935 In a rare public statement, John Scopes claims that he is "not interested" when the Tennessee legislature votes to retain its ban on the teaching of human evolution. In 1950, Scopes told a reporter "my friends who know about [my trial in Dayton] never bring it up in my presence."

1935 Theodosius Dobzhansky (1900–1975; Figure 53) describes the concept of a species as an "actually or potentially interbreeding array of forms," an idea that colleague Ernst Mayr later called the *biological species concept*. Subsequent work by Dobzhansky demonstrated

Figure 53 Theodosius Dobzhansky (left) was a major contributor to the Modern Synthesis. Dobzhansky is shown here with an unnamed assistant and jars containing fruit flies. (*American Philosophical Society*)

how the formation of sterile hybrids between *Drosophila* species kept these species separate, leading Dobzhansky to develop the influential concept of "isolating mechanisms" for the process of speciation.

1935 George McCready Price, Byron Nelson (1893–1972), and others form the short-lived Religion and Science Association to oppose the teaching of evolution. The association claimed that God does not necessarily obey the laws of nature.

1935 Plant geneticist Nikolai Vavilov (1887–1943), a former student of William Bateson and one of the most accomplished biologists in the Soviet Union, amasses a collection of 250,000 seeds of cultivated plants (the most extensive collection in the world). Nevertheless, Vavilov was branded by Trofim Lysenko as a "saboteur" who was doing "destructive" work. Two years later, Lysenko—who was awarded the Order of Lenin (his country's highest honor) eight times—named Vavilov an enemy of the people.

1936 Robert Broom finds more specimens of *Australopithecus* in a cave at Sterkfontein, South Africa. Some of Broom's fossils resembled Raymond Dart's discovery, but others looked more robust and were therefore named *Australopithecus robustus*. Broom's work changed the study of human evolution by showing that australopithecines were some of the earliest hominins. In 1959, paleontologists Louis and Mary Leakey discovered an even more robust *Australopithecus* at Olduvai Gorge that they named *Australopithecus boisei* (see also Appendix C). Sterkfontein is now designated the Cradle of Humanity World Heritage Site.

1936 The University of Heidelberg awards the ERO's Henry Laughlin an honorary degree in recognition of his work regarding eugenics.

1936 English ornithologist Percy Lowe's (1870–1948) paper "The Finches of the Galapagos in Relation to Darwin's Conception of Species" in the journal *Ibis* introduces the term *Darwin's finches*. However, the term was not made famous until 1947 by David Lack's book *Darwin's Finches*.

1936 Germany's Heinrich Himmler (1900–1945) founds an organization named *Lebensborn* (*Fountain of Life*) to produce "Aryan" Germans as the genetic foundation of a "master race."

1936 The renowned Akeley Hall of African Mammals opens in AMNH. Carl Akeley, who died in 1926, was buried in an area depicted in the hall's gorilla diorama.

1936 In the inaugural issue of *Annals of Science*, Ronald Fisher (Figure 50) praises Gregor Mendel (Figure 27) while questioning the validity of Mendel's data, but Fisher does not claim fraud. Fisher argued that Mendel's data fit his conclusions too well, as judged by the chi-square test introduced by Karl Pearson in 1900 (Fisher estimated the probability of getting such good results under standard genetic models to be less than 0.0001). In 1955—the Mendel Centennial—Fisher suggested that a "gardener's assistant" may have deceived Mendel. Few people accepted Fisher's accusations.

1936 The main entrance to AMNH—complete with a skeleton of *Barosaurus* defending her offspring from *Allosaurus*—is completed as a memorial to Theodore Roosevelt.

1936 Oscar Riddle (1877–1968) of the Carnegie Institution of Washington reports to the AAAS that the antievolution movement has persisted because of poor textbooks, the failure of biologists to educate the public about evolution, and poor teaching. Riddle argued "the presumption that for making a teacher of biology there is any substitute for long-continued training under our best college biological departments is an expensive fraud, and the extent to which that presumption is being enforced in one or another guise is now an educational disgrace."

1936 Russian biologist Aleksandr Oparin argues in *The Origin of Life* that life arose through chemical evolution of organic compounds in a reductive environment consisting largely of ammonia, hydrogen, and methane. (*The Origin of Life* was an expansion of a booklet that Oparin first published in the Soviet Union in 1924.) Oparin claimed that a type of natural selection acted on complex molecules in the earth's "primordial soup" and, in the process, eventually produced life. Oparin's work stimulated a flurry of work and debate about the origin of life.

1936 While a Senior Geneticist at the Institute of Genetics of the Academy of Sciences of the Soviet Union, Hermann Muller defends Nikolai Vavilov while denouncing Trofim Lysenko as a fraud. Soon thereafter, Muller had to leave Moscow.

1937 Theodosius Dobzhansky's (Figure 53) *Genetics & the Origin of Species* helps found the Modern Synthesis of evolutionary biology by reconciling the fieldwork of naturalists with the mathematical models of population geneticists. Dobzhansky demonstrated that populations typically harbor a surprising amount of genetic variation, which could allow populations to evolve quickly as environmental conditions change. He also redefined evolution in genetic terms as "a change in frequency of an allele within a gene pool." Dobzhansky remarked "evolutionary plasticity can be purchased only at the ruthlessly dear price of continually sacrificing some individuals to death from unfavorable mutations." Dobzhansky—a theistic evolutionist ("Evolution is God's, or Nature's method of creation")—melded these experimental findings with the theoretical work of Ronald Fisher, Sewall Wright, and J. B. S. Haldane. Dobzhansky's book has been hailed as second only to *On the Origin of Species* in terms of importance to evolutionary biology. In subsequent editions of the book, the importance of nonadaptive evolutionary change (i.e., genetic drift as championed by Wright) was reduced in importance relative to selection. Historians have identified this as a critical change in the transformation of the Modern Synthesis into the "neo-Darwinian" perspective that emphasized selection as the main force producing evolution.

1937 Trofim Lysenko, a former peasant, is appointed deputy to the Supreme Soviet and Director of the Odessa Institute of Genetics and Plant Breeding. Despite his meager education, Lysenko had become well known for his claims that vernalized seeds could pass on their acquired (i.e., Lamarckian-inherited) cold-hardiness to offspring. Lysenko promised rapid advances in agriculture, claiming that he could produce new strains of wheat and other crops within three years (instead of the 12 years required by his mentor Nikolai Vavilov). Lamarckian inheritance continued to be accepted by some geneticists and fit nicely with the Soviet philosophy that the environment determines an organism's traits, in contrast with the supposed genetic determinism of Darwinian evolution. Lysenko eventually became Director of the Lenin Academy of Agricultural Science and dictated to Soviet science a view of genetics that repudiated Mendel and Darwin.

1938 Ella Smith's textbook *Exploring Biology* notes that "no one acquainted with the facts doubts that evolution, or continued change in plants and animals, has taken place. No one has discovered a single fact to disprove the theory of evolution, and the facts that establish its truth are abundant. . . . Evolution is a fact." Most biology textbooks, however, continued to omit or downplay evolution.

1938 Julian Huxley—grandson of Darwin's advocate Thomas Huxley and the then-Secretary of the Zoological Society of London—asks English schoolteacher and amateur ornithologist David Lack (1910–1973) to study Galápagos finches for an entire breeding season. This work allowed Lack and his colleagues to document natural selection in the wild acting through interspecific competition. Huxley later declared the antievolution movement to be "dull, but dead." Like those before and after him to make such claims, Huxley was wrong.

1938 French biologist Edouard Chatton (1883–1947) proposes the names *prokaryote* and *eukaryote* based on the absence and presence, respectively, of nuclei.

1938 Harold Clark, a protégé of young-earth creationist George McCready Price, visits oilfields in Texas and Oklahoma and sees firsthand that older fossils occur in lower sediments than younger fossils. He confessed to Price that "the rocks do lie in a much more definite sequence than we have ever allowed. [Your claims about Flood geology] do not harmonize with the conditions in the field. . . . The same sequence is found in America, Europe, and anywhere that detailed studies have been made." Price, who claimed that Clark had been associating with "tobacco-smoking, Sabbath-breaking" evolutionists and had been influenced by Satan, unsuccessfully tried to have Clark condemned by Adventist leaders.

1938 In an editorial in *Science*, Oscar Riddle rallies scientists fighting for the acceptance of evolution by asking "shall the public that decides the fate of our democracy conceive nature and man as research discloses them or as uninformed and essentially ignorant masses can variously imagine them?"

1938 Off the coast of South Africa, Hendrik Goosen catches a 5-foot, 127-pound coelacanth (*Latimeria chalumnae*), a fish thought to have become extinct more than 65 million years ago. Another living specimen of this so-called "living fossil"—which has limblike fins—was not seen for 14 years.

1938 Sir Arthur Keith dedicates a memorial to Charles Dawson at the Piltdown gravel pit where Piltdown Man was allegedly discovered.

1938 The Deluge Society (formally known as The Society for the Study of Deluge Geology and Related Sciences) is founded by George McCready Price and others, and from 1941 to 1944 publishes 20 issues of the *Bulletin of Deluge Geology and Related Sciences*. Most members of the society, including Price, were Seventh-Day Adventists. By 1945, the Society included more than 600 members, all of whom were required to believe that creation lasted no more than "six literal days."

1938 The National Association of Biology Teachers (NABT) publishes the 1st edition of *The American Biology Teacher*.

1939 The insecticidal properties of DDT are discovered by Swiss chemist Paul Müller (1899–1965). Within a few years after DDT's introduction, several species of insects had evolved resistance to the chemical.

1939 Nikolai Vavilov claims that "Lysenko's position [on heredity] not only runs counter to the group of Soviet genetics, it runs counter to all of modern biology." Lysenko responded that, "I do not recognize Mendelism. . . . I do not consider former Mendelian-Morganist genetics a science" and fired all geneticists having Darwinist views. While on a plant-collecting trip in the Ukraine, Vavilov was arrested for treason, sabotage, espionage, and counterrevolution and was sentenced to death before a firing squad. Although his death sentence was commuted, Vavilov—who had hoped to use genetics to feed the world—was starved to death in a Siberian prison camp.

1939 Fossil-hunter Roland Bird (1899–1978) describes the Paluxy Riverbed fossils, a series of impressions suggested by some as evidence that humans co-occurred with dinosaurs (rather than arising millions of years after dinosaurs went extinct). In reference to human-shaped footprints, Bird noted that "[i]t was ridiculous to think they were human footprints," yet these prints, in strata containing dinosaur prints, were cited by young-earth creationists for years to come. In 1986, John Morris of ICR declared the footprints to be "at best, ambiguous and unusable as an antievolutionary argument at the present time." Today, the dinosaur tracks excavated by Bird are in AMNH on display behind *Apatosaurus*, which in 1905 became the first sauropod dinosaur ever mounted (the giant fossil was remounted in 1992).

1940s Many fundamentalists, tired of the militant attitude of their movement, begin calling themselves "evangelicals."

1940 Will Houghton (1887–1947), president of the Moody Bible Institute, argues in *Moody Monthly* that Nazism is based on evolution and materialism, and that universities are responsible for modernism having replaced orthodox faith with Darwinism and Marxism.

1940 In *Science and Truth*, geologist L. Allen Higley—the first president of the Religion and Science Association—advocates gap creationism while claiming that God's "entire plan of salvation" is written in the constellations. Higley declared bacteria and mountains to be a "perversion of nature."

1940 Eugène Dubois, whose discovery of *Pithecanthropus* was questioned for decades, dies an embittered man. He is buried in an unconsecrated corner of a cemetery in Venlo, Netherlands, under a carving of a skull and crossbones.

1941 American geneticists George Beadle (1903–1989) and Edward Tatum (1909–1975) use vitamin-deficient mutants of the fungus *Neurospora* to show that genes control chemical reactions in cells. Their discovery became known as the "one gene, one enzyme" hypothesis.

1941 Frank Marsh creates a system of "discontinuity systematics" that he calls *baraminology*, from the Hebrew words *bara* ("created") and *min* ("kind"). Baraminologists, who view the *baramin* as a taxonomic rank representing the "created kinds" of Genesis, used baraminology to document boundaries between "microevolution" and "macroevolution," as well as to prove the existence of a designer. Marsh, a student of flood geologist George McCready Price and a self-described "fundamentalist scientist," declared that Satan is a "master geneticist" and that the black skin of African Americans is an "abnormality" resulting from Satan's use of hybridization to destroy the original perfection of life. Like many fellow Seventh-Day Adventists, Marsh believed that the world was the site of "a cosmic struggle between the Creator and Satan."

1941 American historian Howard K. Beale (1899–1959) reports in *A History of Freedom of Teaching in American Schools* that one-third of teachers in the United States are "afraid to express acceptance of evolution." Writer Irving Stone (1903–1989) declares the Scopes Trial a "death blow" to creationism. The following year, a national survey showed that fewer than half of high school biology teachers in the United States teach evolution.

1941 William Houghton, president of Moody Bible Institute, convenes in Chicago a meeting of Christian scientists who are sympathetic to his belief that evolution damages society. The group founded the American Scientific Affiliation (ASA), an organization that provided information to "accurately" interpret "the facts of science and the Holy Scriptures." ASA members had to sign a statement attesting to the inerrancy of the Bible that claimed "I cannot conceive of discrepancies between statements in the Bible and the real facts of science." ASA members have included Henry Morris and John Whitcomb, Jr.; these two young-earth creationists met at an ASA meeting but both later left the organization as it turned away from George McCready Price's Flood geology. In 1961, Morris and Whitcomb published *The Genesis Flood*, and Morris and Duane Gish (also a former member of the ASA) formed the CRS in 1963. ASA continued to be an active organization (e.g., it published *Teaching Science in a Climate of Controversy: A View from the American Scientific Affiliation* in response to publications by NAS promoting evolution) and counted Francis Collins, former Director of the National Institutes of Health Human Genome Research Institute, among its members.

1942 In *Man's Most Dangerous Myth: The Fallacy of Race*, British anthropologist Ashley Montagu (1905–1999) questions the validity of race as a biological concept and rates Charles Darwin's *On the Origin of Species* "as influential, in virtually every aspect of human thought" as the Bible.

1942 American astronomer Harvey Nininger (1887–1986) suggests that asteroid impacts could explain otherwise puzzling gaps in the fossil record.

1942 In *Systematics and the Origin of Species,* evolutionary biologist Ernst Mayr (1904–2005) draws upon his extensive observations of geographic variation to propose a model for how species

arise. Mayr's ideas established the evolutionary significance of geographic isolation and small selective advantages. His observation that most species of birds in New Guinea have nonoverlapping geographic ranges, combined with indications that islands often harbor their own species, led to the concept of allopatric speciation. Mayr's *allopatric model of speciation* required geographic separation of a once cohesive population into discrete units; these now-isolated populations were then no longer a single interbreeding unit, and natural selection and chance effects could cause the separated populations to diverge into new species. Melding Dobzhansky's and Mayr's *biological species concept* with the related allopatric model was enormously influential in evolutionary biology because it treated species as real evolutionary units created by identifiable processes that are open to study (although speciation is now considered a complex process that may operate differently in different types of organisms). Mayr wrote his famous book in part to refute Richard Goldschmidt's 1933 notion of "hopeful monsters."

1942 Georges-Louis Buffon's (Figure 9) influential *Histoire Naturelle* is reissued with 31 illustrations by Spanish artist Pablo Picasso (1881–1973).

1942 At the Wannsee Conference (so named for being held in the Berlin suburb of Wannsee), senior officials of Nazi Germany announce the "final solution to the Jewish question"—namely, the plan for Jewish deportation, enslavement, and annihilation. This conference produced one of the few explicit mentions of an evolutionary principle by the Nazi party: the mass murder of Jews was justified on the grounds that "the possible final remnant will, since it will undoubtedly consist of the most resistant portion, have to be treated accordingly, because it is the product of natural selection and would, if released, act as a seed of a new Jewish revival."

1942 Julian Huxley publishes *Evolution: The Modern Synthesis*, an overview of the integration of Mendelian genetics, population genetics, and paleontology with evolutionary biology. Huxley was not a major intellectual architect of the Modern Synthesis, but his book popularized the phrase *Modern Synthesis*. The last page of Huxley's book introduced the term *evolutionary biology*, an interdisciplinary branch of biology that has developed into a system of theories to explain different aspects of organismic evolution. Huxley's review of evolutionary theory, which included evidence unavailable to Darwin, re-energized Darwin's ideas. Huxley concluded that, with the successful integration of evolution, biology "no longer presents the spectacle of a number of semi-independent and largely contradictory sub-sciences but is coming to rival the unity of older sciences like physics." Huxley's subsequent perspectives on the centrality of evolution extended beyond biology in that he considered evolutionary biology to be the logical basis for a science-based philosophy of life (describing the concept of God as "not only unnecessary but intellectually dubious") and considered the possibility that evolutionary change was necessarily progressive. Huxley supported eugenics, and in the 1960s he advocated financial inducements or penalties (depending on the traits, good or bad, exhibited by individuals), artificial insemination, and the long-term storage of sperm from particularly fit donors to achieve eugenic goals.

1943 Salvador Luria (1912–1991) and Max Delbrück (1906–1981) report that random mutations, not just selection, can confer resistance in bacteria.

1943 The mass production of penicillin prompts many people to believe that infectious diseases will soon be a thing of the past. However, resistant strains of bacteria appeared within three years, and discoveries of resistant bacteria accompanied the use of every subsequent antibiotic. For example, resistance to streptomycin appeared in the same year that it was approved for use (1947), resistance to tetracycline appeared three years after its approval (1952), and resistance to methicillin and cephalothin (the first antibiotic in the cephalosporin class) appeared two years after their introduction (1959 and 1964, respectively). Whereas pneumococcal pneumonia could be cured in the early 1940s by administering 10,000 units of penicillin four times per day for four days, giving today's sufferers 25 *million* units of penicillin every day has no effect. The evolution of bacteria immune to the effects of antibiotics has occurred countless times and, as one would predict, is most common in areas where the use of antibiotics is highest. Providing strong evidence for selection in action, attempts to vanquish pathogenic bacteria with antibiotics have only selected for stronger forms.

1944 Frank Marsh claims in *Evolution, Creation and Science* that the geologic record agrees with the Bible, in particular with the action of the Noachian Flood. Geneticist Theodosius Dobzhansky (Figure 53) cited *Evolution, Creation and Science* as a "sensibly argued defense of special creation," noting that "in rejecting macroevolution, Marsh's book taught the valuable lesson that no evidence is powerful enough to force acceptance of a conclusion that is emotionally distasteful."

1944 In an article in *The Journal of Experimental Medicine* titled "Studies on the Chemical Nature of the Substance Inducing Transformation of Pneumococcal Types," Oswald Avery (1877–1955), Colin MacLeod (1909–1972), and Maclyn McCarty (1911–2005) show that DNA, rather than protein, is the genetic material. This discovery was a founding event in the development of molecular genetics.

1944 George Gaylord Simpson (1902–1984) (Figure 54), heavily influenced by Dobzhansky's *Genetics and the Origin of Species*, brings paleontology fully into the Modern Synthesis with his *Tempo and Mode in Evolution.* Interestingly, *Tempo* was published while Simpson was in the United States Army. Simpson, who preferred the term *post-Darwinian* to *Modern Synthesis*, showed that the fossil record is compatible with branching, nondirectional patterns predicted by Darwin's theory, thereby delivering a death knell for orthogenesis and neo-Lamarckism. Simpson also demonstrated that macroevolution could be explained by the accumulated effects of short-term evolutionary processes. Simpson's claim that there is a "regular absence of transitional forms" in the fossil record has been cited repeatedly by creationists to discredit the theory of evolution.

1944 The National Science Teachers Association (NSTA), an evolution-education advocacy group, is established.

1945 David Lack's (1910-1973) monograph about Galápagos finches argues that differences in bill size are species-recognition signals (i.e., reproductive isolating mechanisms). Two years later, in *Darwin's Finches*, Lack claimed that differences in bill size were adaptations to specific food niches.

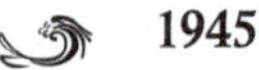

1945 As had occurred earlier in George McCready Price's Religion and Science Association, the Deluge Geology Society becomes

Figure 54 George Gaylord Simpson incorporated paleontology into the Modern Synthesis by showing that the fossil record is compatible with evolution by natural selection. Simpson is shown here in Venezuela in 1938 with a baby guanaco. (*American Philosophical Society*)

influenced by old-earth creationists and abandons biblical literalism. Soon thereafter, membership in the Deluge Geology Society declined rapidly.

1946 Harvard anthropologist Earnest Hooton (1887–1954) publishes the 2nd edition of *Up from the Ape* (initially released in 1931), which advocates learning about humans by studying non-human apes, thereby expanding anthropology from an anatomical to a behavioral science: "If the finds of fossil man hitherto brought to light mean anything at all, they mean that nature has conducted many and varied experiments upon the higher primates, resulting in several lines of human descent." Hooton was influential in training the first generation of professional physical anthropologists in the United States. As a member of the National Research

Council's Committee on the Negro, Hooton used his academic credentials to legitimize negative racial stereotypes, including that blacks are evolutionarily more primitive (and a separate line of descent) than whites.

1946 Seventh-Day Adventist Harold Clark's geology textbook *The New Diluvialism* claims that rising floodwaters successively destroyed ecological zones and thereby produced the predictable arrangement of fossils visible today. Unlike George McCready Price, Clark conceded the validity of the geologic column. Clark's so-called "ecological zonation theory," which remains popular among some creationists, enabled creationists to accept the validity of the geologic column while rejecting an ancient earth. However, Clark's idea was rejected by scientists for many reasons, including the fact that some organisms (e.g., corals and clams) appear in virtually all strata and that organisms living in the same ecological zone (e.g., fish and whales, birds and flying reptiles) appear in different strata. Clark's idea was also distasteful to George McCready Price; *The New Diluvialism* prompted Price to challenge Clark with *Theories of Satanic Origin*.

1946 Famed geneticist Hermann Muller wins the Nobel Prize in Physiology or Medicine for his work on mutagenesis. Despite this recognition, Muller continued to resent his colleague and former mentor Thomas Morgan for not giving him enough credit for his work.

1946 While in graduate school, Southern Baptist Henry Morris (1918–2006) self-publishes *That You Might Believe*, his first attempt to provide a scientific basis for creationism. In this book, the gentle and diplomatic Morris claimed that the teaching of evolution could damage students. The 1st edition of *That You Might Believe* allowed for gap creationism, but this was later expunged. *That You Might Believe* has been in print for more than 60 years.

1946 In *Common-Sense Geology*, George McCready Price brags that he was "the first in modern times to revive the ancient and honorable Flood theory" and notes that his idea is becoming increasingly popular.

1946 In the Ediacara Hills of South Australia, geologist Reginald Sprigg (1919–1994) finds the earliest known communities of animals. These enigmatic animals, known as Ediacarans, lived during the

Precambrian and comprise a strange collection of mostly sessile creatures.

1946 The Society for the Study of Evolution is established and—thanks to a grant from the American Philosophical Society (founded by Benjamin Franklin)—begins to publish its journal, *Evolution*, the next year. Ernst Mayr was the journal's first editor. One of the biggest problems facing the editors was finding first-rate research not involving fruit flies or similar organisms. That same year, the ASA had its first formal meeting, at which it was decided to begin publishing *Modern Science and Christian Faith*.

1947 Young-earth creationist Frank Marsh's *Evolution or Special Creation?* claims that "surely the time is ripe for a return to the fundamentals of true science, the science of creationism." Marsh later claimed that Satan caused disease by the "derangement" of organisms, that the Bible teaches Mendelism, that angels brought animals to Noah, and that evolutionists reject creationism because "evolutionists do not take the time to read the Bible carefully for themselves."

1947 David Lack, director of the Edward Grey Institute of Field Ornithology, publishes *Darwin's Finches* based on his research in the Galápagos Islands. Lack's book (which was reissued in 1961) made Darwin's name synonymous with the finches. (Ironically, Darwin's *On the Origin of Species* mentioned Galápagos mockingbirds, but not finches.) After reading Lack's book, evolutionary biologist E. O. Wilson proclaimed that Darwin's finches "shout the truth of evolution;" soon thereafter, textbooks began including the finches as evidence for Darwin's theory, and the finches became an icon of evolution. (More detailed studies, spanning decades, of the finches have been conducted by a team of biologists led by Peter and Rosemary Grant.) Lack published a variety of books, including *Evolutionary Theory and Christian Belief* (1957) after his conversion from agnosticism to Anglicanism. Unperturbed that a benevolent deity could reign over nature's struggle for existence, Lack—who believed that God "placed man in a special relationship to Himself"—claimed "man is surely unqualified to judge whether this [natural] ordering is in any way evil, or contrary to divine plan."

1947 In *Everson v. Board of Education*, the United States Supreme Court announces that states cannot legitimately help religion in any

form. Hugo Black's (1886–1971) majority opinion reaffirmed the concept of a "wall of separation" between church and state.

1947 Sir Arthur Keith concedes that australopithecines were human ancestors (see also Appendix C).

1947 Harold Clark's *Creation Speaks* presents a rare "positive" treatment of creationism instead of a debunking of evolution.

1947 British author and Christian apologist C. S. Lewis's (1898–1963) *Miracles* foreshadows the modern ID movement by distinguishing naturalism from supernaturalism. Lewis, who also wrote *The Screwtape Letters* (1942) and *The Chronicles of Narnia* (1950–1956), advocated the classic argument from design while hoping that "we may be living nearer than we suppose to the end of the Scientific Age."

1948 Geochemist Claire Cameron Patterson (1922–1995) begins work on his dissertation, in which he tries to determine the earth's age by counting lead isotopes in meteorites. Patterson's work later showed that the earth is approximately 4.55 billion years old (see also Appendix A).

1948 P. J. Wiseman's *Creation Revealed in Six Days: The Evidence of Scripture Confirmed by Archaeology* develops *prophetic-day creationism*, which claims that the six days of creation were ordinary days during which God described the successive creation events to Moses or some other seer. Wiseman's idea generated little interest.

1948 Supreme Court Justice Hugo Black's majority opinion in *McCollum v. Board of Education* claims that government cannot provide religious instruction in public schools. The decision emphasized the importance of public schools to American culture: "The public school is at once the symbol of our democracy and the most pervasive means for promoting our common destiny. In no activity of the state is it more vital to keep out divisive forces than in its schools." Public schools later became a battleground in the evolution-creationism controversy.

1948 Louis (1903–1972) and Mary (1913–1996) Leakey find a partial skull of 18-million-year-old extinct ape *Proconsul* on Rusinga Island in Kenya's Lake Victoria. This was the first skull of a fossil ape ever found. Newspapers claimed "The Leakeys Find

Important Fossil-Man Ancestor," but they were wrong; *Proconsul* was ancestral to great apes that came later.

1948 Trofim Lysenko's suppression of genetics in the Soviet Union peaks when he expunges ideas incompatible with Lamarckism from textbooks. The following year, Lysenko announced that Soviet scientists had successfully transformed wheat into rye by planting wheat in areas specifically favorable to rye. However, by the 1950s, facing continued famines despite assurances from Lysenko that the problem had been solved, Soviet scientists began questioning Lysenko's claims and programs.

1948 The American Scientific Affiliation publishes the first issue of *Modern Science and Christian Faith*. ASA members, who increasingly held advanced degrees in science, rejected George McCready Price's Flood geology. The following year, ASA began publishing a nondoctrinal periodical, the *Journal of the American Scientific Affiliation* (which later became *Perspectives on Science and Christian Faith*).

1948 In his presidential address to the International Congress of Genetics in Stockholm, Hermann Muller denounces Trofim Lysenko and his policies. In response, the Eastern bloc of delegates walked out of the Congress.

1949 British astronomer Fred Hoyle (1915–2001), who advocated the argument from design and claimed that *Archaeopteryx* (Figure 24) was a fake, coins the term *big bang* in a sarcastic comment about Georges-Henri Lemaître's theory that the universe is expanding. Hoyle, who believed that the universe has always looked like it does now, later claimed that *On the Origin of Species* "committed mankind to a course of automatic self-destruction."

1949 In *The Meaning of Evolution*, George Gaylord Simpson (Figure 54) makes his famous observation: "Evolution has no purpose; man must supply this for himself."

1950 Ronald Fisher (Figure 50) argues that Darwinism, and not Lamarckism, can be more easily reconciled with Christianity.

1950 American chemist Willard F. Libby (1908–1980) develops radioactive carbon dating for estimating the ages of materials that came from plant products. With a half-life of 5,730 years, ^{14}C is

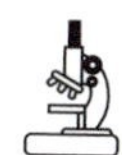

especially useful for dating objects less than 50,000 years old and is therefore of particular use in establishing archaeological chronologies. In 1960, Libby won the Nobel Prize in Chemistry.

1950 British computer scientist Alan Turing (1912–1964) suggests that there is an "obvious connection between machine learning and evolution," thereby originating the discipline of evolutionary computing.

1950 German biologist Willi Hennig (1913–1976) publishes *Grundzüge einer Theorie der phylogenetischen Systematik*, proposing phylogenetic systematics as a way of assessing evolutionary relationships. Hennig's work noted that biological similarities do not necessarily imply close phylogenetic affinity. When Hennig's book was translated into English in 1966 (*Phylogenetic Systematics*), it revolutionized systematics and taxonomy.

1950 Immanuel Velikovsky's (1895–1979) bestseller *Worlds in Collision* claims that mainstream science is concealing evidence that would contradict the accepted view of the earth's history.

1950 Jerome Lawrence (1915–2004) and Robert E. Lee (1918–1994) write the play *Inherit the Wind*, which uses a fictionalized account of the Scopes Trial to expose the threat to intellectual freedom associated with the anti-communist hysteria of the McCarthy era. Although *Inherit the Wind* included some historically accurate moments (e.g., Darrow's condemnations of anti-intellectualism), the play's many historical inaccuracies were noted prominently in the playwrights' note: "*Inherit the Wind* is not history. . . . Only a handful of phrases have been taken from the actual transcript of the famous Scopes Trial. . . . *Inherit the Wind* does not pretend to be journalism. It is theater. It is not 1925." Regardless, the play and subsequent movie strongly influenced perceptions of the relationship between religion and science. *Inherit the Wind* was not performed publicly until 1955.

1950 Plant geneticist George L. Stebbins, Jr.'s (1906–2000) *Variation and Evolution in Plants* applies evolutionary principles to plants and discusses polyploidy as a mechanism for the rapid speciation observed in plants.

1950 Pope Pius XII's (1876–1958) *Humani Generis*—an encyclical about human origins—declares that "the teaching of the Church leaves the doctrine of evolution an open question . . . to be examined and discussed" so long as it recognizes that "souls are immediately created by God." Pius was the first pope to explicitly address evolution at length.

1950 Discoveries in Africa's Olduvai Gorge suggest that hominins used tools at least 1.5 million years ago.

1950 The National Science Foundation (NSF) is established as a federal agency because, in the words of President Harry Truman, "[g]overnment has a responsibility to see that our country maintains its position in the advance of science." Most biology textbooks published in the 1950s did not mention the word *evolution*, and all textbooks placed the coverage of organic change in one of the final chapters. Surveys showed that at least 30% of biology teachers did not discuss evolution in their courses. NSF later funded many evolution-centered projects in basic sciences and education.

1951 American biologist James Watson (b. 1928) joins the Cavendish Laboratory and starts studying DNA with graduate student Francis Crick (1916–2004). Watson and Crick later proposed the double-helix structure of DNA.

1951 Plant geneticist Barbara McClintock (1902–1992) describes "jumping" genes (transposable elements) that can move within an organism's genome.

1951 In a letter to Bernard Acworth (1885–1963) of the Evolution Protest Movement, author C. S. Lewis notes "[w]hat inclines me now to think you may be right regarding [evolution] as the central and radical lie in the whole web of falsehood that now governs our lives is not so much your arguments against it as the fanatical and twisted attitudes of its defenders."

1951 George Gaylord Simpson of AMNH notes that "[t]he history of the horse family is still one of the clearest and most convincing for showing that organisms really have evolved, for demonstrating that, so to speak, an onion can turn into a lily."

1951 British physician and geneticist Henry Bernard Davis (H. B. D.) Kettlewell (1907–1979) begins studying the peppered moth, *Biston betularia*. Kettlewell's lab work showed that dark-winged moths

tended to occur on dark backgrounds, and light-winged moths tended to occur on lighter backgrounds. Field experiments demonstrated that birds could function as selective agents, and that birds had to learn to recognize a type of prey before they could exploit it. Kettlewell confirmed his claims with mark-release-recapture experiments in polluted forests near Birmingham and in pristine forests near Dorset. Kettlewell's work showed that in polluted areas, light-colored moths are more conspicuous and susceptible to predation by birds than are dark moths. Nobel laureate Niko Tinbergen's (1907–1988) movies of the differing predation rates of the moths by birds were shown at science meetings throughout the world, and Kettlewell's work appeared in a *Scientific American* article titled "Darwin's Missing Evidence." Kettlewell's work was described in 1978 by Sewall Wright as "the clearest case in which a conspicuous evolutionary process has actually been observed" and for many years was cited in biology textbooks as the best example of natural selection in action.

1952 Alfred Hershey (1908–1997) and Martha Chase (1927–2003) show that DNA is the genetic material in bacteriophage viruses.

1952 Another attempt to repeal Tennessee's Butler Law fails, thanks largely to the efforts of Bryan College (Figure 49).

1952 Rosalind Franklin (1920–1958) produces the first X-ray diffraction photographs of DNA.

1953 At a meeting of the ASA, Henry Morris meets John Whitcomb, Jr., and they form a partnership that will produce *The Genesis Flood*, a foundation of the modern "creation science" movement. In that book, Morris and Whitcomb dismissed modern science not on the basis of new scientific findings, but instead "on the basis of overwhelming Biblical evidence."

1953 In an article in *Science* titled "A Production of Amino Acids Under Possible Primitive Earth Conditions," Stanley Miller and Richard Urey confirm some of Aleksandr Oparin's claims about the possible impact of the early atmosphere on the origin of life. Miller and Urey passed a spark through Oparin's re-created early atmosphere and produced amino acids.

1953 In an article in *Nature* titled "A Structure for Deoxyribose Nucleic Acid," James Watson and Francis Crick announce the double-helix structure of DNA. Their model, which provided a means to understand mutation, genetic coding, and replication, began the era of molecular genetics, and provided yet more powerful evidence for Darwin's theory.

1953 Famed dinosaur-illustrator Charles Knight's last words to his daughter Lucy are "[d]on't let anything happen to my drawings." She didn't: Knight's iconic work remains on display at AMNH, The Field Museum, the Bronx Zoo, and elsewhere.

1953 Anthropologist Kenneth Oakley (1911–1987) (Figure 55) and his colleagues show that Piltdown Man is a fraud. Many experts now believe that Piltdown Man—one of the most spectacular scientific frauds of the 20th century—was perpetuated by Martin Hinton (1883–1961), the deputy keeper of zoology at NHM.

1954 Theology professor Bertrand Ramm's (1916–1992) *The Christian View of Science and Scripture* accepts the Bible's "divine origin . . . and inspiration," but dismisses young-earth creationism: "Conservative

Figure 55 Anthropologist Kenneth Oakley helped expose Piltdown Man as a fraud. Oakley is the one in the middle. (*Getty Images*)

Christianity is caught between the embarrassments of simple fiat creationism, which is indigestible to modern science, and evolutionism, which is indigestible to much of Fundamentalism." Ramm then proposed a solution, "pictorial-revelatory days," which is the belief that the days of creation were used metaphorically by God to describe creation to the author of Genesis 1. Ramm rejected a worldwide flood, accepted the earth's antiquity, and endorsed progressive creationism, which is the belief that God created life gradually via evolution and in the sequence documented by the fossil record. Ramm, who rejected "narrow bibliolatry" while claiming "creation was *revealed* in six days, not *performed* in six days," accused fundamentalists of being inconsistent and ignorant of science. Although Billy Graham and many Christian biologists embraced Ramm's ideas, fundamentalists such as John Whitcomb, Jr., were outraged by Ramm's claims.

1954 Anthony Allison (b. 1925), who grew up in Kenya, proposes that the high frequency of the form of the hemoglobin gene that confers sickle-cell anemia in some African populations results from selection for that allele due to the natural resistance to malaria it confers. This work, an oft-cited example of "heterozygote advantage," suggested that human populations were still influenced by selective forces.

1954 Biologist John W. Klotz's (1918–1996) *Genes, Genesis, and Evolution* typifies many creationists' demands for capitulation rather than an examination of evidence. Klotz, who was also a Lutheran preacher, knew enough about biology to wish that he did not have to defend creationism and a worldwide flood, but—echoing Augustine's claims 1,600 years earlier—felt he had no choice because "Scripture speaks and that settles it for me." In 1985, Klotz's *Studies in Creation* acknowledged that "the creationist is free to postulate either a young or an old earth."

1955 Biologist Oscar Riddle's *The Unleashing of Evolutionary Thought* explores some implications of evolution. Chapter titles included "Social Inheritance," "The Biological Inequality of Men," and "Religion's Power to Divide."

1955 French paleontologist and Jesuit priest Pierre Teilhard de Chardin's (1881–1955) *The Phenomenon of Man* is published soon after his death (the Catholic Church had blocked earlier publication). Teilhard invoked orthogenesis by proposing that the universe was constantly evolving toward forming the body of Christ (what he referred to as the Omega Point). He believed that early phases (geologic and biological evolution) in this developmental sequence had already occurred, and that the final stage—where all human thought would become fused—was ongoing, as evidenced by global communication systems.

1955 Chemistry professor Homer Jacobson publishes a paper titled "Information, Reproduction and the Origin of Life" in *American Scientist* that describes conditions after the earth's early cooling. Jacobson was shocked when his paper was cited by creationists as evidence for the requirement of divine intervention in the development of life. In 2007—more than five decades after the paper's publication—Jacobson retracted the work.

1955 The American Institute of Biological Sciences (AIBS) forms an Education and Professional Recruitment Committee to develop a program in biology education. The Committee recommended that evolution be included in all biology programs.

1955 The play *Inherit the Wind* opens to favorable reviews in Dallas, Texas. Three months later it opened at Broadway's National Theatre for a three-year run. The play described "the famous 'Monkey Trial' that rocked America" and that was "the most explosive trial of the century." When its Broadway run ended, *Inherit the Wind* was the most successful and longest-running drama in Broadway's history. The success of the play prompted *Encyclopedia Britannica* to first include the Scopes Trial in their 1957 edition. The movie version of *Inherit the Wind*, released in 1960, strongly shaped the public's view of the Scopes Trial and the evolution-creationism controversy.

1956 Britain's Clean Air Acts establish so-called smokeless zones in heavily polluted areas. Soon thereafter, as air quality improved, melanic moths became less common as populations of the lighter-colored moths (those more likely to be camouflaged by lichen-encrusted trees) recovered. These observations provided strong evidence for natural selection.

1956 Claire Patterson's article, "Age of Meteorites and the Earth," is published in *Geochimica et Cosmochimica Acta* (see also Appendix A). Using three different radiometric dating methods, Patterson documented that ocean sediment (derived from a variety of rocks

from many places) and several meteorites (including the Canyon Diablo meteorite, which excavated Barringer Crater in northern Arizona; Figure 32) all had the same age—approximately 4.55 billion years old—and therefore so should the earth. Subsequent determinations of the ages of numerous meteorites, lunar rocks, and other samples have consistently provided ages of 4.5–4.7 billion years.

1956 H. L. Mencken (Figure 43), whose coverage of the Scopes Trial is legendary, dies and is buried in Loudon Park Cemetery in Baltimore, Maryland.

1957 Arthur Garfield Hays, one of Scopes's defenders, notes that he can hardly believe that "religious views of the Middle Ages" could persist in the age of "railroads, steamboats, the telephone, the radio, the airplane, all the great mechanistic discoveries."

1957 In *Why I Am Not A Christian*, British philosopher Bertrand Russell (1872–1970) attacks the argument from design by noting that the world is understandable "as a result of muddle and accident, but if it is the outcome of deliberate purpose, the purpose must have been that of a fiend."

1957 The Soviet Union successfully launches the 183-pound *Sputnik I*, the first orbiting artificial satellite. It took 93 minutes for *Sputnik I* to orbit the earth. This launch, combined with the failure of the American *Vanguard* satellite a few weeks later, triggered widespread concern that science and technology in the United States lagged behind those of the Soviet Union. This concern ultimately prompted NSF to implement the first significant reforms in science education, one of which was creating the Biological Sciences Curriculum Study (BSCS) in 1958 at the University of Colorado. BSCS later produced biology textbooks that "put evolution back into the curriculum" and, in the process, revived the creationism-evolution controversy.

1958 French biologists François Jacob (b. 1920) and Jacques Monod (1910–1976) begin experiments that explain the genetic control of protein synthesis in bacteria. DNA stores and transmits genetic information the same way in all organisms. Jacob and Monod, with French biologist André Lwoff (1902–1994), won the 1965 Nobel Prize in Physiology or Medicine "for their discoveries concerning genetic control of enzyme and virus synthesis."

1958 In a talk presented at the annual meeting of the Central Association of Science and Mathematics Teachers, Nobel laureate Hermann Muller tells attendees "[o]ne hundred years without Darwinism are enough." In a subsequent essay, Muller challenged biology teachers to do a better job of teaching evolution and lamented that biology in public schools was dominated by "antiquated religious traditions."

1958 Julian Huxley notes "Darwin's essential achievement was the demonstration that the almost incredible variety of life, with all its complex and puzzling relations to the environment, was explicable in scientific terms." The following year, Huxley noted that Darwin's theory "is no longer a theory but a fact. . . . Darwinism has come of age so to speak. We are no longer having to bother about establishing the fact of evolution."

1958 Loren Eiseley's (1907–1977) *Darwin's Century* describes the development of evolutionary theory. *Darwin's Century* became a staple on college campuses for the next two decades.

1958 Ronald Fisher's (Figure 50) *The Genetical Theory of Natural Selection* is reissued and influences another generation of biologists, including William Hamilton (1936–2000), who rates the book "only second in importance within evolution theory to Darwin's *Origin*" and "one of the greatest books of the present century." Hamilton's original interest in evolutionary biology, like Fisher's, was its potential use for "enhancing the human stock."

1958 The Geoscience Research Institute, sponsored by the Seventh-Day Adventist Church, establishes the journal *Origins* to reconcile evolution with biblical creation.

1959 Another attempt to repeal the Arkansas antievolution law fails. The leader of the attempt, Willie Oates (1918–2008) (Figure 56), was branded an atheist and withdrew her bill. The following year, Oates lost her bid for reelection.

1959 A symposium at the University of Chicago to celebrate the centenary of Charles Darwin's *On the Origin of Species* is the final gathering of the key architects of the Modern Synthesis. As part of the celebration, NSF convened 63 leading high school biology teachers at the National Conference of High School Biology Teachers. The teachers criticized the poor coverage of

Figure 56 Arkansas legislator Will Etta "Willie" Oates, also known as "The Hat Lady," was a state legislator and community activist who learned the political consequences of publicly supporting the teaching of evolution. In 1959, after leading a failed attempt to repeal Arkansas's ban on the teaching of human evolution, Oates was defeated in her bid for reelection. When Oates died in 2008, she was buried wearing her favorite hat. (*Susan and Jon Epperson*)

evolution in textbooks, but were hesitant to endorse evolution as "fact."

1959 Biologist Oscar Riddle notes that "[b]iology is still pursued by long shadows from the Middle Ages, shadows screening from our people what our science has learned of human origins . . . a science sabotaged because its central and binding principle displaces a hallowed myth."

1959 Ecuador designates most of the Galápagos Islands as a National Park (Figure 17).

1959 Loren Eiseley suggests that Charles Darwin took many of his ideas about natural selection from Edward Blyth (1810–1873), a correspondent of Darwin's who Darwin acknowledged in Chapter 1 of *On the Origin of Species*. In 1979, Eiseley's executors expanded Eiseley's claim in *Darwin and the Mysterious Mr. X* (Blyth was "Mr. X").

1959  Mary Leakey—a direct descendant of John Frere (on her mother's side)—discovers the 1.75-million-year-old *Zinjanthropus boisei* (renamed *Australopithecus boisei*) at Olduvai Gorge in Tanzania (Figure 57; see also Appendix C). Subsequent discoveries by Mary and Louis Leakey convinced biologists that humans originated in Africa and then spread through Asia, Europe, and beyond.

1959  Nobel laureate Joshua Lederberg (1925–2008) coins the term *exobiology* to link origin-of-life studies with the search for extraterrestrial life.

1959 The ASA publishes zoologist Russell Mixter's *Evolution and Christian Thought*, a collection of essays that mark a clear departure with biblical literalists who insist on a young earth and worldwide flood.

1959  The Middle Tennessee State University chapter of the American Association of University Professors petitions the legislature to

Figure 57 This monument at Olduvai Gorge in Tanzania marks where Mary Leakey found fossils of *Zinjanthropus boisei*. Mary's husband Louis judged the discovery "the connecting link between the South African near-man and true man as we know him." (*Don Moll*)

repeal the Butler Law. In response, the Rutherford County Court condemned the "free-thinking" educators and noted that the "God-fearing men" of the court opposed repealing the law.

1959 In an article published in *Systematic Zoology* celebrating the 10th edition of *Systema Naturae* (1758), taxonomist William Stearn (1911–2001) chooses Linnaeus's remains, buried in the Uppsala Cathedral, as the type specimen for *Homo sapiens*.

1959 In *Did God Create a Devil?,* Herbert Armstrong—the founder of the Worldwide Church of God—uses gap creationism to explain the origin and nature of Satan. Armstrong's son, Garner Ted Armstrong, later announced "[t]he BIGGEST false doctrine ever is EVOLUTION."

1960 The Charles Darwin Research Station is established on Santa Cruz in the Galápagos Islands (Figure 17). Today, the Station hosts thousands of visitors every year (Figure 58).

1960 *The Flintstones*, television's first prime-time animated series, debuts on ABC. The show, which portrayed a "modern stone-age family" of humans who lived with dinosaurs, saber-toothed tigers, wooly

Figure 58 The Charles Darwin Research Station resides on Santa Cruz island in the Galápagos archipelago. The station is dedicated to preserving the natural ecosystems of the islands. (*Sehoya Cotner*)

mammoths, and other extinct animals, lasted 166 episodes and indoctrinated millions of children and adults with the notion that humans lived contemporaneously with dinosaurs. This claim was repeatedly rejected by biologists, geologists, and other scientists, but has remained a foundation of young-earth creationism (e.g., that dinosaurs and humans were created on the sixth day of creation). Indeed, the Creation Museum, operated by the antievolution organization Answers in Genesis, includes dioramas depicting humans living with dinosaurs. In 2002, NSF reported in its *Science and Technology Indicators* that 48% of Americans believe that humans and dinosaurs lived at the same time.

1960 The movie version of *Inherit the Wind* has its world premiere at the Berlin Film Festival and its American premiere at the Dayton Drive-In Theater at a celebration commemorating the 35th anniversary of the Scopes Trial. John Scopes attended the Dayton premiere and was honored with a parade and a key to the city. The 127-minute movie was directed by Stanley Kramer (1913–2001) and starred two Oscar-winning Hollywood legends, Spencer Tracy (1900–1967; playing Henry Drummond, the character representing Darrow) and Fredric March (1897–1975; playing Matthew Harrison Brady, the character representing Bryan). Scopes, who promoted the movie at the studio's request, acknowledged that the movie was not historically accurate, but that it "captured the emotions in the battle of words between Bryan and Darrow." *Inherit the Wind* had a tremendous impact; despite its disclaimers to the contrary, most people believed that the movie was an accurate description of the Scopes Trial. Late in 1960, *Inherit the Wind* became the world's first in-flight movie when Trans World Airlines used it to lure first-class passengers. As *Inherit the Wind* played in theaters, young-earth creationists John Whitcomb, Jr., and Henry Morris put their final touches on their upcoming book, *The Genesis Flood*, which popularized young-earth creationism.

1960 In a paper titled "The World into which Darwin Led Us" in *Science*, George Gaylord Simpson (Figure 54) notes that it is unlikely that anything in the world exists specifically for our benefit or ill: "It is no more true that fruits, for instance, evolved for the delectation of men than that men evolved for the delectation of tigers."

1961 In a footnote in *Torcaso v. Watkins*, U.S. Supreme Court Justice Hugo Black lists "Secular Humanism" as a religion that does not

"teach what would generally be considered a belief in the existence of God." Creationists later claimed that the teaching of evolution promoted the religion of secular humanism.

1961 A survey of 1,000 high school teachers reveals that two-thirds of the teachers believe that they can teach biology effectively without accepting evolution. Another survey later in the decade showed that more than 60% of biology teachers believe that evolution is a theory and therefore cannot be said to have definitely occurred.

1961 American geologists Jack Evernden and Garniss Curtis become the first to use radiometric methods to date a fossil of an early human ancestor (*Zinjanthropus*). The fossil's age (1.89–1.57 million years) was four times older than previously thought and changed paleontologists' conception of the rate of human evolution (see also Appendix C).

1961 In *Science and Religion*, Bertrand Russell dismisses creationists' "appearance of age" argument by noting that humans could have appeared five minute ago "with holes in our socks and hair that needed cutting."

1961 John Whitcomb, Jr., and Henry Morris ask Moody Press to publish their book promoting Flood geology and young-earth creationism, but their request is turned down; editors at Moody Press accept day-age creationism and do not believe that the book will be popular. Whitcomb and Morris's book was eventually published by Presbyterian and Reformed Publishers and was titled *The Genesis Flood: The Biblical Record and its Scientific Implications*. The 518-page, footnote-laden book became the most important book (aside from the Bible) in the history of modern creationism. *The Genesis Flood* claimed that geologic strata have been arranged by deceitful, atheistic geologists, despite the fact that the patterns of fossils in geologic strata were documented (long before the publication of *On the Origin of Species*) by geologists indifferent or hostile to the concept of evolution. *The Genesis Flood* also included photos of the Paluxy riverbed footprints to claim that humans lived with dinosaurs, and presented George McCready Price's Flood geology (minus its Adventist trappings) as the only acceptable interpretation of Genesis: "Any true science of historical geology must necessarily give a prominent place in its system to the biblical Flood," and Christians "must certainly accept [the biblical Flood] as unquestionably true." Whereas Flood-geology pioneer Price had avoided pointing out the incompatibility of his ideas with the gap and day-age creationism favored by many fundamentalists (e.g., William Jennings Bryan and William Bell Riley), Whitcomb and Morris denounced these "compromises" by Christians such as Bertrand Ramm while starkly claiming that the geologic evidence used to discredit the biblical Flood is invalid. Whitcomb and Morris presented two complementary ideas: (1) natural phenomena can only be explained by principles of biblical inerrancy (e.g., "the fossil record, no less than the present taxonomic classification system, and the nature of genetic mutation mechanism, shows exactly what the Bible teaches"), and (2) Flood geology is the central paradigm of creationism. Whitcomb and Morris rejected any reconciliation of science and religion ("Our main concern . . . must not be to find ways of making the Biblical narratives conform to modern scientific theories"), instead arguing that people should try "to discover exactly what God has said in the Scriptures, being fully aware of the fact that modern science, laboring under the handicap of non-Biblical philosophical presumptions (such as materialism, organic evolution, and uniformitarianism) are in no position to give us an accurate reconstruction of the early history of the earth and its inhabitants." Whitcomb and Morris insisted that "there is nothing unreasonable" about the claim that the universe "must have had an 'appearance of age' at the moment of creation," but critics charged that such a belief meant that God is a trickster and charlatan who deliberately plants misleading clues about the earth's history. Since there was no death in the world before the Original Sin described in Genesis, the fossil record was evidence of sin and death, not evolution and life. Whitcomb and Morris proclaimed biblical creationism as literal truth by arguing that humans cannot know and should not speculate about the details of creation, adding that scientific claims and geologic uniformitarianism "need to be challenged in the name of Holy Scripture." The authors stressed that "the fully instructed Christian knows that the evidence for full divine inspiration of Scripture is far weightier than the evidences for any fact of science" and excused the scientific and economic successes of modern stratigraphy as being accidental: "The uniformitarian hypothesis and the evolutionary framework of geological ages have been shown to be largely irrelevant to the actual practice of petroleum exploration." *The Genesis Flood*, which challenged the theistic evolution that typified organizations such as the ASA, revived antievolutionary thought from 35 years of obscurity and helped convince half of America that the earth is just a few thousand years old. As of 2009, *The Genesis Flood*—which Stephen Jay Gould judged as "the founding document of the creationist movement"—had sold more than 250,000 copies. It is in its 44th printing, but has never been revised.

1961 Marshall Nirenberg's (b. 1927) discovery that the nucleotide triplet of uracil (UUU) codes for the amino acid phenylalanine begins the challenge of cracking the genetic code. The code was fully deciphered in 1966.

1961 Mary and Louis Leakey discover a 1.8-million-year-old *Homo habilis* (see also Appendix C). Louis claimed that *H. habilis* was on the true evolutionary line to humans and that all australopithecines (including *Zinjanthropus*) were not. Louis held this view until his death in 1972.

1961 In *General Biology*, E. L. Core claims "we do not actually know the phylogenetic history of any group of plants and animals."

1962 American anthropologist Carleton Coon's (1904–1981) controversial *The Origins of Races* argues that *Homo sapiens* evolved five times in five different locations. Coon claimed that the "backwardness" of "Negroes" was due to their being the last of *Homo erectus* to evolve into *Homo sapiens*. Most anthropologists rejected this multiregional model for human evolution, instead favoring the out-of-Africa model, in which modern *Homo sapiens* evolved in one locality in Africa over 150,000 years ago and spread out of Africa approximately 60,000–75,000 years ago. Coon's books are often cited as examples of "scientific racism" used to justify segregation.

1962 American historian Richard Hofstadter's (1916–1970) Pulitzer-Prize-winning *Anti-Intellectualism in American Life* claims "today the evolution controversy seems as remote as the Homeric era to intellectuals in the East."

1962 Francis Crick, James Watson, and Maurice Wilkins (1916–2004; a physicist who worked with Rosalind Franklin on X-ray diffraction studies of DNA) share the Nobel Prize for Physiology or Medicine for determining the structure of DNA. Franklin had died in 1958, at age 37, of ovarian cancer.

1962 American chemist Linus Pauling (1901–1994) and Austrian-born biologist Emile Zuckerkandl (b. 1922) suggest that genetic changes can be used as a kind of "molecular clock" to date the divergence of species.

1962 In *Animal Dispersion in Relation to Social Behavior*, British zoologist Vero Copner Wynne-Edwards (1906–1997) attempts to explain reproductive constraint in a "group selectionist" framework (e.g., that behaviors evolve "for the good of the group"). After being widely accepted, group selection was later discredited, although some biologists continue to advocate the validity of group selection in particular circumstances.

1962 Religious leaders in Phoenix, Arizona, demand that BSCS books be removed from schools because the books come "as close to teaching atheism as one can at the secondary level." Howard Seymour (superintendent of the Phoenix school system) quelled the protest by proclaiming, "students are not expected to believe" what they are taught about evolution.

1962 Supreme Court Justice Hugo Black's majority opinion in *Engel v. Vitale* bans state-sponsored prayer in public schools. Later, Black was the only Justice in *Epperson v. Arkansas* to express any reluctance with the court's unanimous decision.

1962 The preparation of BSCS books goes smoothly despite reduced support from NSF. Early results showed that students who used the BSCS books scored higher on a variety of tests than did students who used other books.

1963 Creationists and self-appointed textbook censors Mel (1915–2004) and Norma Gabler (1923–2007) begin examining textbooks and testifying at meetings of the Texas State Board of Education. A decade later, the Gablers devoted all their time to the creationism cause and founded Education Research Analysts, which tried to ban textbooks that "eliminate coming to Christ for forgiveness of sin."

1963 Ernst Mayr's *Animal Species and Evolution* emphasizes that rapid evolution and speciation can occur in isolated populations on the periphery of species' ranges. Stephen Jay Gould and Niles Eldredge (b. 1943) later used this *peripatric model* to develop the concept of punctuated equilibrium, whereby species were proposed to exist unchanged for long periods of time followed by bursts of diversification.

1963 Duane Gish, Henry Morris, Frank Marsh, and like-minded young-earth creationists organize the Creation Research Society (CRS), which becomes a leading creationist organization of the

late 20th century. The society's founders abandoned the ASA because it had become "soft on evolution" and had "capitulated to theistic evolution." The society espoused a literal interpretation of the Bible—including the reality of a worldwide Noachian flood—and emphasized "publication and research which impinge on creation as an alternate view of origins." By 1973, when Henry Morris claimed that evolution is guided by Satan, the society had almost 2,000 members. Today, CRS is the oldest society continuously publishing creationism-based research.

1963 BSCS publishes its three biology textbooks, each of which emphasizes evolution. The books were described not by the books' foci (ecology, molecular biology, cell biology), but instead by their colors (green, blue, yellow) to avoid the implications that the books were specialized for advanced biology courses or that BSCS was trying to establish a national curriculum for biology (Figure 59). Some states accepted the BSCS textbooks, but education officials in Texas insisted that publishers delete statements such as "To biologists there is no longer any reasonable doubt that evolution occurs." Publishers accepted these revisions, which appeared in subsequent editions of the blue version. Within a few years after being released, BSCS books were being used in nearly half of all high school biology courses in the United States. They were also used in several other countries; Ken Ham, the founder of the antievolution organization AiG, taught from a BSCS textbook when he was a public school teacher in Australia in 1975. Publishers of competing books soon began producing similar books.

1963 In Phoenix, Arizona, Southern Baptist preacher Aubrey Moore claims that the teaching of evolution is "the first step in communism" and demands that it be removed from the curriculum. Moore warned school board members that if they "won't do it, we're going to get people together who will do something about it." When Superintendent Howard Seymour rejected Moore's demands, the bombastic Moore began gathering signatures to force a public referendum about his antievolution message, telling citizens "there is nothing in the Bible about a fish turning into a man." Moore failed to gather enough signatures to force the referendum.

⚖ **1963** In *School District of Abington Township v. Schempp*, the United States Supreme Court declares that it is unconstitutional for states to

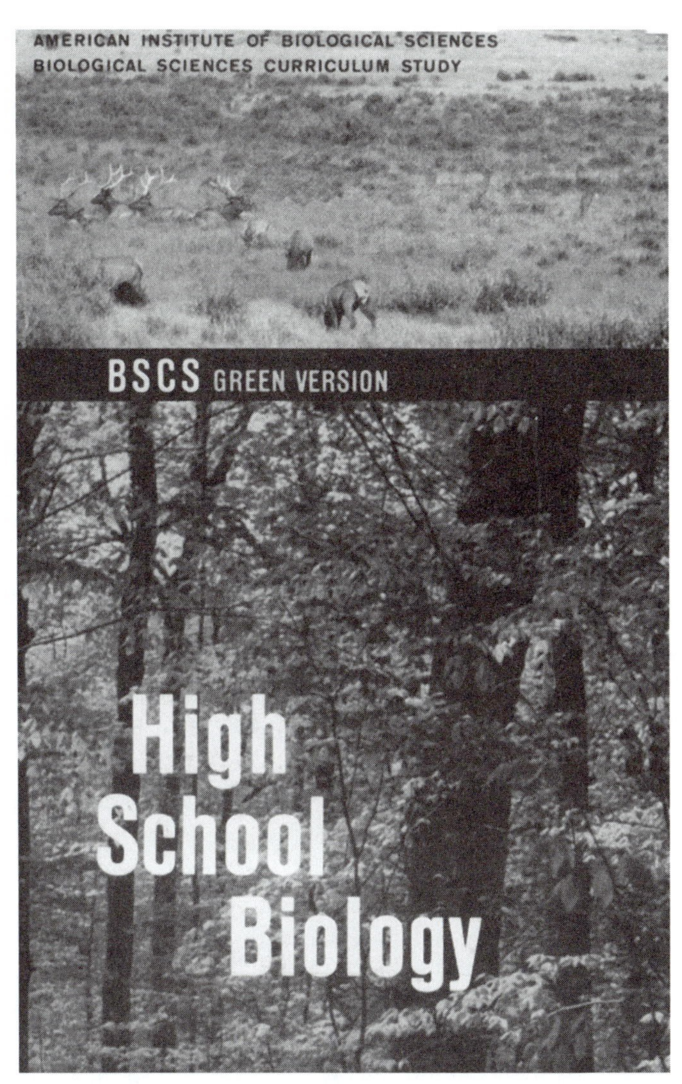

Figure 59 In the early 1960s, BSCS returned evolution to a prominent place in high school biology textbooks. Shown here is the first edition of the BSCS Green Version textbook. (*Reprinted with permission of BSCS*)

require Bible reading and recitation of the Lord's Prayer in public schools. People upset by the decision described it as an attempt to "outlaw God."

1963 John Scopes retires as a geologist and begins granting more interviews about his 1925 trial.

1963 NSF provides what will eventually be $4,800,000 for *Man: A Course of Study (MACOS)*, an introduction to evolution and behavioral/social science for elementary school students.

1963 French novelist Pierre Boulle (1912–1994) publishes *Planet of the Apes*, arguably the most famous science-fiction story of human evolution.

1963 Unable to convince the Orange County Board of Education to ban the teaching of evolution, Nell Segraves and Jean Sumrall (who later helped create the Creation Science Research Center) petition the California Board of Education to require that evolution be taught as theory rather than fact. Segraves and Sumrall believed that the teaching of evolution promoted atheism and was therefore unfair to Christian children. Soon thereafter, conservative Superintendent of Education Max Rafferty ordered that all biology textbooks in California identify evolution as a theory.

1963 Lutheran preacher Walter Lang (1913–2004) founds the Bible-Science Association (BSA) "to stimulate an exchange of ideas on Bible-Science relationships" and begins publishing the mimeographed *Bible-Science Newsletter*. BSA produced a variety of books and sponsored tours, seminars, films, radio programs, and meetings. BSA, which eventually boasted 15,000 members, focused its ministry in southern California through the work of Nell Segraves and Jean Sumrall. Along with the CRS, BSA claimed that people did not have to accept evolution to be scientific and modern. Despite its name, most BSA members were not scientists. BSA publications endorsed "the Bible as the inerrant word of God, true in every subject which it touches, whether the plan be salvation, science, or history" and noted that "talking about dinosaurs, Noah's Ark . . . and UFOs can lead to . . . salvation through Jesus Christ." BSA claimed that dinosaurs became extinct because of sin (*Dinosaurs and Sin,* 1973), that radiometric dating is unreliable (*What is the Age of this Lava Flow?,* 1974), and that Nobel Prizes are given to older scientists because younger ones are evolutionists and not as productive (*Creationists: The Better Scientists,* 1978). BSA is the second-oldest active U.S. creationist organization (after CRS), and its newsletter is the oldest active creation-science publication.

1963 The Francis Galton Laboratory of National Eugenics is renamed the Galton Laboratory of the Department of Human Genetics and Biometry.

1963 BSCS publishes *Biology Teachers Handbook*, which proclaims, "It is no longer possible to give a complete or even a coherent account of living things without the story of evolution."

1964 CRS begins publishing a journal—*Creation Research Society Quarterly*—because its 10 founders had been unable to publish their creationism-based ideas in established scientific journals.

1964 As part of a 57-day filibuster to oppose the Civil Rights Act, West Virginia Senator Robert Byrd (b. 1917) reads Genesis into the Congressional Record, adding, "Noah saw fit to discriminate against Ham's descendants."

1964 Claiming that BSCS textbooks (Figure 59) are the "most vicious attack we have ever seen on the Christian religion," Church of Christ preacher Reuel Gordon Lemmons (1912–1989) of Austin, Texas, campaigns to block the adoption of pro-evolution textbooks in Texas. Like Nell Segraves and Jean Sumrall in California, Lemmons demanded that evolution be taught "as a theory" and was helped by antievolution groups (e.g., the CRS) and Mel and Norma Gabler. The Gablers believed that the humanistic emphasis of the BSCS textbooks would lead to the rejection of traditional morality.

1964 Cambridge University establishes Darwin College in the former home of astronomer George Darwin (Charles and Emma's son). Graduates of Darwin College include zoologist Dian Fossey (1932–1985).

1964 In an essay in the *American Zoologist*, evolutionary biologist Theodosius Dobzhansky (Figure 53) writes "nothing in biology makes sense except in the light of evolution." In 1973, he published an often-cited essay with this title in *The American Biology Teacher*.

1964 Louis Leakey and his associates propose the name *Homo habilus* ("handy person") for the human remains associated with simple tools at Olduvai Gorge in Tanzania (see also Appendix C).

1964 Susan Epperson (Figure 60), the daughter of a biology professor, begins teaching biology at Central High School in Little Rock, Arkansas. The next year, Epperson challenged the Arkansas law banning the teaching of human evolution.

1964 The California Board of Education hears Nell Segraves and Jean Sumrall demand that "atheistic and agnostic" evolution be

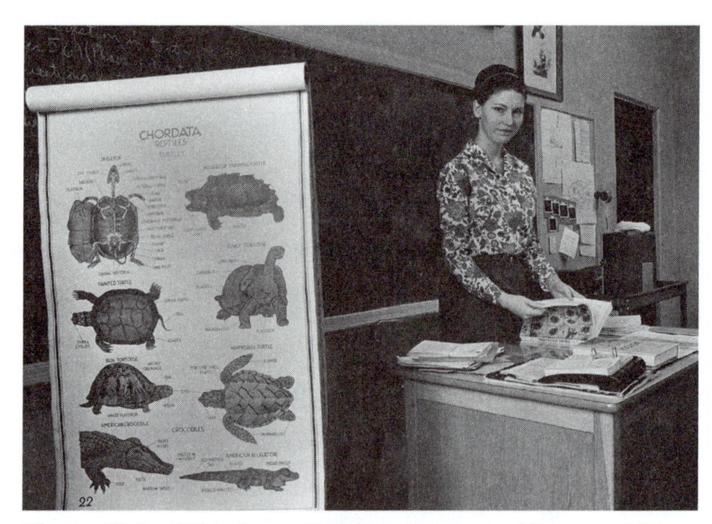

Figure 60 In 1964, Susan Epperson began working as a biology teacher in Little Rock's Central High School. The following year, Epperson's lawsuit to overturn the state's 1928 ban on the teaching of human evolution became the first evolution-related lawsuit since the Scopes Trial. (*Associated Press*)

described as a theory instead of a fact in state-approved biology textbooks. Segraves and Sumrall's argument was based on *School District of Abington Township v. Schempp*, in which the United States Supreme Court declared that mandated Bible-reading in public schools is not religion-neutral because it violates the rights of non-believers. Segraves and Sumrall argued that if it is unconstitutional "to teach God in the school," then it must also be unconstitutional "to teach the absence of God." The Board of Education unanimously rejected the requests that textbooks be rewritten to accommodate creationists' concerns.

1964 The Texas State Textbook Committee rejects claims that the five biology textbooks recommended for adoption in the state's schools—which include the three BSCS books—promote atheism. The committee's decision was supported by state education commissioner J. W. Edgar.

1964 While a graduate student at the University of London, William Hamilton (1936–2000) publishes two articles in *The Journal of Theoretical Biology* that propose a formal theory of kin selection. By taking a "gene's eye" view of selection, Hamilton proposed that behaviors that benefit another individual can arise if the benefit is oriented preferentially toward related individuals (i.e., individuals that are likely to share the same version of the altruist gene). Hamilton's specification of the conditions under which altruistic behavior should evolve encouraged examination of altruism in nature. This work confirmed Hamilton's predictions, opening doors to a new understanding of evolution and social behavior.

1964 The New York World's Fair opens. As had happened at the Chicago World's Fair three decades earlier, an exhibit of life-size models of dinosaurs exhibited by Sinclair Oil triggers a wave of dinosaur-mania in the United States (Figure 61).

1964 Yale University's John Ostrom (1928–2005) names *Deinonychus* ("terrible claw"), a small, bird-like theropod from the Lower Cretaceous of Montana (30 years earlier, the species had been discovered and informally named *Daptosaurus* by Barnum Brown). This discovery resurrected Thomas Huxley's (Figure 23) claim that dinosaurs are closely related to birds.

1964 Ashley Montagu's *The Concept of Race* argues that our racial distinctions are scientifically illegitimate.

1964 E. B. "Henry" Ford's (1901–1988) *Ecological Genetics* establishes the field of the same name and powerfully demonstrates that natural selection is an ongoing force that can be studied in real time.

Figure 61 The 1964–1965 New York World's Fair produced another wave of dinosaur-mania, thanks again to Sinclair Oil. This photo shows some of the nine life-size Sinclair dinosaurs being barged down the Hudson River to the Fair. In the background is Manhattan. Today, the logo of Harry Sinclair's famous company includes an outline of an apatosaur. (© *Sinclair Oil Corporation, 1966. All rights reserved, used by permission*)

1964 Walter Galusha's *Fossils and the Word of God* promotes gap creationism while claiming that there were no carnivores in Eden, that boa constrictors might have swallowed watermelons, that antediluvians had electricity, and that Noah talked animals into helping him build the ark.

1964 In *Essays of a Humanist*, Julian Huxley claims that "God is a hypothesis constructed by man to help him understand what existence is all about" and that "evolution is a process, of which we are products, and in which we are active agents. There is no finality about the process, and no automatic or unified progress."

1965 Another attempt to repeal the Arkansas antievolution law, this one led by Nathan Schoenfeld, fails.

1965 House Bill 301 banning the teaching of evolution in Arizona's public schools is introduced in the Arizona Senate, but it dies in the Senate Education Committee.

1965 NBC produces and broadcasts a new version of *Inherit the Wind*. Meanwhile, John Scopes claimed "restrictive legislation on academic freedom is forever a thing of the past."

1965 Susan Epperson (Figure 60), a high school teacher at Central High in Little Rock, Arkansas, meets with representatives of the Arkansas Education Association (AEA) to discuss challenging Arkansas's antievolution law. Epperson agreed to test the 1928 law because of her "concept of responsibilities as a teacher of biology and as an American citizen." The ACLU offered to help Epperson and the AEA, but the assistance was declined because of negative connotations the ACLU's involvement might produce. Arkansas Attorney General Bruce Bennett (1917–1979), who defended Arkansas's antievolution law against Epperson's challenge, claimed that the law was a simple exercise of administrative control of the curriculum, and that Epperson wanted "to advance an atheistic doctrine" (despite the fact that Epperson was a devout Christian). Hoping to avoid a "Scopes fiasco," Epperson asked for a declaratory judgment that the teaching of evolution represents a constitutional right and that obeying the law would force her to ignore "the obligations of a responsible teacher of biology." Epperson's challenge was later joined by Hubert H. Blanchard, Jr., who argued that the state's antievolution law damaged his sons' educations. Arkansas governor Orval Faubus (1910–1994) insisted that the law—the only such law passed by a public referendum—was still "the will of the people" and declared the Genesis account of creation "good enough for me." Faubus, who supported the antievolution law "as a safeguard to keep way-out teachers in line," prompted the retired John Scopes to urge Arkansas residents to "figure out what [Faubus] is thinking so you can protect yourself from [him]." Epperson's lawsuit was the first challenge of an antievolution law since John Scopes's trial in 1925 (see also Appendix D).

1965 The California State Advisory Committee on Science Education begins drafting new curriculum recommendations for its public schools. The committee's recommendations, which appeared in the Fall of 1969 as *Science Framework for California Public Schools, Kindergarten–Grades One through Twelve*, triggered lawsuits and controversy.

1965 The CRS bans old-earth ideas from its publications and commits itself to publishing a creationism-based biology textbook. Those efforts culminated in 1970 with the publication of *Biology: A Search for Order in Complexity*.

1966 Clifford Burdick (1919–2005) describes his alleged discovery of pollen from conifers in Precambrian rocks as "science-shaking original-pioneer work." This work—like many of his earlier claims—was used by creationists to question standard dating methods, but it was soon discredited.

1966 Encouraged by Max Rafferty (California's Superintendent of Public Instruction), Nell Segraves and Jean Sumrall ask the California Board of Education to mandate equal time for creationism in classrooms and textbooks. The Board refused the request.

1966 American biologist George Williams (b. 1926) responds to group selection arguments (i.e., evolution acting "for the good of the group") with *Adaptation and Natural Selection*, a forceful argument for selection at the level of the individual organism ("group-related adaptations do not, in fact, exist"). However, Williams also warned against "unwarranted uses of the concept of adaptation," such that natural selection "should be used only as a last resort." Stephen Jay Gould later described Williams's book—one of the first publications to reference the work of then-graduate student William Hamilton—as "the founding document for Darwinian fundamentalism."

1966 Henry Morris describes his strict creationist ideas in his book *Studies in the Bible and Science*. Morris claimed that if anyone wants to know about creation, the "sole source of true information is . . . divine revelation. God was there when it happened . . . [the Bible] is our textbook on the science of Creation!"

1966 John Wright, the last of the Scopes Trial jurors, dies at age 84.

1966 *Epperson v. Arkansas* begins on April 1 in a packed courtroom and lasts for less than a day. Bruce Bennett invoked the memory of William Jennings Bryan (Figure 42) by claiming that evolution is atheistic and that Arkansas has the right to determine what is taught in Arkansas's public schools. Two months later, Judge Murray Reed—noting "this Court is not unmindful of the public interest in this case"—issued a nine-page decision declaring that the Arkansas antievolution law was unconstitutional because it tends to "hinder the quest for knowledge, restrict the freedom to learn, and restrain the freedom to teach." The ruling by Reed (an appointee of Governor Orval Faubus) made it legal—at least temporarily—to teach human evolution in Arkansas's public schools (see also Appendix D).

1966 Willi Hennig's *Phylogenetic Systematics* is translated into English and soon thereafter becomes the foundation for cladistics, a hierarchical classification of species based on evolutionary ancestry. The vast impact of cladistics made Hennig, with Linnaeus, among the most influential systematists.

1966 Arkansas voters oust Attorney General Bruce Bennett and several other segregationists.

1966 Pope Paul VI (1897–1978) abolishes the *Index of Forbidden Books*.

1967 In *The Naked Ape*, English zoologist Desmond Morris (b. 1928) describes humans as another species of ape.

1967 American biologist Marshall Nirenberg and his colleagues show that the same genetic code is shared by bacteria, plants, and animals, and that the same mRNA encodes the same protein in all biological systems.

1967 Encouraged by a group of fundamentalist preachers, Archie Cotton—a local coalmine operator and member of the Campbell County Board of Education—charges science teacher Gary Lindle

Scott with unprofessional conduct, neglect of duty, and violation of Tennessee's Butler Law. At a closed meeting, Cotton convinced the Board of Education to vote unanimously to fire Scott. The next day, Scott's story appeared on the front page of *The New York Times*. Although Scott received little local support, he was later helped by the ACLU and the National Education Association, who arranged for Scott to be represented by civil-rights attorney William Kunstler (1919–1995). Scott's attorneys used a class-action suit in federal district court to charge that Tennessee's Butler Law was unconstitutional; the suit included Scott, two of his students, 59 Tennessee teachers, and NSTA. Scott asked for a permanent injunction to restrain state and local officials from enforcing the Butler Law and offered to drop his lawsuit if the Tennessee Senate repealed the Butler Law. Facing bad publicity and high expected costs for defending their decision to fire Gary Scott, the Campbell County Board of Education voted 7 to 1 to reinstate Scott.

1967 Knoxville attorney Martin Southern files a lawsuit in the name of his 14-year-old son in Knox County Chancery Court to test the validity of Tennessee's Butler Law. Southern claimed that the law had "limited" his son's education and asked for a declaratory judgment to overturn the law. Chancellor Len G. Broughton took the matter under advisement, and the Tennessee House of Representatives began debating whether to repeal the Butler Law. When Memphis Democrat D. O. Smith (cosponsor of House Bill 48, introduced to repeal the Butler Law) asked for the repealer to be brought before the House, the House's Sergeant-at-Arms brought a caged monkey to Smith's desk. The debate included numerous proclamations of faith, after which the House supported the bill by a vote of 59 to 30. The Tennessee Senate's two-hour debate about repealing the Butler Law was broadcast on television. The Senate's vote was a tie (16 to 16), which defeated the bill. The Senate then voted 23 to 10 to support Senate Bill 536, which amended the Butler Law to ban the teaching of evolution as "fact." However, the Tennessee House rejected the amendment, thereby leaving uncertain the fate of the Butler Law. Tennessee Governor Buford Ellington (1907–1972) later signed legislation to repeal the Butler Law, and Southern dropped his lawsuit.

1967 *The New Scofield Reference Bible* stops advocating gap creationism and claims that Genesis provides no information about the earth's age.

1967 In *Did Man Get Here by Evolution or by Creation*, the Watchtower Bible & Tract Society argues that Satan invented evolution and that the end of the world is imminent. More than 20 million copies of this booklet have been distributed.

1967 Henry Morris announces the Bible contains "all the known facts of science" and that "there neither is, nor can be, any proof of evolution. . . . Only the Creator—God himself—can tell us what is the truth about the origin of all things. And this He has done, in the Bible, if we are willing simply to believe what He has told us. . . . If we expect to learn anything more than this about the Creation, then God above can tell us."

1967 Lynn Sagan (later Margulis) proposes in the *Journal of Theoretical Biology* the endosymbiotic theory for the evolution of the eukaryotic cell. After encountering a decades-old suggestion that a symbiotic relationship among bacterial cells could be the origin of eukaryotes, Sagan proposed that the cytoplasmic organelles of eukaryotes that house extranuclear genetic material—mitochondria and chloroplasts—were at one time free-living bacteria that had started living within other bacterial cells. Over time, this symbiotic relationship evolved into a single unit, the eukaryotic cell. Sagan expanded her discussion of endosymbiosis in the *Origin of Eukaryotic Cells* (1970), and the endosymbiotic theory became a well-established concept. Margulis embraced James Lovelock's (b. 1919) Gaia Hypothesis and later proposed that symbiosis is the major evolutionary force on the planet: evolutionary novelties (e.g., species) arise when new symbiotic relationships are created, after which these new associations are "edited" by natural selection.

1967 Richard Leakey, Louis and Mary's youngest son, finds Omo I and Omo II, skulls of *Homo sapiens* that are 195,000 years old. These are some of the oldest examples of *Homo sapiens* (see also Appendix C).

1967 Using Emile Zuckerkandl and Linus Pauling's "molecular clock" technique, Vincent Sarich (b. 1934) and Allan Wilson (1934–1991) argue in an article in *Science* ("Immunological Time Scale for Hominid Evolution") that humans are more closely related to chimpanzees than chimps are to gorillas. Sarich and Wilson challenged the accepted belief of an ancient human lineage by claiming that "man and African apes shared a common ancestor five million years ago." Louis Leakey dismissed the claim as "not in accord with the facts available today."

1967 The Arkansas Supreme Court issues an unsigned, two-sentence *per curiam* ruling that reverses Judge Murray Reed's decision by declaring the Arkansas antievolution law "a valid exercise of the state's power to specify the curriculum in its public schools." The court's decision again made it a crime to teach human evolution in Arkansas, but did not address larger issues or comment on the validity of evolution. Susan Epperson (Figure 60) decided to appeal the decision.

1968 Biologists complete deciphering the genetic code. The Human Genome Project—which provided the actual DNA sequence of a human—was completed 33 years later.

1968 *Epperson v. Arkansas* is argued before the U.S. Supreme Court. Epperson's brief to the Supreme Court closed with dramatic references to "the famous Scopes case" and the "darkness in that jurisdiction" that followed it. The state countered by appealing to the authority of the Scopes case and closed with excerpts from the Tennessee Supreme Court's opinion in that case. The U.S. Supreme Court ruled unanimously in *Epperson v. Arkansas* that banning the teaching of evolution is unconstitutional because the First Amendment to the Constitution does not allow a state to require that teaching and learning must be tailored to the principles or prohibitions of any particular religious sect or doctrine. John Scopes—by this time 68 years old—noted that "[t]his is what I've been working for all along. . . . I'm very happy about the decision. I thought all along—ever since 1925—that the law was unconstitutional." The *Epperson* decision was subsequently applied in a variety of other decisions, including those involving censorship of textbooks, Bible readings in public schools, and academic freedom. As a result of *Epperson*, all laws banning the teaching of human evolution in public schools were overturned by 1970 (see also Appendix D).

1968 Japanese biologist Motoo Kimura (1924–1994) proposes the neutral theory of evolution in which genetic drift is the main force that changes allelic frequencies. Kimura argued that most mutations are selectively neutral (i.e., do not influence fitness), and his ideas became a basis for the refinement of "molecular clock" methods used to estimate the time passed since species diverged from a common ancestor. Kimura's theory, which remains controversial, was claimed by some creationists to refute Darwinian evolution, but this claim was unfounded.

1968 The first exobiology journal—now titled *Origins of Life and Evolution of the Biosphere*—is founded.

1968 Michael Polanyi's article in *Science* titled "Life's Irreducible Structure" compares machines and living organisms. This article inspired many of the proponents of ID.

1969 Acting on behalf of her daughter Frances, Mrs. Arthur G. Smith of Jackson, Mississippi, sues in state court for an injunction against the enforcement of the state's ban on teaching human evolution. Smith argued that her daughter was being denied a proper education and, citing *Epperson v. Arkansas,* that the antievolution law was unconstitutional. The lower court dismissed Smith's lawsuit and claimed that the state's antievolution law was constitutional, despite the U.S. Supreme Court's ruling to the contrary on the Arkansas law in *Epperson v. Arkansas*. Smith appealed the decision to the Mississippi Supreme Court (see also Appendix D).

1969 Following a heated discussion, the California State Board of Education sends the *Science Framework* guidelines to the State Department of Education for revision. Soon thereafter, State Department of Education Superintendent Max Rafferty announced that the guidelines would be rewritten to include creationism. These efforts resulted in evolution being treated as a "theory," not as a fact. Publishers changed their textbooks to accommodate the new recommendations, and years of debate and legal wrangling about the teaching of evolution and creationism in California ensued.

1969 Led by aerospace engineer and ASA member Vernon Grose (b. 1928), creationists file a 13-page memorandum that convinces the California Board of Education (of which seven members had been appointed by Governor Ronald Reagan) to vote unanimously to include scientific creationism in the state's 205-page *Science Framework*. Grose claimed, "creation in scientific terms is not a religious belief." Grose's claims represented a change in creationists' tactics—instead of asking states to ban evolution, they now asked that teachers teach creationism. Outraged members of the California State Advisory Committee on Science Education issued a statement drafted by Stanford's Paul DeHart Hurd opposing the changes in *Science Framework*.

1969 Famed Southern Baptist preacher W. A. Criswell's (1909–2002) *Why I Preach That the Bible Is Literally True* promotes young-earth creationism and claims that if evolution is true, humans are without salvation.

1969 More than 40 years after his famous trial, John Scopes receives an average of one letter per day about his trial.

1969 The $10,000,000 provided by NSF and other agencies to the BSCS has contributed significantly to the improvement of biology education in the United States. Project Director Arnold Grobman announced "[i]t appears now that the major storms are over. There is every indication that the teaching of evolution is generally accepted in America and will become far more commonplace that it ever was before." Grobman, like those before and after him who made similar predictions, was wrong.

1969 The Texas Board of Education, under pressure from Mel and Norma Gabler (who claimed that the evolution-based BSCS biology textbooks had been written by communists), removes two BSCS books from the list of textbooks approved for use in the state's public schools.

1969 As eugenics becomes increasingly unpopular, the journal *Eugenics Quarterly* changes its name to *Social Biology*.

1970 While on a sabbatical from Virginia Polytechnic Institute, Henry Morris meets Baptist preacher Tim LaHaye (b. 1926) at the Torrey Memorial Bible Conference in Los Angeles. Morris later resigned his faculty position in Virginia and moved to California to help LaHaye establish Christian Heritage College (now San Diego Christian College). Students attending the conservative college were required to take six semester hours of "scientific creationism."

1970 During a two-day visit to Peabody College in Nashville, Tennessee, John Scopes (Figure 62) makes his final public appearance and his first appearance in a Tennessee classroom in 45 years. *Time* magazine reported that "[h]istory has treated Scopes well and he was greeted like a returning hero."

1970 Lester Showalter publishes *Investigating God's Orderly World*, a Bible-based science textbook claiming "the theory of the evolution of man . . . is wrong because it is contrary to God's Word—the Bible." Although "ungodly men" may doubt the biblical

Figure 62 After his famous trial in 1925, John Scopes worked in the oil industry and shunned publicity. This photo shows Scopes in 1969, a year before his death. (*Jerry Tompkins*)

Flood, "we must accept and believe the Bible and not the scientists." This book was intended for science teachers wanting to "uphold faith in the six-day Creation as given in Genesis 1" and who believe the Bible "even in the face of unexplainable information."

1970 Henry Morris unveils the phrase *creation science* in a course at fundamentalist Christian Heritage College. The creation science movement became immensely popular. Thanks largely to the work of Morris, strict creationism—once a minority point of view among creationists—became the movement's dominant view.

1970 In its "The Devil's Advocate" column, *The American Biology Teacher* publishes an article by young-earth creationist—and famed debater—Duane Gish (b. 1921) titled "A Challenge to Neo-Darwinism." Gish's article stressed familiar antievolution arguments (e.g., lack of transitional forms) while claiming "there is very little evidence, if any, to support the general theory of evolution." Gish pleaded for biology teachers to give a "balanced presentation" of the evidence. These calls by creationists for fairness and "equal time" became increasingly popular in the decades that followed.

1970 The Orthodox Union, which determines the quality of being kosher (i.e., which foods are acceptable for Orthodox Jews to consume), publishes *A Science and Torah Reader*, declaring that evolution is scientifically invalid and incompatible with Orthodoxy. Although ancient Jewish scholars such as Maimonides (1135–1204), Nahmanides (1194–1270), and Gersonides (1288–1344) had suggested that the creation described in the Torah should not be read literally, influential Orthodox rabbis such as Moshe Feinstein (1895–1986) denounced evolution as heretical.

1970 An effort to repeal Mississippi's antievolution law fails, leaving the fate of the law to the state's courts. In a ruling on Mrs. Arthur Smith's appeal of a lower court's decision to uphold the state's antievolution law, the Mississippi Supreme Court—citing *Epperson v. Arkansas*—ruled in *Smith v. State* that Mississippi's antievolution law—the last surviving law of its kind—was void. The state did not appeal. Within a span of three years, legislatures and courts had rejected all of the existing antievolution laws (see also Appendix D).

1970 John Scopes dies at age 70 and is buried in a family plot in Oak Grove Cemetery in Paducah, Kentucky, not far from Lone Oak School where he first learned about evolution. Scopes was buried beneath the inscription "A Man of Courage" (Figure 63).

1970 Leona Wilson sues the Houston Independent School District in federal district court on behalf of her daughter and other students because the school has violated the students' constitutional rights by teaching evolution as fact and without referring to other theories of origin. The plaintiffs also repeated an increasingly common theme among antievolutionists, arguing that the state was using the teaching of evolution to establish the "religion of secularism."

Figure 63 John Scopes's 1925 trial for allegedly teaching human evolution remains the most famous event in the history of the evolution-creationism controversy. When Scopes died in 1970, he was buried beneath the inscription "A Man of Courage." (*Randy Moore*)

The resulting lawsuit (*Wright v. Houston Independent School District*) was the first lawsuit initiated by creationists, but it was later dismissed before reaching trial when Judge Woodrow Seals ruled that the free exercise of religion is not accompanied by a right to be insulated from scientific findings incompatible with one's beliefs (see also Appendix D).

1970

Prompted by the California textbook controversies in the late 1960s, Nell Segraves, Kelly Segraves, and Henry Morris found the Creation Science Research Center (CSRC) to produce teaching materials and "extension ministries" (e.g., seminars and radio programs) for schools wanting to include Flood-based creationism in their science curricula. Soon thereafter, Morris's CRS produced John Moore's *Biology: A Search for Order in Complexity*, a pro-creationism textbook that claimed "there is no way to support the doctrine of evolution" and promoted what came to be known as the "two-model approach"—that is, a "creation model" (which mirrors Genesis and no other creation myth) and an "evolution model." Moore's creationism-based biology textbook claimed "the most reasonable explanation for the actual facts of biology as they are known scientifically is that of Biblical creationism." The textbook, which was published by Zondervan (a Christian publishing house) after more than five years in development, was used at several public and many private schools. Creationists claimed that the book "would be of interest to public school systems desiring to develop a genuine scientific attitude in their students." The use of *Biology: A Search for Order in Complexity* in public schools was hindered by legal challenges—most notably *Hendren v. Campbell*—for more than a decade.

1970 Swiss author Erich von Däniken's (b. 1935) *Chariots of the Gods?* claims that early humans were produced by "deliberate 'breeding' by unknown beings from outer space." In 1977, von Däniken attacked the theory of evolution and claimed that the sudden appearance of new animal species throughout history was due to genetic engineering by extraterrestrials. Richard Steeg of the BSA responded to von Däniken by agreeing that it is "actual historical fact" that the earth was visited by aliens who bred with humans, but claimed that the aliens "were angels, not spacemen." According to Steeg, this breeding was evil and produced degeneration.

1970 The Education Development Center publishes the NSF-backed *Man: A Course of Study (MACOS)*, an introduction to evolution and behavioral/social science for elementary school students. Within four years, 1,700 school districts in 47 states had adopted the award-winning program, but nationwide opposition by parents and teachers later reduced sales dramatically. Several members of Congress, led by Representative John Conlan (b. 1930) and Senator Jesse Helms (1921–2008), pressured NSF to limit funding of pro-evolution projects. To avoid further investigations, NSF became cautious in its treatment of evolution-related projects. In 1980, presidential candidate Ronald Reagan used *MACOS* as an example of the federal government's endorsement of subversive values.

1970 The Institute for Creation Research (ICR) is founded by Henry Morris and Tim LaHaye to promote "creation science, biblical creationism, and related fields" (Figure 64). ICR was originally the research division of LaHaye's Christian Heritage College, but became autonomous in 1981. ICR claimed that "all things in the universe were created and made by God in the six literal days of the creation week described in Genesis 1:1–2:3," that the Bible is "factual, historical, and perspicuous," that "all theories of origins or development which involve evolution in any way are false,"

Figure 64 The Institute for Creation Research (ICR) is one of the most influential and well-funded antievolution organizations in the world. Although ICR was originally headquartered in California, today the ICR's offices (shown here) are in Dallas, Texas. (*Darrell and Donna Vodopich*)

and that evolutionary thinking produced abortion, promiscuity, drug abuse, and homosexuality. ICR sells books, videos, and "Bibleland Cruises" with celebrity creationists, and one of its radio programs—*Science, Scripture, and Salvation*—airs on more than 600 stations. ICR's Museum of Creation and Earth History in Santee, California, includes a Noah's Ark diorama that documents how animals lived aboard the ark. The museum, which is visited by more than 25,000 people per year, explains the age of the earth, the meaning of life, what happens when we die, and why there is suffering in the world.

1971 "Lonesome George," the last member of a doomed race of Galápagos tortoises, is discovered on Pinta Island.

1971 Norman Macbeth's *Darwin Retried* claims that the flaws of evolution are an "open secret" among scientists who conspire to hide them. Macbeth, who predicted "it will be a long time before the public is given the full dark picture," urged biologists to "confess."

1971 The TransNational Association of Christian Colleges and Schools is founded by Henry Morris to promote and accredit

Christian schools. The association required that accredited members affirm "the literal existence of Adam and Eve as the progenitors of all people" and the "special creation of the existing space-time universe and all its basic systems and kinds of organisms in the six literal days of creation week."

1971 Arthur Koestler's (1905–1983) *The Case of the Midwife Toad* speculates that Paul Kammerer's (Figure 34) faked nuptial pads in midwife toads were planted by Nazi sympathizers or Darwin supporters intent on discrediting Lamarckian inheritance.

1971 Duane Gish joins fellow creationist Henry Morris at Christian Heritage College as Professor of Natural Science. Gish later moved with Morris to ICR to serve as Associate Director. When ICR started offering graduate degrees, Gish became Senior Vice President.

1971 Fundamentalist preacher Jerry Falwell (1933–2007)—an ardent opponent of evolution—opens Liberty University, which is based upon "an inerrant Bible," "a Christian worldview beginning with belief in biblical Creationism," and "an absolute repudiation of political correctness." In the early 1980s, Liberty University sought state accreditation for its biology department so its graduates could teach in public schools (or, in the words of Falwell, so that Liberty's "hundreds of graduates [could] go out into the classrooms teaching creationism"). Because the course "History of Life" included creationism and was a required part of the science curriculum, the ACLU opposed accreditation. When the course was moved to the humanities department, the biology program was accredited.

1971 In *Chance and Necessity*, French biologist Jacques Monod (1910–1976) dismisses the notion of purpose in the universe.

1971 Niles Eldredge (Figure 65) suggests that the fossil record indicates that most species do not change gradually and continually, but rather go through long periods of stasis followed by rapid but infrequent periods of diversification. Eldredge was influenced by the geographic speciation model developed by Theodosius Dobzhansky (Figure 53) and elaborated by Ernst Mayr, which argued that speciation is most likely at the edges of species' ranges, while the core of the population remains unchanged over long periods. Eldredge's idea initially generated little interest. However, a year later, Stephen Jay Gould and Eldredge published "Punctuated

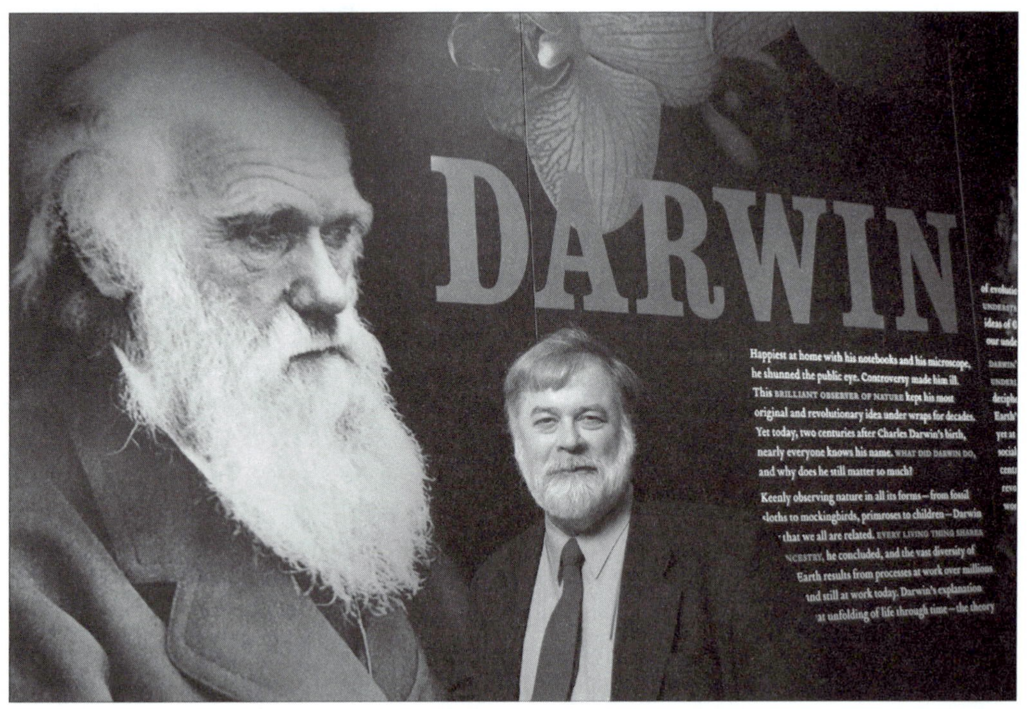

Figure 65 Niles Eldredge, who with Stephen Jay Gould advocated what came to be known as "punctuated equilibrium," stands beside an exhibit about Darwin that opened at the American Museum of Natural History in late 2005. (*Niles Eldredge*)

Equilibria: An Alternative to Phyletic Gradualism" in *Models in Paleobiology*, a book edited by Thomas Schopf. Gould and Eldredge challenged the traditional view of gradualism by arguing that the fossil record demonstrates that evolutionary change is generally nonexistent in species composed of large populations due to the homogenizing effect of interbreeding among individuals. The lack of smooth transitional sequences in the fossil record does not, therefore, represent the unavoidably imperfect (due to the capricious process of fossilization) chronicle of slow gradual change, but rather accurately reflects the actual operation of the evolutionary process. Gould maintained that the idea was generated by Eldredge, but Eldredge credited Gould's "knack for catchy phrases" as being responsible for attracting interest. Eldredge and Gould, who were labeled by some biologists as "traitors to the Darwinian tradition," spent years trying to demonstrate that punctuated equilibrium is consistent with contemporary understanding of evolution and that even Darwin suggested that "gradualism" (often portrayed as the mutually exclusive alternative to punctuated equilibrium) is not the only possible pattern of evolution. Creationists used the debate about "punk ek" to question the status of evolution within science: If scientists themselves cannot agree about how evolution works, then evolution cannot be a well-established concept as has been claimed. Creationists also argued that punctuated equilibrium is an example of scientists' flip-flopping (evolution was first proposed to work slowly, now it is proposed to work quickly) when confronted with contradictory evidence (i.e., lack of transitional forms in the fossil record). After being repeatedly misquoted by antievolutionists, Eldredge published *The Monkey Business: A Scientist Looks at Creationism* (1982) to confront "scientific creationism" and *The Triumph of Evolution and the Failure of Creationism* (2001) to confront ID.

1971 The Board of Education in Columbus, Ohio, passes a resolution encouraging teachers to present both creationism and evolution in their science classes.

1971 The discovery of restriction enzymes enables scientists to build recombinant DNA by combining DNA fragments from two different viruses.

1971 The Michigan legislature considers a bill requiring "equal time" for creationism. After references to the Bible were removed from the bill, the legislation was supported by the House of Representatives. However, the bill died when it reached the Senate too late to be considered.

1971 In *Lemon v. Kurtzman* (a case concerning public support of private schools), the U.S. Supreme Court establishes the so-called "Lemon Test" detailing the requirements for legislation concerning religion. The test's three "prongs" require that a government's action (1) have a legitimate secular purpose; (2) not have the primary effect of either advancing or inhibiting religion; and (3) not foster excessive entanglement of the government with religion. If any of these three prongs is violated, the government's action is unconstitutional under the Establishment Clause of the First Amendment to the United States Constitution. The "Lemon Test" resulted from three separate cases (*Lemon v. Kurtzman*, *Earley v. DiCenso*, and *Robinson v. DiCenso*), which were joined because they involved similar issues. The Lemon Test has been cited repeatedly in court cases associated with the evolution-creationism controversy.

1972 NAS, NSTA, and NABT—spurred into action by creationists' victories in California—begin campaigns to oppose requirements that biology teachers include creationism in their courses. NABT, publisher of *The American Biology Teacher*, established its Fund for Freedom in Science Teaching to combat creationists. The fund received $12,000 in donations, but to placate its many members who were creationists, NABT sponsored a creationism session at its annual meeting. The session was the best-attended session at the meeting. BSCS's Bill Mayer later denounced NABT for acquiescing to creationists' demands.

1972 CSRC splits from Christian Heritage College, primarily because of conflicts between Henry Morris and Kelly Segraves. The Segraves family had kept the CSRC name when Morris founded ICR (Figure 64) to continue his creationism-based research. Although CSRC went into debt, it continued to fight evolution education, along with legalized abortion, women's rights, and gay rights.

1972 AIBS passes a resolution deploring efforts by biblical literalists to interject creationism and religion into science courses. That same year, AAAS denounced creationism as "neither scientifically grounded nor capable of performing the roles required of scientific theories." Most people disagreed; for example, a poll showed that 75% of students at Rhea County (TN) High School—the school at which John Scopes taught in 1925—were creationists who believed that evolution produced corruption, lust, greed, drug addiction, war, and genocide.

1972 Geneticist Richard Lewontin (b. 1929) argues that there is more genetic variation within human populations than between them, no matter where in the world those populations are from.

1972 Artist Jack Chick (b. 1924) publishes *Big Daddy?*, a small, 24-page comic book written with the help of young-earth creationist Kent "Dr. Dino" Hovind. *Big Daddy?* begins with an arrogant biology professor humiliating a student who dares to question evolution. In a classic David-versus-Goliath confrontation, the enraged professor screams that the student is a fanatic and threatens him with jail, but the tide turns when the student tells the professor about "amazing findings which are rarely made public." These "amazing findings" expose evolution as a "big lie" and force the professor to admit that the student is "destroying me." The humiliated and repentant professor then pleads for the student to answer the questions that science cannot answer. The story ends when everyone becomes a creationist and the professor—admitting that he can no longer teach evolution—is fired by heathen administrators. Other Chick publications claimed that Russia sabotaged efforts by Christian heroes to find Noah's ark (*The Ark*) and that evolution is the religion of scientists who laugh at God (*In the Beginning*). *Big Daddy?* is the most widely distributed antievolution publication in history.

1972 Colorado state legislators introduce House Concurrent Resolution 1011 requiring "equal time" for creationism as an amendment to the state constitution. Thanks to the efforts of William Mayer (Director of BSCS) and others, the bill died in the Judiciary Committee.

1972 Duane Gish promotes creationism at the NABT annual convention that bears the theme "Biology and Evolution." At this same convention, Theodosius Dobzhansky (Figure 53) gave his famous "Nothing in Biology Makes Sense Except in the Light of Evolution" presentation. Gish became famous for his skills in debating evolutionary biologists, beginning with an impromptu confrontation with biologist G. L. Stebbins at the University of California at Davis in 1972. By 2005, Gish had participated in almost 300 debates throughout the United States and in more than 40 countries.

1972 *Footprints in Stone*, Stanley Taylor's film about the alleged evidence that humans and dinosaurs lived contemporaneously, is distributed throughout the United States.

1972 In an article titled "The Nature of the Darwinian Revolution" in *Science*, Ernst Mayr notes that the Darwinian "revolution began when it became obvious that the earth was very ancient rather than having been created only 6,000 years ago. This finding was the snowball that started the whole avalanche."

1972 ICR's Henry Morris emphasizes his belief that evolution is the anti-God conspiracy of Satan by noting "the peculiar rings of Saturn, the meteorite swarms . . . reflect some kind of heavenly catastrophe associated either with Satan's primeval rebellion or his continuing battle with Michael and the angels."

1972 Still smarting over how creationists used his 1944 comment about the "absence of transitional forms" to discredit evolution, George

Gaylord Simpson (Figure 54) points out that biologists have discovered "literally thousands of transitional forms" and that "anyone who cites me or my work in opposition [to evolution] is either woefully ignorant or willfully misrepresenting the facts."

1972 The California State Board of Education accepts the texts recommended by the Curriculum Development and Supplemental Materials Commission. When Reverend David Hubbard of Pasadena complained about the dogmatism of scientists, the board voted 7 to 1 to remove dogmatism in explanations of origins in state-approved textbooks. The California Baptist Convention and other creationist groups passed resolutions calling for the California Board of Education to implement its original decision requiring the inclusion of creation science in textbooks. After the board appointed a committee of four creationists to oversee implementation of the new policy, *Science* publisher William Bevan warned that creationists' victories could politicize the nation's classrooms, noting that "if the state can dictate the content of a science, it makes little difference that its motivation is religious rather than political." The board agreed on a compromise that emphasized the speculative nature of Darwinian evolution but did not mention God or Genesis. The board ignored the pro-science recommendations of the state's Curriculum Committee and restored creationist textbooks to the list of approved books. Soon thereafter, the board dissolved the Curriculum Committee and replaced it with a Curriculum Development and Supplemental Materials Commission that was staffed with creationists. The commission's subcommittee outraged many members of the State Board of Education when it refused to recommend creationist textbooks.

1972 Philosopher Karl Popper (1902–1994), who made falsifiability a hallmark of scientific studies, publishes *Objective Knowledge: An Evolutionary Approach*. This book compared the advance of scientific knowledge and evolution by natural selection, noting, "our theories . . . suffer in our stead in the struggle for survival of the fittest."

1972 Masatoshi Nei (b. 1931) develops a method for estimating genetic distances between populations from electrophoresis. Nei's method became a standard for studying the evolutionary relationships of populations. In 1979, Nei developed a method for estimating the number of nucleotide substitutions per site between two DNA sequences from restriction enzyme data.

1972 In *The Remarkable Birth of Planet Earth*, Henry Morris—who believed that Christian salvation requires the acceptance of a worldwide flood that destroyed virtually all life on earth—argues that God would not preside over evolution's "cruel spectacle" and "appalling inefficiency and barbarity." Morris also claimed that humans are "a unique creation of God, entirely without evolutionary relation to the animals," that "Adam was the first man" and that the earth's age can be learned only by studying the Bible: "The only way we can determine the true age of the earth is for God to tell us what it is. And since he has told us, very plainly, in the Holy Scriptures that it is several thousand years of age, and no more, that ought to settle all basic questions of terrestrial chronology." Morris purported to present evidence for divine creation while blaming evolution "for our present-day social, political, and moral problems."

1972 William Willoughby, religion editor of the *Washington Evening Star*, sues H. Guyton Stever (director of NSF) and the Board of Regents of the University of Colorado "in the interest of forty million evangelistic Christians in the United States." NSF provided funds for the development of BSCS's textbooks, and Willoughby wanted NSF to spend the same amount of money "for the promulgation of the creationist theory of the origin of man."

1973 Russell Artist convinces Tennessee state senator Milton Hamilton and several of his colleagues to introduce Senate Bill 394, which calls for "balanced treatment" of evolution and creationism. The bill, which became known as the Genesis Bill, required (1) all textbooks to claim that discussions of origins are not scientific facts and (2) equal numbers of words, space, and emphasis for "other theories, including, but not limited to, the Genesis account in the Bible." Despite the Tennessee attorney general's opinion that the Genesis Bill legislation was probably unconstitutional, the Tennessee Senate began debating the bill. When a group of Tennessee professors claimed that the legislation is "utterly repugnant to the American idea of democracy," Senator Milton Hamilton responded with "[t]his is not a Ph.D. bill; it's a people's bill." The Senate voted 29 to 1 in favor of the Genesis Bill, after which the House of Representatives accepted the Senate bill in lieu of its own. The legislation passed, thereby mandating an equal emphasis on evolution and the Genesis version of creation. The law, which excluded "all occult or Satanical beliefs of human origin," defined the Bible not as a textbook, but instead as "a reference

work," and therefore not subject to the law. This bill—the first requiring "equal time" for evolution and creationism—became law when Governor Winfield Dunn (b. 1927) refused to sign or veto the bill. The law was overturned two years later (*Daniel v. Waters*; see also Appendix D).

1973 A coalition of seven representatives, including Speaker of the House Ned McWherter (governor of Tennessee from 1986 to 1994), introduces the innocuously titled "An Act to Amend Tennessee Code Annotated, Section 49-2008, Relative to Selection of Textbooks." This bill was similar to an "equal time" bill introduced the previous week in the Tennessee Senate.

1973 Creationist Duane Gish appears at Seattle-area churches and schools in hopes of mobilizing support for an "equal time" bill in the state legislature. That bill (House Bill 1021) was killed by the House Education Committee. When the Committee for the Initiative on Creation and Evolution tried to gather enough signatures to force a referendum on the issue (Initiative 47) the following year, it fell far short of the 118,000 signatures needed to transmit the measure to the legislature.

1973 Australian physicist Brandon Carter (b. 1942) coins the term *anthropic principle* to explain the "just right" conditions for life to exist and evolve in the universe and to note that if the universe had been different in any important way, humans would not be here. Anthropic reasoning, which attempts to explain the observation that the universe is fine-tuned for life, was later used by creationists to justify their claims about nature.

1973 In his article "A New Evolutionary Law," evolutionary biologist Leigh Van Valen (b. 1935) proposes the "Red Queen" hypothesis to describe the selective pressures organisms face in a constantly changing environment. Van Valen used the analogy of the Red Queen, a living chess piece from *Alice in Wonderland* who must constantly run to stay in the same place. Similarly, populations of organisms constantly evolve in response to environmental change.

1973 Mel and Norma Gabler found the influential Education Research Analysts in Longview, Texas, to finance their campaign against "secular humanism" and textbooks they consider anti-American. The Texas Board of Education often removed books the Gablers found offensive from the list of state-approved textbooks. Because the textbook market was so large in Texas, these moves affected textbook use nationwide.

1973 Nobel Laureate François Jacob (b. 1920) notes "there are many generalizations in biology, but precious few theories. Among these, the theory of evolution is by far the most important, because it draws together from the most varied sources a mass of observations which would otherwise remain isolated; it unites all the disciplines concerned with living beings; it establishes order among the extraordinary variety of organisms and closely binds them to the rest of the earth; in short, it provides a causal explanation of the living world and its heterogeneity."

1973 Peter (b. 1936) and Rosemary (b. 1936) Grant begin studying finches on Daphne Major, a 100-acre islet in the Galápagos archipelago. After the Grants' first visit to the islands, they described the archipelago as a "gold mine" of a natural laboratory. By 1977 the Grants had banded more than half of the island's birds, and thereafter this percentage remained near 100%. The thoroughness of the Grants' more than 30 years of work enabled them to document first-hand the mechanism of natural selection and observe evolution occurring faster than Darwin had estimated. Because such long-term monitoring of natural populations is rare due to funding and logistical challenges, the Grants' work is important because of the continuous record they and their students have generated, having studied more than 25 generations and 20,000 birds.

1973 In a famous article titled "Nothing in Biology Makes Sense Except in the Light of Evolution" published in *The American Biology Teacher*, Theodosius Dobzhansky (Figure 53) notes "[l]et me try to make crystal clear what is established beyond reasonable doubt, and what needs further study, about evolution. Evolution as a process that has always gone on in the history of the earth can be doubted only by those who are ignorant of the evidence or are resistant to evidence, owing to emotional blocks or to plain bigotry. By contrast, the mechanisms that bring evolution about certainly need study and clarification. There are no alternatives to evolution as history that can withstand critical examination." Dobzhansky also argued that creationists' "appearance of age" argument implies a deceitful creator who planted false evidence as if "deliberately to mislead sincere seekers of truth." *The American Biology Teacher* also published articles by Duane Gish and John N. Moore promoting creationism; these articles were prefaced by a statement saying that most biologists reject creationism.

1973 The Fifth Circuit Court of Appeals upholds Judge Woodrow Seals's dismissal of *Wright v. Houston Independent School District*, declaring that Wright's request of equal time for creationism is "an unwarranted intrusion into the authority of public school systems to control the academic curriculum" (see also Appendix D).

1973 The Georgia Board of Education approves *Biology: A Search for Order in Complexity* for adoption in the state's public schools. Soon thereafter, ICR (Figure 64) sponsored a symposium at an Atlanta Baptist church at which Henry Morris urged citizens to organize to pass an antievolution law. Citizens for Another Voice in Education (CAVE) was formed, but the group was not influential.

1973 The Kanawha County (West Virginia) School Board adopts John Moore's *Biology: A Search for Order in Complexity* and other publications produced by creationist Nell Segraves. The following year, residents of the county threatened violence when they discovered that the English and biology textbooks used in local schools questioned fundamentalist beliefs. Coal miners went on strike to protest the books, preachers vilified the books from their pulpits, teachers were threatened, snipers fired on school buses and police escorting the teachers, and protesters destroyed three cars, attacked school buses, and vandalized the Board of Education building. The board responded by adopting creationist materials for the entire school district, adding that students could be excused from reading any book that their parents found objectionable.

1973 NABT and several science teachers join the challenge to Tennessee's Genesis Law in federal district court in Nashville.

1973 In *The World That Perished*, John Whitcomb, Jr., claims that once animals were aboard Noah's Ark, God "imposed a year-long hibernation [that] removed the burden of their care completely from the hands of Noah and his family."

1973 The Oregon School Board requires school libraries to include creationist materials and tells teachers to urge students to "weigh the information and arrive at their own conclusions."

1973 *Willoughby v. Stever* is dismissed by the U.S. District Court in Washington, DC, on the grounds that (1) books supported by taxes allocated to NSF disseminate scientific findings, not religion, and (2) the First Amendment does not allow the state to require that teaching be tailored to particular religious beliefs. Willoughby appealed his case to the U.S. Supreme Court (see also Appendix D).

1974 A three-judge panel hears the challenge to Tennessee's Genesis Bill. The judges accepted the state's argument for abstention, after which attorneys for the plaintiffs appealed the decision to the circuit court of appeals. As the federal lawsuit spawned by Tennessee's Genesis Bill moved toward circuit court, Chancellor Ben Cantrell of Tennessee state court declared the Genesis Bill to be unconstitutional because it established religion (see also Appendix D).

1974 Famous young-earth creationist Henry Morris, several members of ICR, Scott Memorial Church, and Heritage College establish Creation-Life Publishers in San Diego. One of the company's first books was *Scientific Creationism*, a "reference book" that appeared in two editions: one for public schools, containing no references to the Bible, and another for Christian schools that included a chapter titled "Creation According to Scriptures." Morris, reviving the claims of George McCready Price, argued that "the long geological ages of evolutionary history never really took place at all" and denounced uniformitarianism while claiming "the *real* message of the fossils" is that "there is no truly objective time sequence to the fossil record." Morris proclaimed "[t]he Bible is a book of science" while claiming "[i]f the Bible is the Word of God—and it is—and if Jesus Christ is the infallible and omniscient Creator—and He is—then it must be firmly believed that the world and all things in it were created in six natural days and that the long geological ages of evolutionary history never really took place at all." Later, Morris claimed that Satan invented evolution at the Tower of Babel.

1974 In a remote desert region of Ethiopia's Afar Triangle, Donald Johanson (b. 1943) discovers one of the most complete australopithecine skeletons. Its owner, nicknamed Lucy, was a little bigger than a chimp and had a brain about one-third the size of that of modern humans. [Lucy, a member of the species *Australopithecus afarensis*, died about 3.2-million years ago.] Johanson and his

colleagues placed Lucy at the base of the human lineage. For the next 20 years, textbooks put Lucy's species as ancestral for all humankind. Lucy became an ancestral ambassador; although researchers have discovered older and more complete fossil hominins, Lucy remains a landmark to which others are compared (see also Appendix C).

1974 The Darwin Correspondence Project is founded by American scholars Frederick Burkhardt (1913–2007) and Sydney Smith (1911–1988) to locate, study, and publish summaries of all letters written by Charles Darwin. By 2009, the project—the largest scholarly project about one person—included 15,000 letters.

1974 John Ostrom's paper titled "*Archaeopteryx* and the Origin of Flight" revives Thomas Huxley's (Figure 23) century-old claims that birds are the descendants of dinosaurs.

1974 Stephen Jay Gould begins his monthly "This View of Life" column in *Natural History* magazine. Gould's essays were usually about evolutionary biology, but he was masterful at weaving history, politics, and popular culture (including his beloved game of baseball) into engaging discussions of the sometimes-arcane aspects of science. Many of the essays were collected into several popular books (e.g., *The Panda's Thumb*, *Ever Since Darwin*, and *Hen's Teeth and Horse's Toes*) that provided accessible discussions of evolutionary topics to general readers. Gould continued writing the column until 2001 when a recurrence of cancer forced him to curtail his activities.

1974 American social scientist Donald Campbell (1916–1996) coins the term *evolutionary epistemology* to describe how epistemology sometimes involves natural selection.

1974 The Atlanta Board of Education accepts its textbook committee's recommendation to reject *Biology: A Search for Order in Complexity*, noting that the book "contains numerous errors."

1974 The California Board of Education reverses its 1969 decision and eliminates creationism from state-adopted textbooks. That same year, California Attorney General Evelle Younger (1918–1989) ruled that the California Board of Education's elimination of creationism from the state's guidelines was constitutional.

1974 The Tennessee legislature refuses to remove the most blatant aspects of religion from its Genesis Bill. The fate of the law rested with the higher courts.

1974 Engineer Harold Hill's *How to Live Like a King's Kid* tells how NASA scientists discovered Joshua's Long Day, which allegedly occurred when Joshua commanded the sun to stand still (Joshua 10:11–13). Hill's story prompted many citizens to contact NASA for more information.

1974 The Texas Education Policy Act adopts Mel and Norma Gabler's recommendation that all biology textbooks state that evolution is a theory rather than a fact and is only one of several explanations of human origins. Despite the protests of some scientists, the Texas Board of Education used the new policy to reject all three BSCS textbooks.

1974 The U.S. Supreme Court refuses to review *Wright v. Houston Independent School District*, thereby ending the first lawsuit initiated by creationists (see also Appendix D).

1974 Virginia repeals its forcible sterilization law, which had been used in 1927 to sterilize Carrie Buck.

1975 Antievolutionist Kelly Segraves's *Sons of God Return* links demons, Flood geology, and UFOs. Segraves claimed that UFO pilots are "fallen angels and followers of Satan" and that "God sent the animals to Noah."

1975 Controversy created by the NSF-sponsored *MACOS* project prompts the U.S. House of Representatives to pass the Bauman Amendment giving Congress direct supervision and veto power over every project funded by NSF. The bill died in the Senate.

1975 E. O. Wilson's (b. 1929) *Sociobiology: The New Synthesis* argues that human behavior is a product of evolutionary forces. Most of the book dealt with ant colonies, but Wilson's two chapters about human evolution evoked controversy.

1975 Henry Morris again reminds his followers "Satan himself is the originator of the concept of evolution." In the preface to Morris's *The Troubled Waters of Evolution*, Tim LaHaye—who became famous as a coauthor of the *Left Behind* series of books—claimed that evolution "has wrought havoc in the home, devastated morals, destroyed man's hope for a better world, and contributed to the political enslavement of a billion or more people."

1975 In a 2-to-1 decision, the Sixth Circuit Court of Appeals announces in *Daniel v. Waters* that Tennessee's Genesis Bill (passed in 1973) is "patently unconstitutional" because it is little more than a revised edition of the Butler Law that led to the Scopes Trial in 1925. Circuit Court Judge George Edwards noted that the law was intended to displace evolution with biblical teachings. Tennessee did not appeal the decision (nor that of *Steele v. Waters*, a closely related case), and the demise of the Genesis Bill caused other "equal time" bills to die quietly in several state legislatures (see also Appendix D).

1975 In Texas, one of the nation's largest purchasers of textbooks, 80% of biology textbooks in use do not mention evolution.

1975 William Dankenbring's (b. 1941) *The First Genesis* endorses gap creationism and claims that the K-T extinction was caused by a cosmic battle between God and Satan. Three years later, in *Beyond Star Wars*, Dankenbring claimed that the Great Pyramid is a memorial to the Flood and that UFOs are meant to convince people to reject Jesus.

1975 New Zealand-born biologist Allan Wilson and his American student Mary-Claire King (b. 1946) report that human and chimpanzee DNA sequences differ by only 1.5%. King and Wilson suggested that the primary difference between human and chimp DNA involves turning genes on and off.

1975 Robert Kofahl and Kelly Segraves publish the popular and overtly religious *The Creation Explanation: A Scientific Alternative to Evolution*. The authors lauded biblical catastrophism while dismissing evolution and much of science (e.g., asserting that Einsteinian relativity is "under strong criticism"). After learning that ice floats because "the Creator designed it that way," readers were told that they must choose between creation and evolution. Faith in biblical literalism would "redeem an individual . . . from the destructive effects of the evolutionary faith."

1975 The California Board of Education, with several of its members having been replaced by appointees of Governor Jerry Brown

(b. 1938), approves no science textbooks that include creationist explanations.

1975 The U.S. Supreme Court refuses to hear the appeal of *Willoughby v. Stever*, thereby ending the case (see also Appendix D).

1976 Arizona Republican Congressman John Conlan sponsors an amendment to the National Defense Education Act that would "prohibit federal funding of any curriculum project with evolutionary content or implications." The amendment passed the House by a vote of 222-to-174, but was narrowly defeated by the Senate.

1976 Philosopher Karl Popper's (1902–1994) *Unended Quest: An Intellectual Autobiography* claims that "Darwinism is not a testable scientific theory, but a *metaphysical research programme*—a possible framework for testable scientific theories." Creationists soon began using Popper's claim to argue that evolution is not a valid scientific theory. Popper later recanted his claim.

1976 Indiana's West Clark Community Schools adopts *Biology: A Search for Order in Complexity* as the only approved textbook for its biology classes. This led to *Hendren v. Campbell* the following year.

1976 In *Gods of Aquarius*, Brad Steiger (b. 1936) claims that UFOs have guided evolution and that future evolution will be guided by technology as humans "literally become as gods under God."

1976 Electrical engineer Harold Hill's *From Goo to You by Way of the Zoo* (later retitled *How Did It All Begin?*) claims that biologists are hiding evidence that would undermine their cause. Hill's book, which includes a foreword by German rocketeer Wernher von Braun (1912–1977), claimed that shoe leather can be turned into gold, aliens are telling us to "turn on to Jesus," and that professors ignore evidence against evolution because they do not want to give up their "fat royalty incomes" and "pretty coeds."

1976 Kentucky enacts a law stipulating that "no teacher in a public school may stress any particular denominational religious belief" and that teachers who cover evolution in their classes can also teach "the theory of creationism as presented in the Bible." The law encouraged teachers to read the Bible to students as part of their "instruction on the theory of creation" and stipulated that students who adhere to the biblical account should get credit on all exams. In 1990, a similar version of the law (Kentucky Revised Statute 158.177) was enacted.

1976 Richard Dawkins's (b. 1941) provocative *The Selfish Gene* popularizes and extends the proposal by George Williams that the unit of selection is the gene rather than the individual. Starting with Darwin, it had become well accepted that natural selection causes adaptive evolutionary change through differential reproductive success of individuals. Dawkins proposed, however, that individuals are "vehicles" for the true "replicators," the genes; that is, individuals are a gene's way of copying themselves. Dawkins also proposed that replicators can exist in a variety of forms and introduced the term *meme* to describe ideas, concepts, and practices that are replicated culturally (e.g., songs, tools, fashions, religious beliefs). Some biologists were uncomfortable with gene-centered selection because it implied an extreme form of genetic determinism. Harvard paleontologist Stephen Jay Gould was especially critical, labeling Dawkins a "Darwinian fundamentalist."

1976 Mary Leakey, Paul Abell (1924–2004), and coworkers discover sets of footprints made by at least two upright hominins (probably *Australopithecus afarensis*) in volcanic ash 3.6 million years old at Laetoli, Tanzania (about 25 miles southwest of Olduvai Gorge; see also Appendix C). After fears that they would be damaged by erosion or animals, the footprints were buried in the 1990s.

1976 Protests against *Man: A Course of Study (MACOS)*, an award-winning and pro-evolution program for elementary school students, cause sales to drop by 70%.

1976 In *The Ark on Ararat*, Tim LaHaye and John Morris claim that the end of the world is near, that heat of the Great Tribulation will melt snow on Mount Ararat and reveal Noah's ark, and that readers should "climb aboard the Ark" and "receive Jesus Christ as your personal Savior." Morris, who believed that the "present-day animal distribution must be explained on the basis of migrations from the mountains of Ararat," later claimed "[l]ike it or not, if you're going to be a Bible-believer, the earth is young."

1977 Trustees of the Dallas (Texas) Independent School District vote 6 to 3 to approve *Biology: A Search for Order in Complexity* for adoption, to purchase 60 copies as reference books, and to train teachers to use the book.

1977 Irwin Ginsburgh's *First Man, Then Adam!* presents a modified day-age creation story that involves Adam and Eve crash landing in a spaceship from Pleiades in Eden 5,700 years ago. Adam and Eve then reproduced with Stone Age people, after which they repaired their spaceship and left. Ginsburgh claimed that UFOs are the vestiges of this history and that the Tree of Knowledge described in Genesis was the space aliens' computer.

1977 Robert Kofahl's *Handy Dandy Evolution Refuter* claims that God created the Y chromosome and that evolution is evil because it contradicts the Bible.

1977 In a series of booklets titled *Connections*, Marshall and Sandra Hall claim that evolution is a lie, that miracles are real, that theistic evolution is destructive, and that evolution is to blame for problems such as astrology, the United Nations, NASA, homosexuality, women's rights, and the metric system. The Halls demanded that Congress investigate and convict of treason the supporters of evolution.

1977 American biologist Carl Woese (b. 1928) redefines the tree of life by dividing bacteria into two classes—bacteria and archaebacteria (archaea). Although archaea—many of which live in extreme conditions—resemble bacteria, genetic analyses showed that they are an evolutionarily distinct group. Later, Woese proposed that all organisms can be divided into three domains: eukaryota, eubacteria, and archaea.

1977 Stephen Jay Gould's *Ontogeny and Phylogeny*—which is dedicated to D'Arcy Thompson—foreshadows the development of evolutionary developmental biology ("evo-devo").

1977 Nobel laureate Fred Sanger (b. 1918) and his colleagues sequence the DNA of a bacteriophage.

1977 The Indiana Civil Liberties Union (ICLU) sues in an Indianapolis court on behalf of parents and students (including ninth-grader Jon Hendren) to ban the use of *Biology: A Search for Order in Complexity* in the state's public schools because, among other things, the book's use would establish religion. Despite the testimonies of scientists that creationism is a scientifically useless idea, the State Textbook

Commission upheld its original recommendation that the book is acceptable. Soon thereafter, the ICLU returned to court to challenge the decision. In *Hendren v. Campbell*, Marion, Indiana, Superior Court Judge Michael Dugan II ruled that the use of *Biology: A Search for Order in Complexity* in the state's public schools violated constitutional bans because it "advanced particular religious preferences and entangled the state with religion." The Textbook Commission removed *Biology: A Search for Order in Complexity* from its list of approved texts and did not appeal Dugan's decision. Soon thereafter, school officials in Dallas, Texas, also removed the textbook from classrooms (see also Appendix D).

1977 The Rhea County Courthouse, built in 1891, is designated a National Historic Landmark by the National Park Service. Two years later, a $1 million grant restored the courthouse, which currently houses the Scopes Trial Museum in its basement. Today, the courtroom contains the original judge's bench, four tables, dais rail, jury chairs, and spectator seats.

1978 After being denied tenure at Bowling Green State University, young-earth creationist Gerald "Jerry" Bergman (b. 1946) sues the university, alleging that he was fired for his religious views. Bergman's case was dismissed in 1985, and his appeal was rejected in 1987 when the court ruled that Bergman's denial of tenure was not due to his religious views, but instead to concerns about ethical issues. Bergman later inflated evolution into a political philosophy, noting that "[i]f Darwinism is true, Hitler was our savior and we have crucified him. . . . If Darwinism is not true, what Hitler attempted to do must be ranked with the most heinous crimes of history and Darwin as the father of one of the most destructive philosophies of history."

1978 After Congress provides almost $500,000 for an exhibit titled "The Emergence of Man: Dynamics of Evolution" at the Museum of Natural History at the Smithsonian Institution, Dale Crowley, Jr., sues to either cancel the exhibit or force the Smithsonian Institution to provide equal space and money for an exhibit discussing the biblical story of creation. In *Crowley v. Smithsonian Institution*, District Judge Barrington Parker, Jr. (b. 1944) rejected the claim that the Smithsonian should give equal time to the biblical story of creation, noting that "[t]he plaintiffs can carry their beliefs into the Museum with them, though they risk seeing science exhibits contrary to that faith." Parker ruled that providing the remedy requested by Crowley would violate the Establishment Clause, as noted in *Epperson v. Arkansas* and *Daniel v. Waters*. *Crowley* ended in 1980 when the U.S. Supreme Court refused to hear Crowley's final appeal. *Crowley v. Smithsonian Institution* established that the federal government (1) can fund public exhibits that promote evolution, and (2) is not required to provide money to promote creationism.

1978 Creation-Life Publishers produces a public-school edition of Duane Gish's *Evolution? The Fossils Say NO!* Gish repeated creationists' criticisms of evolution ("Evolution theory is indeed no less religious nor more scientific than creation") while claiming that creationism is not a religious doctrine. Gish condemned theistic evolution as "bankrupt," adding that "not for a moment do I believe that the theory of evolution can be reconciled with the Bible. . . . You really cannot believe the Bible and the theory of evolution both." Gish believed that he had avoided religious statements by using the word *Creator* instead of *God*.

1978 Karl Popper recants his earlier claim that Darwinism is not a testable scientific theory. His recantation was largely unnoticed by many creationists.

1978 In ICR's *Age of the Earth*, Harold Slusher and Thomas Gamwell argue that the earth is thousands, rather than billions, of years old.

1978 Panspermia advocates Sir Fred Hoyle and Chandra Wickramasinghe (b. 1939) propose an extraterrestrial influence for evolutionary change in their book *Lifecloud—The Origin of Life in the Universe*.

1978 Paul Ellwanger founds Citizens for Fairness in Education, which bases its legislative efforts for equal time for creationism on Wendell Bird's upcoming *Impact* resolution. Ellwanger's first bill was defeated in the South Carolina legislature, but within a year it appeared in legislatures of eight other states.

1978 Developmental biologist Edward Lewis (1918–2004) proposes that specific genes in fruit flies are responsible for guiding the development of whole segments of an animal's body and that the linear order of these genes on a chromosome corresponds to the anterior-posterior order of action of the genes in the organism. Soon thereafter, Christiane Nüsslein-Volhard (b. 1942) and Eric

Wieschaus (b. 1947) discovered similar genes that regulate overall body development in fruit flies. All three researchers shared the Nobel Prize in Physiology or Medicine in 1995 for this work. Study of homeotic—"master control"—genes now is a foundational concept in evolutionary developmental ("evo-devo") biology, which includes the study of how evolutionary novelties can arise by modification in the timing of the activation of these "developmental toolkit" genes.

1978 In *Yale Law Journal*, Wendell Bird publishes a strategy for incorporating the teaching of creationism into public schools. The article, which won the Egger Prize for the best student article, claimed that teaching only evolution denies "academic freedom" and violates the free exercise of religion by forcing students to learn heretical views. Bird argued "treatment of either the theory of evolution or the theory of scientific creationism must be limited to scientific evidences and must not include religious doctrine." Bird also justified his resolution by noting that evolution "is contrary to the religious convictions and moral convictions of many students and parents," that creationism is as scientific as evolution, and that evolution is as religious as creationism. Soon thereafter, Bird's proposal became the foundation for creationists' demands for "equal time" and "balanced treatment" in classrooms. Bird's four-page "Resolution for Balanced Presentation of Evolution and Scientific Creationism," which was published in ICR's *Impact* series, was prefaced by an editorial indicating that the resolution should be used with local school boards, not as a model for legislation. Legislation based on Bird's resolution was later ruled unconstitutional in *McLean v. Arkansas Board of Education* and *Edwards v. Aguillard*.

1979 The Cobb County (Georgia) school board allows students to ignore the biology requirement for graduation if they have religious objections to evolution.

1979 Pope John Paul II requests that the Catholic Church reexamine the Galileo case with an eye toward reconciliation.

1979 In the popular television show *In Search of Noah's Flood*, actor Leonard Nimoy (b. 1931) notes that "no written word has survived as much skepticism as the story of Noah's Ark."

1979 Kelly Segraves and the Creation Science Research Center sue in California to stop distribution of *Science Framework* and to end presentations of evolution as fact in California's public schools. The petition was denied, but *Segraves v. California* was quickly dubbed "Scopes II" by the media. Although Segraves demanded equal time for creationism, he later backtracked by asking only that the state "stop posing the theory that man and all life on Earth developed from a common ancestor, as a fact." The judge—claiming that "everybody won"—ordered California to distribute its 1972 "anti-dogmatism" policy to school officials, textbook publishers, and science teachers and to include the policy in all future versions of *Science Framework*.

1979 Kenneth Ham (b. 1951) founds, with John Mackay, the Creation Science Foundation in Australia, which later merges with the Creation Science Association. Ham then moved to the United States to work with ICR and in 1994 opened a U.S. branch of his Australian organization now called Answers in Genesis (AiG). By 2005, AiG-U.S. had separated from other branches of AiG. Ham, a young-earth creationist, wrote several books and produced a radio program that supported the inerrancy of the Bible, claiming that "if something disagrees with the Bible, it is wrong, regardless of the evidence." In 2007, AiG opened the sprawling $27,000,000 Creation Museum near Cincinnati, Ohio.

1979 An author using the pen-name John Woodmorappe (b. 1954) publishes *Radiometric Dating Reappraised*, which accuses geologists of "fudging" their data to document an ancient earth.

1979 Legislatures in nearly a dozen states begin considering equal-time bills. One such bill, introduced in Georgia, would have helped ICR reap $2,000,000 in sales of creationist textbooks. All of the bills failed.

1979 Stephen Jay Gould and Richard Lewontin argue that much in biology has little or no direct connection to adaptive advantage (i.e., there can be "adaptation without selection"). Gould and Lewontin, who claimed that developmental constraints often mold organisms in nonadaptive ways, appropriated the term *spandrel* (the curved area above an arch in Renaissance architecture) to describe traits developed during evolution as a side-effect of a true adaptation rather than directly as a result of natural selection. (Spandrels, although often elaborately decorated, have no function or purpose; they are simply a byproduct produced by other structures). Gould and Lewontin triggered a debate about the importance of evolutionary adaptations versus mere byproducts of evolution.

1979 The Smithsonian Institution's National Museum of Natural History opens its exhibit titled "The Dynamics of Evolution," which was the subject of the 1978 lawsuit *Crowley v. Smithsonian Institution*.

1980 A poll finds that almost half of the readers of the *American School Board Journal* favor teaching both evolution and creationism; 25% believe that evolution should be the only explanation that is presented; 19% want only biblical discussions to be presented; and 8% want to avoid teaching anything about the origin of humans.

1980 *Science* reports that textbook commissions in most states are under "heavy pressure" to include creationists' materials on their lists of approved books.

1980 In Cobb County, Georgia, teachers threaten to go on strike unless the local school board rescinds an earlier order to include creation science in the curriculum.

1980 Louisiana State Senator Bill Keith—who claims that evolution is the "greatest hoax of the 20th century . . . a fairy tale like the Easter Bunny"—introduces a bill requiring the inclusion of creationism in the state's public schools. When the bill died in committee, Keith—in the name of "academic freedom"—introduced a new bill, drafted by creationist Wendell Bird, requiring equal time for evolution and creationism. After Keith distributed materials from ICR proclaiming that creationism is "pure science" having "no missing links," Keith's modified bill passed the state House of Representatives by a vote of 71 to 19. Keith then announced that "evolution is no more than a fairy tale about a frog that turns into a prince," and the Senate also passed the legislation. When Governor David Treen (b. 1928) signed the bill, it became law. Keith's bill was later overturned by the U.S. Supreme Court in *Edwards v. Aguillard*. Keith claimed that God had delayed *Edwards* so that President Ronald Reagan could appoint more federal judges who supported "creation science" (see also Appendix D).

1980 In a bold, multidisciplinary paper published in *Science*, American physicist Luis Alvarez (1911–1988; Figure 66) and his geologist son Walter (b. 1940) suggest that the Cretaceous-Tertiary extinction was

Figure 66 Louis (left) and Walter Alvarez argued that the extinction of dinosaurs 65 million years ago was caused by a meteor impact. In this photo taken at a limestone outcrop near Gubbio, Italy, Walter (right) touches the top of the Cretaceous limestone at the K-T boundary that marks the abrupt disappearance of non-avian dinosaurs from the fossil record. (*Lawrence Berkeley National Lab*)

caused by a giant asteroid that hit the earth 65 million years ago (see also Appendix B). The Alvarezs' claim was initially ridiculed, but subsequent evidence (including the presence of global iridium anomalies and the discovery in 1991 of the impact crater) supported their argument. In 2007, Jack Chick's *There Go the Dinosaurs* dismissed the Alvarezs' explanation as "a story told by people who don't trust God."

1980 In a speech to a conservative religious group in Dallas, presidential candidate Ronald Reagan (1911–2004) cites *Man: A Course of Study (MACOS)* as an example of the federal government's support of subversive values and asks why NSF did not instead develop curriculum materials supporting Christian values. Needing votes from fundamentalists, Reagan told reporters that he questioned evolution and wanted "the Biblical theory of creation" to be taught in public schools.

1980 In an article published in the journal *Paleobiology*, Stephen Jay Gould declares that the Modern Synthesis is "effectively dead, despite its persistence as textbook orthodoxy." Creationists cited this quote as evidence of the inability of scientist to agree about evolution.

⚖ **1980** Teachers in junior high schools in Little Rock, Arkansas, begin reviewing science textbooks for adoption. Creationist and math teacher Larry Fisher, wanting the alleged shortcomings of evolution to be included in science textbooks, sent ICR's *Impact* resolution (coauthored by Wendell Bird) to the superintendent of the Pulaski County Special School District, noting that the district would reap a public relations bonanza if the resolution was implemented because "about 80%" of people support the resolution. Fisher's activities ultimately led to *McLean v. Arkansas Board of Education*, in which Judge William Overton ruled that creation science has no scientific merit (see also Appendix D).

1980 Texas declares that evolution must be presented as "only one of several explanations of the origin of mankind." Several other states later did the same thing.

1980 The American Humanist Association begins producing *Creation/Evolution*, a journal intended to counter the political activities of creationists. That same year, Tim LaHaye's *The Battle for the Mind* warned readers that humanists were planning "a complete world takeover by the year 2000."

⚖ **1980** Various state legislatures continue to debate "equal time" legislation based on Paul Ellwanger's model. Iowa's equal-time bill, which was opposed by the Iowa Academy of Science, was referred to the House Finance Committee, where it died.

1980 At a conference of Orthodox Jewish scientists, physicist Lee Spetner declares *Archaeopteryx* (Figure 24) a deliberate fraud. Spetner's 1997 book, *Not By Chance! Shattering the Modern Theory of Evolution*, continued his attack on evolutionary biology.

1981 ICR (Figure 64), recently separated from Christian Heritage College, establishes a graduate college to educate creationists. All of ICR's graduate programs (including astronomy, biology, geology, and science education) were approved by California educational authorities. In 1990, the California Superintendent of Public Instruction revoked ICR's license to award degrees, but two years later the license was reinstated by a federal judge.

1981 In "Evolution as Fact and Theory" published in *Discover* magazine, Stephen Jay Gould notes that "it is infuriating to be quoted again and again by creationists—whether through design or stupidity, I do not know—as admitting that the fossil record includes no transitional forms."

1981 In *Evolution: Fact or Fiction*, Georgia Court of Appeals Judge Braswell Deen, Jr., denounces evolution as the cause of societal ills, especially crime. Deen, a self-proclaimed expert on "human origins from a law-science perspective," advocated teaching young-earth creationism in public schools. In *Time* magazine, Deen asserted that "this monkey mythology of Darwin is the cause of permissiveness, promiscuity, prophylactics, perversions, pregnancies, abortions, pornotherapy, pollution, poisoning and proliferation of crimes of all types."

1981 American writer John McPhee's (b. 1931) *Basin and Range* first uses the phrase *deep time* to describe earth's history.

1981 A poll shows that half of all Californians accept the teaching of creationism with evolution in public schools.

1981 After an Associated Press-NBC News poll shows that almost 75% of parents and teachers accept the teaching of creationism in public schools, the Tampa, Florida, school board requires equal time for creationism and evolution. This decision, opposed by many science teachers, made the study of creationism mandatory for its 115,000 students. Soon thereafter, a debate in Tampa involving ICR's Henry Morris was attended by 1,700 spectators and covered by seven radio stations, six television stations, and numerous newspapers.

1981 Sidney Jansma's *UFOs and Evolution* blames Satan for evolution, ESP, astrology, Islam, karate, and Ouija boards. Four years later, in *Six Days*, Jansma advocated young-earth creationism and noted "God says creation—Satan says evolution. . . . I do not see evolution simply as the beastly actions and language of Satan but as his lying and blasphemy as spoken through the possessed."

1981 Arkansas State Senator Jim Holsted introduces his "balanced treatment for scientific creationism" bill (Act 590) without consulting scientists, science educators, the Arkansas Attorney General, or the Arkansas Department of Education. Holsted admitted, "this bill favored the view of the Biblical literalists, of which I am one." Holsted's bill was referred to the Judiciary Committee, which included Holsted as a member and was chaired by born-again Christian Max Howell. Howell gained the committee's endorsement by claiming that the bill stresses "fairness" and "freedom of choice." In the meantime, Arkansas's Moral Majority, led by Reverend Roy McLaughlin, formed Arkansas Citizens for Balanced Education in Origins to promote Act 590. When the ACLU announced plans to challenge the law, creationists formed the Creation Science Legal Defense Fund to raise money to hire Wendell Bird and John Whitehead to defend the law. After voting to suspend parliamentary rules, the Arkansas House of Representatives, lobbied by the Moral Majority, passed the "balanced treatment" bill by a vote of 69 to 18 a day before adjourning. Then, after a 20-minute debate and no hearings, the Arkansas Senate passed the bill by a vote of 20 to 2. Arkansas governor Frank White paid his debt to the Moral Majority by signing Act 590 into law, despite the fact that he had not read the legislation. White justified his actions by asking "[i]f we're going to teach evolution in the public school system, why not teach scientific creationism? Both of them are theories."

1981 Before the ACLU can challenge Louisiana's balanced treatment law, Wendell Bird sues in Baton Rouge federal district court (on behalf of Senator Bill Keith and 54 other plaintiffs) for a declaratory judgment to force Louisiana's schools to comply with the newly passed law. The lawsuit (*Keith v. Louisiana*) was dismissed seven months later when Judge Frank Polozola ruled that it did not raise a federal question (see also Appendix D).

1981 Francis Crick's *Life Itself: Its Origin and Nature* promotes "directed panspermia," the concept that life, having arisen elsewhere in the universe, had been dispersed to earth by aliens. By the 1990s, however, discovery of enzymatic RNA had fostered development of the "RNA world" hypothesis for the origin of life, and Crick acknowledged that he had underestimated the probability of life arising on earth without outside influence. Regardless, Crick's statements (e.g., "[t]he probability of life originating at random is so utterly minuscule as to make it absurd") were repeatedly cited by antievolutionists as evidence that science required a supernatural explanation for life. During the *Kitzmiller v. Dover* trial in 2005, ID-advocate Michael Behe used Crick's writings to suggest that scientists believe the universe is designed.

1981 Henry Morris announces that more than 1,000,000 books produced by ICR are in circulation.

1981 Jerry Falwell (1933–2007), leader of the Moral Majority, tells his followers not to read books other than the Bible.

1981 In *The Star People*, Brad Steiger (b. 1936) and Francie Steiger claim that the extraterrestrials who produced humans arrived on earth 40,000 years ago and that Jesus' father was an extraterrestrial.

1981 *McLean v. Arkansas Board of Education* opens in federal district court in Little Rock, Arkansas, with Judge William Overton (1939–1987) presiding (see also Appendix D).

1981 Republican Congressman William Dannemeyer (b. 1929) of California asks Congress to review and limit funding for the Smithsonian Institution's exhibit titled "The Dynamics of Evolution," which had been a topic of an earlier lawsuit (*Crowley v. Smithsonian Institution*). Dannemeyer claimed that the exhibit promoted the religion of secular humanism, but his request gathered little support.

1981 Sir Fred Hoyle endorses "directed panspermia" in which an extraterrestrial "intelligence which assembled the enzymes" bombarded earth with comets. These comets were purportedly loaded with viruses responsible for the origin and diversity of life on earth.

1981 The ACLU and other groups sue in federal district court to challenge the constitutionality of Arkansas's "balanced treatment" law. When Arkansas Attorney General Steve Clark (b. 1947) excluded Wendell Bird from the state's defense team, Bird claimed that Clark did "an inadequate job" of defending the law, and television preacher Pat Robertson told his followers that Clark was "crooked," "biased," and had supported the ACLU.

1981 At a hearing before an Oklahoma legislative committee debating a bill requiring balanced treatment of evolution and creationism in the state's public schools, a superintendent from a rural school district tells legislators to "leave us alone—we know what we're doing. We're not teaching evolution—we're teaching biblical creationism."

1981 The Alabama House Education Committee supports an equal-time bill, but the bill is killed by a filibuster by Representative Robert Albright, a former biology teacher.

1981 AAAS, meeting in Toronto, sponsors a session to discuss creationism and warn of its damage to science education. The following year, AAAS issued a resolution titled "Forced Teaching of Creationist Beliefs in Public School Science Education" supporting evolution and denouncing "Creationist Science" as a scientifically invalid "threat to the integrity of education and the teaching of science." Also, the American Association of University Professors passed a resolution asking state governments to "reject creation-science legislation as utterly inconsistent with the principles of academic freedom." NAS and NABT passed a joint resolution urging scientists to challenge creationists at the local level, but stressed the futility of public debates with creationists.

1981 The not-for-profit and religiously neutral National Center for Science Education (NCSE) is founded and becomes the only national organization devoted primarily to opposing antievolution activities. Today, NCSE has about 4,000 members.

1981 Famed British paleontologist Colin Patterson (1933–1998) asks a crowd at the AMNH, "[c]an you tell me anything you know about evolution, any one thing, any one thing that is true?" Although Patterson claimed that he was "speaking only about systematics" and "off the record," his comments were cited by creationists alleging that many top scientists reject evolution. Patterson later announced, "I do not support the creationist movement in any way."

1981 The Pro-Family Forum distributes thousands of copies of "Can America Survive the Fruits of Atheistic Evolution?" The pamphlet, which claimed that teaching evolution produced Nazism, communism, abortion, divorce, and venereal disease, was used by several groups trying to force creationism into public schools.

1981 Wayne Moyer, Executive Director of NABT, admits, "[w]e have done a botched job of teaching evolutionary theory." Chemist Russell Doolittle delivered a stronger indictment, noting that science education in the United States is "simply, sadly, awful."

1981 With the help of Richard Turner, a former legal aid to Ronald Reagan, Kelly Segraves alleges in a lawsuit that California is violating his children's religious rights because evolution is being taught in a dogmatic way. The nonjury trial involving Kelly Segraves's petition, often referred to as "Scopes II," opened in Sacramento, California with Judge Irving Perluss presiding. By the second day of the trial, creationists abandoned their demand for equal time and their claim that the teaching of evolution was a state-sponsored establishment of religion. Having thereby changed the trial to a discussion of phrasing in *Science Framework*, Segraves told Judge Perluss that he would accept the removal of dogmatic statements about evolution from *Framework*. In *Segraves v. State of California*, the Sacramento Superior Court did not order major changes to *Framework*, but did require the California Board of Education to recommend a less dogmatic presentation of evolution (by emphasizing that scientific explanations focus on "how" and not on "ultimate cause"). The judge praised the nondogmatism policy as necessary for a pluralistic society and ordered that copies of the policy be sent to all school districts in the state. (In 1989, the antidogmatism policy was extended to all areas of science, not just those involving evolution). Soon after Perluss's decision, Nell Segraves demanded that creationists get "50% of the curriculum and the content. . . . We want 50% of the tax dollars used for education. . . . We have a lot to undo."

1982 Books such as Philip Kitcher's (b. 1947) *Abusing Science: The Case Against Creationism*, Norman Newell's (1909–2005) *Creation and Evolution: Myth or Reality?*, and Niles Eldredge's *The Monkey Business: A Scientist Looks at Creationism* refute creationists' claims.

1982 Writing in *Christianity Today*, geologist Edwin Olsen labels creationists as being intolerant, simplistic, and less honest than evolutionists about their own failings, adding that "[i]n its isolation and inflexibility, 'creationism' . . . is doing more harm than good."

1982 In an article titled "Creationism and the Age of the Earth," *Science* editor Philip Abelson writes that scientific creationists "have no substantial body of experimental data to back their prejudices. Truth is not on their side. In the end their activities must bring only harm to their cause." At that time, 44% of the American public believed that living organisms were created in their present form within the last 10,000 years. In 2009, that percentage remained unchanged.

⚖ **1982** In Arizona, State Republican Representative Jim Cooper (chair of the House Education Committee) introduces legislation to ban the teaching of evolution in ways that "foster the belief in a religion or cause disbelief in religion." Cooper wanted violators to be fined $10,000 and face up to a year in prison, but his bill received little support.

1982 In his second book, *The Extended Phenotype* (1982), Richard Dawkins argues that an individual's phenotype is not restricted to the corporal entity typically identified as the "individual." Beaver dams, termite mounds, and human-built structures, for example, favor replication of the genes that led to the creation of such entities. The true phenotype, therefore, is anything the "vehicle" has, does, or creates that promotes replication of the genes it houses.

⚖ **1982** In *McLean v. Arkansas Board of Education*, federal judge William R. Overton rules that (1) Arkansas's "equal time" law is unconstitutional; (2) creation science is religion, not science; and (3) scientific creationism has no scientific significance (see also Appendix D). Overton, the son of a biology teacher, issued his decision on the same day that Mississippi enacted its "Balanced Treatment for Creation Science and Evolution Science Act." Overton described Wendell Bird's resolution as a "student note" whose "argument has no legal merit." Overton's blunt and devastating decision, which included a definition of the word *science*, destroyed creationists' hopes of using creation science as a means of forcing religion into the science classes of public schools. The nation's first balanced-treatment law had lasted less than one year. Arkansas Attorney General Steve Clark, who described the law's religious overtones as an "insurmountable problem," did not appeal the ruling, and his handling of the case was denounced by Wendell Bird, Pat Robertson, Jerry Falwell, and other creationists. Thanks to the *McLean* decision and local opposition, legislatures soon defeated creationist bills in several other states, including Florida, Georgia, Kansas, South Dakota, Maryland, and West Virginia. Afterward, *Science* published Judge William Overton's entire opinion from *McLean*. Famed young-earth creationist Duane Gish claimed that Overton's decision had established that "secular humanism will now be our official state-sanctioned religion," and Christian apologist Norman Geisler (b. 1932)—an expert witness for the state—charged that Overton's decision would have "devastating consequences for the pursuit of truth in public schools." During the trial, Geisler claimed that he knew "at least twelve persons who were clearly possessed by the devil," that UFOs are "the Devil's major, in fact, final attack on the earth," and that UFOs are real because he "read it in the *Reader's Digest*."

1982 In the Royal Institution's "Omni Lecture," Sir Fred Hoyle claims "that biomaterials with their amazing measure of order must be the outcome of intelligent design."

1982 Jorge Yunis and Om Prakash confirm earlier findings that chromosomes from humans, chimps, gorillas, and orangutans are remarkably similar and can be aligned with one another. Every human chromosome, except one, matches a chromosome in chimps. The exception is chromosome 2, which matches two chromosomes in chimps and other great apes, thereby suggesting that the two chromosomes have fused since humans diverged from other apes.

1982 Judge Adrian Duplantier (1929–2007) rules that Louisiana's balanced treatment law usurps the authority of the Louisiana Board of Elementary and Secondary Education. Louisiana Attorney General William Guste (b. 1922) appealed Duplantier's decision to the Louisiana Supreme Court. Thereafter, the lawsuit was known as *Edwards v. Aguillard*, since the state and its governor (Edwin Edwards [b. 1927]) were now the plaintiffs.

1982 In its "Evolution and Creationism" resolution, the United Presbyterian Church condemns legislation requiring the teaching of creationism. Similar resolutions were passed by the General Convention of the Episcopal Church ("Resolution on Evolution and Creationism") and the Unitarian Universalist Association ("Resolution Opposing 'Scientific Creationism'").

1982 In *Darwinism Defended*, philosopher Michael Ruse (b. 1940) critiques creationism and creationists: "Creationism is wrong;

totally, utterly, and absolutely wrong. I would go further. There are degrees of being wrong. The creationists are at the bottom of the scale. They pull every trick in the book to justify their position. Indeed, at times they verge right over into the downright dishonest. Scientific creationism is not just wrong, it is ludicrously implausible. It is a grotesque parody of human thought, and a downright misuse of human intelligence. In short, to the believer, it is an insult to God."

1982 Lynn Margulis and Karlene Schwartz propose, in *Five Kingdoms*, a revised classification system for life that separates fungi from plants. Although controversial at the time, evidence now indicates that fungi and animals are more closely related than are plants and fungi.

1982 Mississippi enacts its "Balanced Treatment for Creation Science and Evolution Science Act."

1982 Norman Geisler, A. F. Brooke, and Mark Keough publish *The Creator in the Courtroom,* which describes a creationist interpretation—including Geisler's testimony—of *McLean v. Arkansas Board of Education*. Geisler claimed that the Arkansas law mandating equal time for creationism and evolution did not introduce religion into schools because God is not a religious concept. Geisler was one of the first to advocate ID as an alternative to evolution.

1982 Texas State Senator Oscar Mauzy (1926–2000; Chair of the Senate Jurisprudence Committee) asks Texas Attorney General Jim Mattox (1943–2008) for an opinion about the constitutionality of the state's 1974 pro-creationism textbook guidelines (i.e., that evolution be identified as a theory rather than as a fact and that evolution be identified in textbooks "as only one of several explanations of the origins of humankind"). Mattox used Judge William Overton's decision in *McLean v. Arkansas Board of Education* to rule that the guidelines were unconstitutional because they were motivated by "religious sensibilities, rather than a dedication to scientific truth."

1982 The Arizona House and Senate pass a bill similar to that proposed the preceding year by Republican State Representative Jim Cooper, which bans the teaching of evolution in ways that promote or cause disbelief in religion. That same year, the Illinois State Board of Education approved what they touted as world-class standards for science education, despite the fact that the standards did not mention evolution. The standards, which were strongly influenced by the Illinois Christian Coalition, also did not mention human sexuality or multicultural studies.

1982 The Louisiana Supreme Court, in a vote of 4 to 3, overturns Judge Adrian Duplantier's decision, concluding that the legislature can mandate the teaching of creation science. This decision returned the case challenging Louisiana's balanced treatment law to Duplantier's court (see also Appendix D).

1982 Two days after beginning excavations along the Paluxy River in Glen Rose, Texas, Baptist preacher Carl Baugh (b. 1936) announces that his discoveries of human and dinosaur tracks have "unparalleled historic significance." Two years later, Baugh opened the Creation Evidence Museum just outside of Dinosaur Valley State Park in Glen Rose. The popular museum consisted of a small group of trailers and a larger building identified as a "scientifically chartered museum." Baugh used the museum to discredit evolution by showing that people lived contemporaneously with dinosaurs. Baugh later hosted a weekly show on Trinity Broadcasting Network titled "Creation in the 21st Century," where he was referred to as the "foremost doctor on creation science."

1982 Henry Morris claims in *The Troubled Waters of Evolution* that Satan invented evolution during a meeting with Babylonian ruler Nimrod at the Tower of Babel.

1982 Richard Alexander's (b. 1929) *Darwinism and Human Affairs* describes cultures as products of social interactions among individuals who have evolved to maximize their inclusive fitness.

1983 In *The Relevance of Creationism*, Ken Ham claims that creationism is the foundation of Christianity. Ham also complained "many Christian girls go bra-less and wear clingy t-shirts or wear clingy clothes to show off their breasts or sexual parts."

1983 Carl Woese dismisses theoretical accounts of the transformation from nonliving to living matter as "little more than *Just-So Stories*."

1983 The Minnesota Twin Family Study begins registering twins "to identify the genetic and environmental influences on the development of psychological traits." The research program became well-known for its demonstration of the influence of genes on personality traits, illuminating the "nature-versus-nurture" debate.

1983 Victor Kachur's *The World that Was* argues that animals helped Noah build the Ark (e.g., that 600-pound beavers helped control rivers) and that Noah left mammoths behind to freeze. Kachur claimed that the Ark remains on Ararat, inscribed with the names of people who built it.

1983 Using the Benjamin Waterhouse Hawkins-inspired ape-to-man progression on its cover, *Mad* magazine wishes Charles Darwin a happy birthday (in its April, not February, edition).

1983 Nobel laureate P. D. Medawar (1915–1987) notes in *Aristotle to Zoos: A Philosophical Dictionary of Biology* that "a man who believes that fossils are the remains of organisms inundated by Noah's flood can believe anything; no effort of credulity would be too much."

1984 *Science and Creationism*, a collection of essays (including chapters such as *To Hell With Evolution*) edited by anthropologist Ashley Montagu (1905–1999), exposes creationism as pseudoscience.

1984 In its "On Creationism in School Textbooks," the Central Conference of American Rabbis advocates the teaching of evolution as being "basic to understanding science" while claiming that the teaching of creationism will "clearly distort the integrity of science."

1984 Henry Morris denounces the ASA, claiming that it has "capitulated to theistic evolution." The same year, Morris's *History of Modern Creationism* vilified scientists while claiming that "all the real facts of science" support creationism and that the creationism movement is "far too widespread . . . for the evolutionists ever to regain the obsequious submission of the public which they used to enjoy and abuse."

1984 In *The Mystery of Life's Origin*, chemist Charles Thaxton, along with Walter Bradley and Roger Olson, reintroduces the term *intelligent design*. The authors claimed that it is "fundamentally implausible that unassisted matter and energy organized themselves into living systems." Although *Mystery* relegated discussions of design to its epilogue, it is revered among creationists as a founding document of the modern ID movement.

1984 Tennessee Governor Ned McWherter (b. 1930) announces that he will "vigorously pursue" the superconducting supercollider. At the same news conference, McWherter advocated "equal time" for evolution and creationism in science classes. Texas was chosen as the site for the supercollider, but the project was eventually shelved.

1984 Kamoya Kimeu (b. 1940), a member of Richard Leakey's research team, discovers a nearly complete skeleton of a 1.5-million-year-old boy near Lake Turkana, Kenya. The so-called "Turkana Boy" (*Homo ergaster* or *H. erectus*; see also Appendix C) later (2007) became the focus of a protest by Kenyan religious leaders.

1984 Benjamin Zarr publishes *Evolution*, which claims to contain revelations from God. Zarr argued that life was created by God from crystals, that bacteria swallowed infants to produce sperm, that these sperm produced earthworms, and that "[m]an consists of two, separate, four-segmented earthworms." Although Zarr rejected the story of Adam and Eve, he claimed that "CREATIONISTS ARE RIGHT" and that "a newly adapted Bible can be accepted as the LEGAL BASIC THEORY OF EVOLUTION!" Zarr sent copies of this book to libraries to "make a biological improvement in your library."

1984 When the Texas Board of Education repeals the state's 1974 pro-creationism guidelines for textbooks, state education officials demand better coverage of evolution in textbooks. Publishers, hoping to secure some of the state's $80,000,000 textbook market, quickly complied.

1984 The NAS distributes more than 40,000 copies of its *Science and Creationism: A View from the National Academy of Sciences*. *Science and Creationism* strongly advocated the teaching of evolution while rejecting the teaching of creationism in science classrooms.

1984 In *War in the Heavenlies*, popular televangelist and faith-healer Toufik "Benny" Hinn (b. 1952) advocates gap creationism, an ongoing battle between God and Satan, and demons as spirits of pre-Adamic creatures who want to inhabit humans.

1985 Adrian Duplantier rules that Louisiana's law requiring "balanced treatment" for evolution and creationism is unconstitutional. Duplantier considered the case against the state so convincing "that whatever that evidence would be, it could not affect the outcome." Duplantier noted that the concepts of creation and a creator originate in religious conviction and that Louisiana's law was unconstitutional "because it promotes the beliefs of some theistic sects to the detriment of others" (see also Appendix D). Citing the Scopes Trial, the Fifth Circuit Court of Appeals later voted 8 to 7 to support Judge Duplantier's ruling that the Louisiana "balanced treatment" law is unconstitutional. Louisiana Attorney General William Guste announced that Louisiana would appeal the decision to the U.S. Supreme Court. Judge Thomas Gee's (1925–1994) dissenting opinion that "there are two bona fide views" of origins was the first published judicial support for creationist claims since the Scopes Trial. *Edwards v. Aguillard* then went to the U.S. Supreme Court (see also Appendix D).

1985 In *Creation and the Modern Christian*, Henry Morris again urges his followers to fight the spread of evolution. Morris claimed that "evolution is a religion," "evolution is atheistic," "evolution promotes racism," creationism produces "Americanism," and evolution produces communism. The following year, in *Science and the Bible*, Morris again stressed the Bible's "scientific accuracy," rejected evolution in favor of young-earth creationism, and hoped that his book would "win people to a genuine faith in Jesus Christ."

1985 Chemist Graham Cairns-Smith's (b. 1931) *Seven Clues to the Origin of Life* suggests that clay played an important role in the emergence of self-replicating forms.

1985 Kurt Vonnegut's (1922–2007) novel *Galápagos* describes how survivors of "The Nature Cruise of the Century" evolved into a peaceful race of humans after war, disease, and starvation had killed everyone else. Other recent "Darwinian" works of fiction include *The French Lieutenant's Woman* (1969), *The Evolution of Jane* (1998), *Mr. Darwin's Shooter* (1998), and *The Peppered Moth* (2000).

1985 In *The Religion of Evolution*, Henry Morris describes the concept of evolution as "an insult to common sense." That same year, creationist Homer Duncan (1913–2006)—the author of *Evolution:*

The Incredible Hoax (1977)—argued that evolution is "the most stupid thought to enter the mind of man."

1985 Fred Hoyle claims in the *British Journal of Photography* that the impressions of feathers on the *Archaeopteryx* fossils (Figure 24) are forgeries.

1985 The cover of *Time* magazine depicts *T. rex* with the headline "Did Comets Kill the Dinosaurs?"

1985 The California Curriculum Committee begins hearings on science textbooks. The Committee recommended that all of the science textbooks submitted for grades 7 and 8 be rejected. This recommendation was supported by the State Board of Education, which asked publishers to improve their discussion of evolution. Although the 10 revised textbooks were criticized by scientists, they were adopted by a vote of 7 to 2 by the Board.

1985 In *Evolution: A Theory in Crisis*, Michael Denton (b. 1943) claims that "contrary to what is widely assumed by evolutionary biologists today, it has always been the antievolutionists, not the evolutionists, in the scientific community who have stuck rigidly to the facts and adhered to a more strictly empirical approach." Denton dismissed evolution as "a highly speculative idea for which there is no really hard scientific evidence." Denton's book strongly influenced biologist Michael Behe, who became a leader of the ID movement.

1986 An article titled "Exploding the Myth of the Melanic Moth" in *New Scientist* magazine criticizes what has been a main textbook example of natural selection—namely, the increase in moth populations of the dark (melanic) form of the peppered moth (*Biston betularia*) in industrial areas. Although the role of natural selection in increasing the frequency of dark moths in polluted areas (and light moths becoming more abundant after pollution was controlled) was not questioned, some biologists doubted the importance of bird predation as well as other specifics of the story. Antievolutionists cited scientists' questioning of the peppered moth story as evidence that one of the "icons" of evolution was, in fact, a fraud. This spurred several scientists to reinvestigate the original research and to conclude that the basic explanation for the change in frequency of dark forms of the moth was as originally proposed.

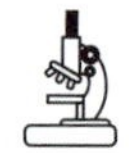

1986 At a scientific symposium at New York's Cold Spring Harbor laboratories, eccentric biochemist (and future Nobel laureate) Kary Mullis (b. 1944) explains the polymerase chain reaction (PCR), a mechanism for amplifying DNA. PCR provided scientists with a quick way to make genetic comparisons between different types of organisms and thus contributed to more accurate phylogenetic studies.

1986 ASA responds to NAS's *Science and Creationism* (1984) by distributing their *Teaching Science in a Climate of Controversy: A View from the American Scientific Affiliation* to more than 60,000 teachers. The 48-page ASA booklet advocated an old earth and theistic evolution, and was described by Lynn Margulis as a publication whose "purpose is to coax us to believe in the ASA's particular creation myth."

1986 Humanlike footprints near dinosaur tracks at Glen Rose, Texas, are shown to be either "inept carvings" from the 1930s or dinosaur tracks that have not weathered like other tracks. These findings convinced some creationists to stop showing the film *Footprints in Stone*.

1986 New Mexico's State Board of Education votes to retain its "evolution is a theory" disclaimer and encourages local school boards to consider presenting "multiple theories of origin."

1986 Geophysicist Glenn Morton's *The Geology of the Flood* claims "if evolution is true, then the Bible is wrong."

1986 Richard Dawkins's *The Blind Watchmaker* counters claims about the implausibility of evolution alone producing complex organisms. The book's title referred to the famous proposition by 19th-century theologian William Paley—in *Natural Theology* (1802)—that if one were to encounter a complex entity with an obvious function, like a watch, it would be logical to conclude that there is a "watchmaker" responsible for its creation. Similarly, life must have a designer. Dawkins confronted Paley's argument by discussing how the nonrandom and cumulative features of natural selection can produce organisms with complex traits. Dawkins enraged many Christians when he noted that "[n]early all peoples have developed their own creation myth, and the Genesis story is just the one that happened to have been adopted by one particular tribe of Middle Eastern herders. It has no more special status than the belief of a particular West African tribe that the world was created from the excrement of ants." Dawkins also famously noted that Darwin "made it possible to be an intellectually fulfilled atheist" and, while lamenting the public's poor understanding of evolution, commented that "[i]t is almost as if the human brain were specifically designed to misunderstand Darwinism, and to find it hard to believe."

1986 The U.S. Supreme Court announces that it will hear the appeal of Judge Duplantier's decision about Louisiana's balanced-treatment law. A dozen *amicus curiae* briefs supporting the lower courts' decisions were filed by 24 scientific organizations and 72 Nobel laureates. When the Court heard opening arguments in *Edwards v. Aguillard*, Wendell Bird argued the creationists' case; ACLU attorney Jay Topkis argued that the law was unconstitutional.

1986 Leroy Hood (b. 1938) and his colleagues at the California Institute of Technology develop a prototype of an automated DNA sequencer. Descendants of this device ultimately enabled biologists to decipher the human genome.

1986 Paleontologist Stephen Jay Gould tells a New Zealand audience that creationism is "peculiarly American." In 2007, approximately 24% of New Zealanders surveyed were biblical creationists.

1986 Pope John Paul II (Figure 67) paraphrases Galileo's 1615 letter to the Grand Duchess Christina when he tells his followers that the Bible "does not wish to teach how heaven was made but how one goes to heaven."

1986 Vertebrate paleontologist Jacques Gauthier produces the first cladistic study to include birds and dinosaurs, thus supporting John Ostrom's claim that theropods such as *Deinonychus* are the closest extinct relative of birds.

1987 Ray Webster, a social science teacher in New Lenox, Illinois, is told by his superintendent to stop teaching creationism. Webster and his student Matthew Dunne sued the school district, claiming that it had violated Webster's right to free speech and Dunne's right as a student to hear about creation science in school. In 1989, in *Webster v. New Lenox School District #122*, U.S. District Judge George Marovich ruled that (1) a teacher does not have a

Figure 67 Pope John Paul II, shown here in 1984 with biologist Stephen Jay Gould at the Vatican, described evolution as "more than a hypothesis." Although the Pope's claims acknowledged the vast amount of evidence supporting evolution by natural selection, many creationists condemned the Pope's conclusion as being heretical. (*Associated Press*)

First Amendment right to teach creationism in a public school, (2) a school district can require a teacher to teach its established curriculum, and (3) Dunne's desires to learn about creation science in school are outweighed by the district's responsibility to avoid violating students' First Amendment rights (see also Appendix D). Webster's appeal ended in 1990 when the Seventh Circuit Court of Appeals upheld Marovich's decision.

1987 In the apocalyptic *The Lie: Evolution*, antievolutionist Ken Ham warns readers "[t]he Bible prophetically warns that in the last days false teachers will introduce lies among the people. Their purpose is to bring God's Truth into disrepute and to exploit Believers by telling them made-up and imagined stories. Such a Lie is among us. That Lie is Evolution." Ham's book, which opened with a drawing depicting evolution as the basis of pornography, abortion, homosexuality, and lawlessness, linked the teaching of evolution to Nazism, drugs, and racism, and included a chapter titled "The Evils of Evolution."

1987 In "Mitochondrial DNA and Human Evolution," published in *Nature*, Allan Wilson, Rebecca Cann, and Mark Stoneking conclude that modern humans share a common ancestor who lived in Africa as recently as 150,000 years ago. This ancestor is now called "Mitochondrial Eve."

1987 Carl Baugh's *Dinosaur: Scientific Evidence that Dinosaurs and Men Walked Together* claims that tracks in the Paluxy riverbed show that humans and dinosaurs lived contemporaneously. Baugh's evidence of "Glen Rose Man" was later shown to be a tooth of a Cretaceous fish.

1987 *Edwards v. Aguillard* is heard by the U.S. Supreme Court (see also Appendix D). Although Wendell Bird (representing Louisiana) claimed that the Louisiana "balanced treatment" law had a primary secular purpose based on "fairness" and "academic freedom," the Supreme Court ruled in a 7 to 2 decision that (1) Louisiana's "balanced treatment" law was unconstitutional because it impermissibly endorsed religion by advancing the religious belief that a supernatural being created humankind; (2) it is unconstitutional to mandate or advocate creationism in public schools, for creationism is a religious idea; (3) requiring the teaching of creation science as "a basic concept of fairness" is "without merit"; (4) banning the teaching of evolution when creation science is not also taught undermines a comprehensive science education; and (5) the law was "facially invalid as violative of the Establishment Clause of the First Amendment, because it lacks a clear secular purpose." Although the Court did not comment on whether creation science was actually science, Justice William Brennan (1906–1997) wrote in the majority opinion that creationism could not be taught as an alternative to evolution because of its religiosity, but that "teaching a variety of scientific theories about the origins of humankind to schoolchildren might be validly done with the clear secular intent of enhancing the effectiveness of science instruction." Creationists interpreted this as an invitation and legal foundation for developing scientific alternatives and presenting them in public schools. (The most popular "alternative" became intelligent design.) The two dissenting votes were by William Rehnquist (1924–2004), an active Lutheran, and Antonin Scalia (b. 1936), a conservative Catholic whose son was a priest. Scalia claimed that Louisiana residents, "including those who are Christian fundamentalists," were entitled "to have whatever evidence there may be against evolution presented in their school, just as Mr. Scopes was entitled to present whatever scientific evidence there was for it." Soon thereafter,

this statement was cited by creationists as a license to teach the alleged "evidence against evolution" and the "strengths and weaknesses" of evolution. *Edwards v. Aguillard* signaled the end of most laws demanding "equal time" and "balanced treatment" for creationism in public schools. In response, some creationists began to abandon Henry Morris's young-earth creationism and, in its place, repackaged creationism as ID. The American Federation of Teachers hailed the *Edwards* decision as a rescue of the nation's schools "from narrow-minded fanatics trying to impose their beliefs on others." However, television preacher Pat Robertson (b. 1930) called the *Edwards* decision an "intellectual scandal," and Henry Morris warned that America was "in imminent danger of judgment from the Creator we are now brazenly repudiating."

1987 *In Mozert v. Hawkins*, the Sixth Circuit Court of Appeals rules that students do not have a right to be excused from classroom activities that expose students to competing ideas, even if some of those ideas are contrary to the students' religious beliefs.

1987 English paleontologist Jenny Clack discovers *Acanthostega*, a Devonian tetrapod that has evidence of functional gills and legs. This discovery suggested that animals evolved legs while still living in water.

1987 Paul MacKinney, chairperson of the Midwest Creation Fellowship, urges that in the wake of *Edwards v. Aguillard*, creationists begin to advocate teaching "arguments against evolution" and portray themselves as victims of discrimination.

1987 University of California-Berkeley law professor Phillip Johnson (b. 1940) reads Richard Dawkins's *The Blind Watchmaker*. Johnson admired Dawkins's rhetorical abilities, but was not convinced by Dawkins's argument for the power of unguided evolution. Johnson believed that evolutionary biologists did not consider all the evidence (i.e., especially that for design) and instead were confusing assumptions and conclusions. Johnson, a respected legal scholar, claimed that scientists accept evolution only because they have a dogmatic, *a priori* commitment to naturalism. According to Johnson, science has not shown that God does not exist; it merely assumes there is no God because to do otherwise is not "science." Accepting evolution necessarily means adhering to an atheistic worldview because one cannot simultaneously accept evolution and believe in God. These conclusions spurred Johnson to write *Darwin On Trial* (1991), the first widely-read modern book about ID.

1988 Another of the many remakes of *Inherit the Wind* is broadcast on network television.

1988 Nicolas Steno, who established the foundation of modern geology, is beatified by Pope John Paul II (Figure 67) on October 23.

1988 California's Superintendent of Public Instruction Bill Honig announces that ICR can no longer offer graduate degrees in science. Honig scheduled a second review of ICR's programs for the following summer.

1989 In *The Long War Against God*, antievolutionist Henry Morris again demonizes evolution: "The very first evolutionist was not Charles Darwin or Lucretius . . . but Satan himself. He has not only deceived the whole world with the monstrous lie of evolution but has deceived himself most of all. He still thinks he can defeat God because, like modern 'scientific' evolutionists, he refuses to believe that God is really God." Morris also claimed "the modern opposition to capital punishment for murder and the general tendency toward leniency in punishment for other serious crimes are directly related to the strong emphasis on evolutionary determinism that has characterized much of this century." Morris, who believed that young-earth creationism is the most certain truth of science, blamed evolution for Nazism, racism, abortion, and other global evils.

1989 Harvard paleontologist Stephen Jay Gould's (Figure 67) *Wonderful Life* argues that the Cambrian "explosion" produced many more basic body-plans than exist today (see also Appendix B).

1989 Biologist Richard Dawkins again enrages creationists when he claims in *The New York Times* that "it is absolutely safe to say that if you meet someone who claims they do not believe in evolution, that person is ignorant, stupid, or insane (or wicked, but I'd rather not consider that)."

1989 Percival William Davis and Dean Kenyon publish the 166-page *Of Pandas and People: The Central Question of Biological Origins*, the first biology textbook to promote ID. The original title of the book was *Creation Biology*, and its definition of *intelligent design* was the same as that of *creation* in earlier editions. *Pandas* sold 23,000 copies in its first five years, and Davis and Kenyon published a 2nd edition in 1993. Both editions of *Pandas*, which ignored the age of the earth and promoted the alleged lack of transitional fossils, were widely criticized by scientists and educators as being "incompetent" and a "wholesale distortion of modern science." In 2005, the Dover, Pennsylvania, school board's attempt to use *Pandas* as an auxiliary textbook in biology classes led to *Kitzmiller et al. v. Dover Area School District*, which established that ID is religion, not science.

1989 The California Board of Education, despite a previous announcement that it will strengthen the teaching of evolution, deletes the claim that evolution is a "scientific fact" from its guidelines. The Board also removed from its guidelines quotes from *Edwards v. Aguillard* and the NAS book *Science and Creationism*.

1989 Bryan College (Figure 49) in Dayton, Tennessee, establishes its Center for Origins Research to "develop a new way of looking at biology that honors the Creator" rather than "trying to discredit evolution." The first director was Kurt Wise (b. 1959). In 2009, the director was Todd Wood (b. 1972).

1989 NSF funds a BSCS project titled "Advances in Evolution: Biological and Geological Perspectives," but—fearing a backlash—asks BSCS Executive Director Joseph McInerney (b. 1948) to remove the word *evolution* from the project's name. McInerney refused the request.

1989 The Texas Education Agency proposes in Proclamation No. 66 that biology textbooks include the "scientific theory of evolution" and "scientific evidence of evolution." The recommendation did not mention scientific creationism or alternative theories. A slightly modified proclamation was approved two months later by a vote of 12 to 3.

1989 While working in Egypt, Philip Gingerich discovers *Basilosaurus*, a fossil whale that has small legs (each with five toes). In 1994, Gingerich discovered an older ancestor of whales (*Rodhocetus*) having even larger legs.

1990 A survey reports that 57% of newspaper editors disagree with the statement that "every word in the Bible is true," and 41% disagree with the statement that "Adam and Eve were actual people," roughly the same proportions that occur in the general population. When editors devoted space to the evolution-creationism controversy, 75% gave equal or more space to creation science than to evolution. Meanwhile, of the roughly 1,600 daily newspapers published in the United States, about 1,400 published daily astrology columns, fewer than 50 of which printed disclaimers stating that the columns were not scientific and were only for entertainment.

1990 California's Superintendent of Public Instruction Bill Honig revokes the license of ICR to offer graduate science degrees. ICR sued in state and federal courts, claiming that its First Amendment rights had been violated. Honig later restored ICR's license and gave the decision about ICR's accreditation to the newly formed Council for Private Postsecondary and Vocational Education.

1990 Conservative philosopher Larry Azar claims in *Twentieth Century in Crisis: Foundations of Totalitarianism* that "evolutionism" was the force behind the totalitarianism of Adolf Hitler.

1990 Country music legend Merle Haggard (b. 1937) tells his fans "evolution is a laughing matter for anybody that's got a rational mind."

1990 Computer consultant Josh Greenberger of the Orthodox Union publishes *Human Intelligence Gone Ape*, which claims to "disprove the theory of evolution in more ways than one" and "show how evolution is genetically impossible."

1990 Pat Robertson founds the American Center for Law and Justice (ACLJ) to counter the work of the ACLU.

1990 In *Evolution and the Myth of Creationism*, American biologist Tim Berra (b. 1943) claims "[f]undamentalists long for the return of a more moral America, an America that may never have been. All around them they see what they perceive as declining morality and spirituality. They reason that if humans share ancestry with the other animals, we have no reason to behave as anything other than animals. This view neglects the fact that humans are the only known animals with the ability to contemplate the consequences of their own actions. It also fails to recognize that there is a great deal of good in the world, the nightly news notwithstanding. Crime existed long before the theory of evolution, even before the writing of the Bible, and biologists do not like crime any more than the creationists do. Evolutionary theory is not a license to run amok, and neither is a belief in the literal interpretation of the Bible a guarantor of moral behavior."

1990 Readers of *Scientific American* learn of the clash between the magazine and contributor Forrest M. Mims III, a science writer and supporter of ID. Although Mims did not advocate ID in his columns for *Scientific American,* editor Jonathan Piel (b. 1938) did not hire him to edit the magazine's "The Amateur Scientist" column. Mims claimed that he did not get the job because of his religious beliefs, but he declined an offer by the ACLU to sue on his behalf.

1990 Self-taught paleontologist Sue Hendrickson (b. 1949) discovers the largest and most (more than 90%) complete *Tyrannosaurus rex* fossil ever found (Figure 68). That fossil, later named "Sue" in her

Figure 68 Sue Hendrickson with "Sue," a 67-million-year-old *Tyrannosaurus rex* that she discovered just outside Faith, South Dakota, in the summer of 1990. Bidding for Sue began at $500,000, but the final price was $7.6 million (the sales commission raised the price to $8.4 million). It took six fossil-hunters from the Black Hills Institute 17 days to excavate Sue, and 10 museum preparators more than two years (30,000 hours of work) to prepare Sue's more than 250 teeth and bones. To see a more famous assemblage of Sue, see Figure 70. (© *1990, Black Hills Institute of Geological Research, Photographer: Peter L. Larson*)

honor, was purchased in 1997 at a Sotheby's auction for $8.4 million by The Field Museum (Chicago, Illinois), where it was unveiled in 2000. McDonald's Corporation, which has a restaurant in the museum, helped provide funds for Sue's purchase in exchange for the rights to reproduce Sue's image in selected venues. The discovery of Sue and her auction set off another wave of dinosaur-mania that included movies (e.g., *Jurassic Park*) and the popular children's show *Barney and Friends.*

1990 The Discovery Institute is founded as a Seattle branch of the Hudson Institute, a conservative think-tank based in Indianapolis, Indiana. Although the Institute originally focused on regional economic, communications, and transportation policy, ID soon became a major emphasis. The Discovery Institute claimed that it rejects efforts to prevent the teaching of evolution or to require the teaching of any particular theory, including ID. Although it publicly disagreed with the "misguided policies" of the Dover (Pennsylvania) School Board to include ID in ninth-grade biology, the Discovery Institute also condemned Judge John Jones III's (b. 1955) decision in *Kitzmiller et al. v. Dover Area School District* (2005).

1990 The Human Genome Project officially begins, with Nobel laureate James Watson as its first director. The project was completed in 2001.

1990 The Kentucky legislature makes effective Kentucky Revised Statute 158.177, which allows teachers to teach biblical creationism and give students credit on exams for answers based on biblical creationism. Nonbiblical beliefs about creation were not covered by the law. This law defied the 1987 U.S. Supreme Court decision in *Edwards v. Aguillard*, but has not been challenged. Meanwhile, education officials in Louisiana grouped evolution with the occult, witchcraft, and drug use as topics inappropriate for the state exit exams.

1990s A series of Gallup Polls shows that, as in previous decades, the public supports the inclusion of creationism in science classes of public schools.

1991 Young-earth creationist Marshall Hall's *The Earth is not Moving* rejects the Copernican model of a rotating and orbiting earth. In 2007, state legislators in Georgia and Texas distributed a memo drafted by Hall blaming evolution on the Jews.

1991 In *The Third Chimpanzee—The Evolution and Future of the Human Animal*, biologist Jared Diamond (b. 1937) provocatively describes humans as a third species of chimpanzee (along with the common chimpanzee and pygmy chimp, or bonobo).

1991 *Bishop v. Aronov* challenges University of Alabama physiology professor Phillip Bishop's use of class meetings to promote ID and his religious faith ("evidence of God in human physiology"). The court ruled that (1) academic freedom is not an independent First Amendment right, (2) a university can control the style and content of speech in school-sponsored events, and (3) a university can control its curriculum, provided the university's "actions are reasonably related to legitimate pedagogical concerns."

1991 Sequence data from large sections of whole chromosomes supports the hypothesis that humans have one fewer chromosome than chimpanzees, a result of the fusion of two chromosomes at some point in the hominin lineage. Specific sequences indicative of the ends (telomeres) of chromosomes were found in the middle of the human chromosome 2 sequence, as expected by the end-to-end fusion of two once-separate chromosomes. Later research found evidence of a now-defunct second centromere in human chromosome 2. Together, these data support the close evolutionary relatedness of humans and chimpanzees.

1991 Geologists working in the Yucatán Peninsula discover the Chicxulub Crater, the site of the meteor impact that may have doomed non-avian dinosaurs to extinction. The crater is more than 110 miles in diameter.

1991 A national study concludes "over a quarter—and perhaps as many as half—of the nation's high school students get educations shaped by creationist influences."

1991 In Morton, Illinois, the school board decides that there is too much evolution in the curriculum and orders staff to develop a creationism-based curriculum for balance. Administrators at nearby schools advocated teaching creationism. Many biology teachers responded by not mentioning evolution in their classes.

1991 In Orange County, California, parents and teachers complain that San Juan Capistrano Valley High School biology teacher John Peloza is teaching creationism and proselytizing for conservative

Christianity. When Peloza was later reprimanded and told by the district to stop teaching creationism and follow California's guidelines for teaching evolution, Peloza sued. The school district argued that Peloza was creating "curriculum anarchy," and Peloza argued that the district was violating his right to free speech by forcing him to teach the religion of evolution. The following year, U.S. District Judge David Williams dismissed Peloza's lawsuit while describing it as "frivolous and unreasonable," ruling that the district acted appropriately in prohibiting teachers from teaching creationism. Williams's 10-page decision emphasized that a teacher cannot teach his or her own curriculum if it violates the state's educational guidelines. Peloza appealed Williams's decision (see also Appendix D).

1991 Phillip Johnson publishes *Darwin on Trial,* which argues that evolution fails to explain the observed patterns we see in nature. *Darwin on Trial*—which marked the maturation of the ID movement into an organized entity—was rejected by scientists. Harvard paleontologist Stephen Jay Gould's (Figure 67) review in *Scientific American* was typical and claimed that Johnson's book was "full of errors, badly argued, based on false criteria, and abysmally written." Although Johnson emphasized keeping the discussion purposely vague about who or what the "designer" is (even claiming that it could be an alien that brought life to earth), he also admitted that he wanted to "get the issue of intelligent design, which really means the reality of God, before the academic world and into the schools." In a review of *Darwin on Trial* published in *Nature*, philosopher David Hull (b. 1935) noted "[w]hat kind of God can one infer from the sort of phenomena epitomized by the species on Darwin's Galápagos Islands? The evolutionary process is rife with happenstance, contingency, incredible waste, death, pain and horror. . . . The God of the Galápagos is careless, wasteful, indifferent, almost diabolical. He is certainly not the sort of God to whom anyone would be inclined to pray."

1992 Creationists in Minnesota's Anoka-Hennepin School District ask that the curriculum include "evidence against evolution." This strategy, which became increasingly popular in the ensuing years, prompted even more teachers to avoid the topic of evolution.

1992 Ian Campbell and his coworkers suggest that volcanic activity of the Siberian Traps caused the Permian-Triassic mass extinction 251 million years ago (see also Appendix B).

1992 The popular children's television show *Barney and Friends* stars an optimistic, anthropomorphic, purple *Tyrannosaurus rex* who sings and dances. Barney's friends include Baby Bop (a green triceratops), B. J. (a yellow protoceratops), and Riff (an orange hadrosaur)—as well as several *Homo sapiens* juveniles.

1992 Joe Kirschvink's article titled "Late Proterozoic Low-latitude Glaciation: The Snowball Earth" suggests the controversial Snowball Earth Hypothesis. Kirschvink's article attracted little attention for several years, but in 1998 Paul Hoffman and his colleagues supported Kirschvink's claim when they suggested that the earth underwent world-wide glaciations followed by greenhouse gas-based heating, dramatic changes that spurred the evolution of multicellular life.

1992 Pope John Paul II (Figure 67) admits that 17th-century Italian scientist Galileo Galilei was treated "unfairly" by the Catholic Church. The pope regretted the "tragic mutual incomprehension" of the case.

1992 Principals of the nascent ID movement—Phillip Johnson, Michael Behe, Stephen Meyer, and William Dembski—meet at Southern Methodist University to develop a plan for advancing ID as a social and political force. It was here that the "wedge strategy" was developed: "A log is a seeming solid object, but a wedge can eventually split it by penetrating a crack and gradually widening the split. In this case the ideology of scientific materialism is the apparently solid log."

1992 San Francisco State University biology professor Dean Kenyon is censured for teaching creationism in his classroom. Kenyon had written *Biochemical Predestination* in 1969, a book that advocated biochemical evolution (the preface of the Russian edition was written by A. I. Oparin), and he was inspired to study evolution by Julian Huxley. However, in 1976, Kenyon read *The Genesis Flood* and other neocreationism books and became a creationist.

1992 Jerome Barkow and husband-and-wife team Leda Cosmides and John Tooby publish the edited volume *The Adapted Mind: Evolutionary Psychology and the Generation of Culture*, which forms the basis of the new discipline of evolutionary psychology. Evolutionary psychologists view human behavior and mental processes from an adaptive perspective and propose that the human brain consists of a large collection of genetically-determined modules that have evolved to solve particular problems faced by individuals. Such conclusions about human behavior have garnered attention from the popular media as well as condemnation by some evolutionary biologists.

1992 The California Board of Education and the Institute of Creation Research announce an out-of-court agreement that gives ICR $225,000 and allows ICR to continue to teach creation science.

1993 American biologist William Schopf reports in *Science* his discovery of 11 types of microbes—all less than 0.01 inches long—in a rock formation in Australia. The microscopic fossils—which resemble modern cyanobacteria—were embedded in a rock 3.485 billion years old. Schopf's fossils are more than 1.3 billion years older than any other comparable group of fossils ever found.

1993 An Israeli dairy—hoping to capitalize on the dinosaur craze triggered by *Jurassic Park*—sells milk with free dinosaur stickers and is threatened with loss of its kosher status. As a critic noted, "[t]his is like seeping sacrilege. . . . Dinosaurs symbolize a heresy of the creation of the world because they reflect Darwinistic theories."

1993 *Utahraptor* is discovered in Grand County, Utah, by Jim Kirkland (b. 1954) and his colleagues. The following year, this giant velociraptor became famous when it appeared in the movie *Jurassic Park*.

1993 In an article published in *Proceedings of the National Academy of Sciences*, biologists Jeffrey Palmer and Sandra Baldauf conclude that animals and fungi are evolutionarily more closely related than are fungi and plants. This conclusion upended the long-accepted evolutionary relationship among the three groups.

1993 The AAAS publication *Benchmarks for Science Literacy* promotes evolution as an integral part of the science curriculum.

1993 In *Natural History* magazine, biologist Jared Diamond notes "[a]s for the claim that evolution is an unproved theory, that's nonsense. Evolution is a fact, established with the same degree of confidence as our 'theory' of the round earth, our 'germ theory' of disease, and the 'atomic theory' of matter. Yes, there is lively debate about the particular evolutionary mechanisms that caused particular changes, but the existence of evolutionary change is not in doubt. Our own bodies provide walking evidence."

1993 In *Beyond the Culture Wars*, English professor Gerald Graff (b. 1937) coins the phrase *teach the controversy* to encourage the teaching of conflicts surrounding academic issues to help students understand how knowledge is established. Graff's phrase was later used by creationists trying to get their religious beliefs into biology curricula. Graff, a self-described secularist, lamented, "I felt as if my pocket had been picked when the ID crowd appropriated my slogan."

1993 *Maclean's*, "Canada's Weekly Newsmagazine," reports that "even though less than a third of Canadians attend a religious service regularly . . . 53% of all adults reject the theory of scientific evolution."

1993 A Gallup poll reports that 47% of Americans believe that "God created man pretty much in his present form at one time within the last 10,000 years" and an additional 35% believe that evolution is guided by God. Only 11% accepted naturalistic evolution, and 58% favored teaching creationism in public schools.

1994 AiG opens for business in Kentucky and begins denouncing evolution and promoting religious fundamentalism by focusing on questions related to the book of Genesis. AiG proclaimed that evolution undermines Christianity and that "in six 24-hour days God made a perfect creation" in 4004 BCE (Figure 8). AiG became the most dominant of the many antievolution organizations; by 2008, AiG had hundreds of employees, operated the $27-million Creation Museum, sent its *Answers* magazine to more than 50,000 readers, sponsored thousands of events each year, and had a Web site visited by more than 50,000 people every day. AiG controls almost 60% of the more than $20 million in donations to the 10 largest antievolution organizations.

1994 In *The Young Earth*, John Morris claims "only Scripture gives *specific* information about the age of the earth and the timing of its unobserved events. Rocks, fossils, isotopic arrays, and physical systems do not speak with the same clarity as Scripture."

1994 In *The Bell Curve*, psychologist Richard Herrnstein (1930–1994) and political scientist Charles Murray (b. 1943) argue that intelligence is highly heritable and that there are fundamental differences among races in terms of intelligence. The book created a firestorm of protest that fueled debate on the respective roles of nature and nurture in determining intelligence.

1994 In *Evolution: Fact, Fraud, or Faith*, former Indiana state legislator Don Boys declares war on evolution and proclaims he will "bare-knuckle it with any evolutionist." Boys advocated young-earth creationism, dismissed evolutionary biology as laughable "drivel," and claimed that evolutionary biologists are "pathetic," "venomous," "unscholarly skunks." Later, Boys—who believed that "God is angry with evangelicals and mushy fundamentalists"—claimed that gap creationism is endorsed by "foolish Christians" and harms other Christians because it leads to disbelief and disobedience.

1994 In *Peloza v. Capistrano Unified School District*, the Ninth Circuit Court of Appeals declares that requiring a teacher to teach evolution does not violate the Establishment Clause of the U.S. Constitution (see also Appendix D). The court described as "patently frivolous" Peloza's claim that his freedom of speech had been denied." The case ended in 1995 when the U.S. Supreme Court refused to hear Peloza's appeal.

1994 The AMNH reports that 45% of Americans agree "human beings evolved from earlier species of animals."

1994 American psychologist Steven Pinker's (b. 1954) *The Language Instinct* argues that the human capacity for language is an adaptive instinct crafted by natural selection.

1994 The Tangipahoa (Louisiana) Parish Board of Education adopts a resolution by a vote of 5 to 4 disclaiming the endorsement of evolution. The resolution, which required elementary and high school teachers to read a disclaimer to students before teaching evolution, stated that the teaching of evolution is "not intended to influence or dissuade the Biblical version of Creation or any other concept." Parents of children in Tangipahoa Parish began a protracted legal battle when they sued in U.S. District Court for the Eastern District of Louisiana, challenging the validity of the disclaimer under the U.S. and Louisiana constitutions barring laws "respecting an establishment of religion."

1994 The Tennessee Senate again considers legislation to restrict the teaching of evolution in its public schools. The Senate Education

Committee approved the legislation by a vote of 8 to 1, but the legislation was defeated in the Senate by a vote of 20 to 13.

1994 The discovery of *Australopithecus ramidus* is announced by American paleontologists Tim White (b. 1950) and Gen Suwa on the cover of *Nature* with the headline "Earliest Hominids." The *Times* of London described the fossil as "The Bone that Rewrites the History of Man." This fossil was the first in 20 years to challenge Lucy for her status as the earliest human ancestor (see also Appendix C).

1994 Physician Randolph Nesse (b. 1948) and evolutionary biologist George Williams publish *Why We Get Sick: The New Science of Darwinian Medicine*, building upon earlier suggestions by parasitologist Paul Ewald of the importance of an evolutionary perspective in understanding human health and medicine.

1995 Tufts University philosopher Daniel Dennett's *Darwin's Dangerous Idea*: *Evolution and the Meaning of Life* claims that what is considered to be indicative of the human soul—consciousness—is merely another product of adaptive evolution. Dennett considered natural selection a "universal acid" that, when invoked as an explanatory framework, dissolves all other possible explanations, lays bare how the world actually operates, and leaves in its wake a changed worldview.

1995 Scientists from AMNH and the Mongolian Academy of Science report in *Nature* their discovery of an *Oviraptor* that died 80 million years ago in Mongolia's Gobi Desert while sitting on its nest of eggs. This was the first definitive evidence of modern avian brooding among dinosaurs.

1995 In *Dinosaur in a Haystack*, Stephen Jay Gould (Figure 67) notes that "[o]ur creationist detractors charge that evolution is an unproved and unprovable charade—a secular religion masquerading as science. They claim, above all, that evolution generates no predictions, never exposes itself to test, and therefore stands as dogma rather than disprovable science. This claim is nonsense. We make and test risky predictions all the time; our success is not dogma, but a highly probable indication of evolution's basic truth."

1995 In *In the Beginning . . . A Catholic Understanding of the Story of Creation and The Fall*, Joseph Ratzinger (b. 1927)—now better known as Pope Benedict XVI—argues that the biblical creation story is symbolic and not constrained by physics and biology.

1995 During an appearance before the Alabama State Board of Education, Forrest Hood "Fob" James, Jr. (b. 1934)—a former professional football player who succeeded George Wallace as governor of Alabama from 1979 to 1983 (as a Democrat) and from 1995 to 1999 (as a Republican)—ridicules evolution by slowly crossing the stage, beginning in a crouch and then ending erect. The following year, James used discretionary funds to send all of Alabama's biology teachers a copy of Phillip Johnson's *Darwin on Trial.* James also vowed to use state troopers and the National Guard to keep the Ten Commandments on display in an Alabama courtroom. The Alabama Board of Education later voted 6 to 1 to require that all 40,000 biology textbooks used in its public schools include a "Message from the Alabama State Board of Education" that claimed, among other things, that evolution is "a controversial theory." Future versions of the sticker were, however, weaker.

1995 Paleontologist Martin Pickford's (b. 1943) *Richard E. Leakey: Master of Deceit* depicts Leakey and other paleoanthropologists as greedy, manipulative "parasites." Pickford's book, which described numerous sexual indiscretions in embarrassing detail, portrays a ruthless, competitive side of anthropology unknown to the public. Pickford, who had attended high school with Richard Leakey, later discovered *Orrorin,* which he proposed as the earliest known hominin.

1995 NABT's "Statement on the Teaching of Evolution" describes evolution as "an unsupervised, impersonal, unpredictable, and natural process of temporal descent with genetic modification that is affected by natural selection, chance, historical contingencies, and changing environments." The words *unsupervised* and *impersonal* were challenged by critics (notably philosopher Alvin Plantinga [b. 1932] and religion scholar Huston Smith [b. 1919]), claiming the statement was proof that evolution was a materialist ideology and not a theologically neutral scientific theory. The ensuing furor prompted NABT to delete *unsupervised* and *impersonal* from the organization's statement on evolution, a decision claimed as a victory by antievolutionists.

1995 Phillip Johnson's *Reason in the Balance: The Case Against Naturalism in Science, Law, and Education* continues Johnson's criticism of naturalistic evolution, specifically, its tendency to exclude theistic factors as possible explanations of natural events. Johnson equated evolutionary naturalism with the tolerance of abortion, homosexuality, pornography, genocide, and other "social evils."

1995 In *The Demon-Haunted World: Science as a Candle in the Dark*, famed astronomer Carl Sagan (1934–1996) notes that "I meet many people offended by evolution, who passionately prefer to be the personal handicraft of God than to arise by blind physical and chemical forces over aeons from slime. They also tend to be less than assiduous in exposing themselves to the evidence. Evidence has little to do with it. . . . The clearest evidence of our evolution can be found in our genes. But evolution is still being fought, ironically, by those whose own DNA proclaims it—in the schools, in the courts, in textbook publishing houses, and on the question of just how much pain we can inflict on other animals without crossing some ethical threshold."

1995 Tim LaHaye and Jerry Jenkins (b. 1949) begin publishing the *Left Behind* series of novels. These books, whose sales have exceeded 60 million copies, popularized biblical end-times prophecy and spawned an industry of movies, television specials, video games, and associated materials, all proclaiming premillenial dispensationalism. The *Left Behind* series is the fastest selling adult fiction series in history.

1995 In response to a complaint, Tulsa zoo administrators remove a reference about human-chimp ancestry from a chimp exhibit.

1995 Craig Venter (b. 1946), Hamilton Smith (b. 1931), and their colleagues determine the first genomic sequence of a free-living organism (the bacterium *Haemophilus influenzae*).

1995 Richard Dawkins notes in *River Out of Eden* that "[t]he universe we observe has precisely the properties we should expect if there is, at bottom, no design, no purpose, no evil and no good, nothing but blind, pitiless, indifference."

1996 CRS begins publishing the journal *Creation Matters*, which contains "somewhat lighter" articles than some of its other periodicals. In 2009, CRS has approximately 1,700 members.

1996 The National Science Education Standards present evolution as one of the "unifying concepts and processes" and list it prominently in the Content Standards for grades 9–12.

1996 Edward Larson (b. 1953) and Larry Witham (b. 1952) repeat psychologist James Leuba's landmark studies of the religious beliefs of American scientists that Leuba conducted in the early 20th century. The following year they reported in *Nature* that the rate of belief among U.S. scientists had not changed since 1933 (still averaging around 40% across those surveyed). Larson and Witham noted that 65 years earlier this number was "shocking" to many because of how low it was, but in 1996 this same number surprised many people because of how large it was within a discipline often viewed as opposing religion. A survey of biologists in NAS found almost 95% of respondents to be atheists or agnostics. Two years later, however, NAS published *Teaching About Evolution and the Nature of Science*, in which they noted "whether God exists or not is a question about which science is neutral."

1996 Guillermo Gonzalez (b. 1963) and Jay Richards of the Discovery Institute publish *The Privileged Planet (The Search for Purpose in the Universe*), which claims that the universe is "fine-tuned" to allow life to exist and that the chance that such a universe could happen on its own through random processes is negligible. This, the authors claimed, indicates design.

1996 Michael Behe's (b. 1952) (Figure 69) iconoclastic *Darwin's Black Box: The Biochemical Challenge to Evolution* (named Book of the Year for 1996 by *Christianity Today*) proposes that it is impossible to imagine how the intricate "molecular machines" of a cell could have evolved by natural selection because only the final version with all the necessary parts in place produces a working system. Cellular components (like Behe's favorite example, the flagella) are therefore "irreducibly complex" and are scientific evidence for the action of an intelligent designer. Although Behe claimed that ID is "one of the greatest achievements in science," virtually all scientists have dismissed it as religion, not science.

1996 In his "Message to the Pontifical Academy of Sciences," Pope John Paul II (Figure 67) describes evolution as "more than a hypothesis" and announces that he sees no conflict between religious teachings and the theory of evolution. The Pope's statement enraged many creationists.

1996 In the ICR periodical *Back to Genesis*, Henry Morris proclaims "that all true science supports Biblical creationism."

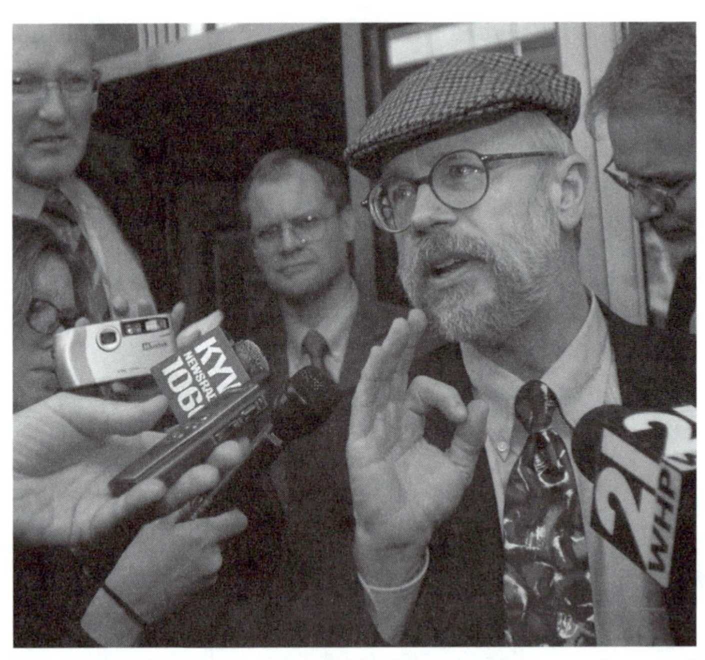

Figure 69 Michael Behe is a Lehigh University biology professor and creationist who promotes ID. Behe's *Darwin's Black Box*—which was condemned and ridiculed by mainstream scientists—became a foundation for ID's "scientific" opposition to evolution. This photo shows Behe talking to reporters after testifying in *Kitzmiller v. Dover*, a 2005 lawsuit in which ID was ruled to be religion, not science. (*Associated Press*)

1996 Republican parties in seven states adopt platforms that endorse the teaching of creationism in public schools.

1996 School administrators in Draffenville, Kentucky, recall textbooks from classrooms and glue together pages that discuss the "big bang."

1996 Tennessee State Senator Tommy Burks introduces legislation (Senate Bill 3229) requiring that evolution be taught as a scientific theory and that anyone who teaches evolution as a fact could be fired. Zane Whitson, the bill's sponsor in the state's House of Representatives, proclaimed that people who opposed the legislation "aren't believers like I am. . . . The atheists are opposed to this bill." The bill passed committee votes in both the House and Senate, but was rejected by the full Senate.

1996 The Discovery Institute announces the establishment of the Center for the Renewal of Science and Culture—later renamed the Center for Science and Culture (CSC)—to serve as the home base for research involving ID.

1996 ICR (Figure 64) publishes *Noah's Ark: A Feasibility Study*, by a young-earth creationist using the pen-name John Woodmorappe. Woodmorappe argued that Noah and his family could have fed all 16,000 animals aboard the Ark if they had used modern production techniques. One reviewer, after reading Woodmorappe's claim that Noah trained all of the animals beforehand to defecate and urinate on demand into buckets, noted that "[t]his, of course, makes Noah the greatest animal trainer in history."

1996 The Education Committee of the Ohio Senate rejects (by a vote of 12 to 8) legislation requiring that the teaching of evolution be accompanied by discussions of the "arguments against evolution."

1996 Dolly the sheep becomes the first mammal cloned from an adult cell. When Dolly died in 2003, her obituary in *The New York Times* noted that she was survived "by six lambs, produced in the customary way, with a ram."

1996 Los Alamos scientist and fundamentalist Christian John Baumgardner creates the computer program *Terra* with the aim of proving that Earth is less than 10,000 years old.

1996 A feathered dinosaur—the turkey-size *Sinosauropteryx prima*, which was discovered near Liaoning, China—is announced by Chen Pei Ji at the annual meeting of the Society of Vertebrate Paleontology.

1996 The Georgia legislature rejects an amendment to an education bill that would "provide that local boards of education may establish optional courses in creationism."

1996 NSF reports that 44% of Americans accept that "human beings as we know them today developed from earlier species of animals.'"

1996 While seeking the Republican Party's nomination for president, political commentator Pat Buchanan (b. 1938) reignites the evolution-creationism debate by claiming that "I don't think it is demonstrably true that we have descended from apes. I don't believe it. I do not believe all that . . . I think [parents] have a right to insist that Godless evolution not be taught to their children or their children not be indoctrinated in it."

1997 BSA becomes Creation Moments, Inc., and shifts its focus to producing short radio programs.

1997 Baylor University President Robert Sloan, Jr., establishes the Institute for Faith and Learning to unify faith and science.

1997 Young-earth creationist Larry Vardiman of ICR claims that the earth is covered with "about a mile of sediments which appear to have been formed by the Genesis Flood." Two years later, ICR's John Morris claimed "[t]he Flood would have totally restructured the surface of the globe."

1997 The Illinois State Board of Education quietly deletes the word *evolution* from its education standards and science test. The "evolutionless" standards were approved by the state's Superintendent of Education, many educators, and an official representing a teachers' union.

1997 In the *New York Review of Books*, biologist Richard Lewontin confesses what proponents of ID have often claimed: "We take the side of science in spite of the patent absurdity of some of its constructs, *in spite* of its failure to fulfill many of its extravagant promises of health and life, *in spite* of the tolerance of the scientific community for unsubstantiated just-so stories, because we have a prior commitment, a commitment to materialism."

1997 A Gallup Poll reports that 68% of Americans believe that "creationism should be taught along with evolution" in public schools.

1997 In a controversial article titled "Why They Kill Their Newborns" published in *The New York Times*, Harvard psychologist Steven Pinker (b. 1954) invokes Darwinian ideas to explain infanticide, adding that killing one's newborn baby is not as serious as killing one's child later in life.

1997 Biology teacher Roger DeHart of Burlington, Washington, is reassigned to teaching earth science after it is discovered he is using the pro-ID textbook *Of Pandas and People*. DeHart had been discussing ID in his classroom for nearly 10 years, with tacit approval from the administration and with no registered complaints. The ACLU's threat of legal action finally caused the school to relieve DeHart of teaching biology.

1997 In *The Science of God*, physicist Gerald Schroeder explains an old earth by claiming that Genesis 1 was written by a being who lived in "logarithmic time" (in which the "first day" was 8 billion years, the "second day" was 4 billion years, the "third day" was 2 billion years, and so forth), and that the writing of Genesis was assumed by someone living in "nonlogarithmic" time after Genesis 1. Schroeder also claimed that there is no evidence for Noah's flood and that God first gave human souls to Adam and Eve.

1997 In *Freiler v. Tangipahoa Parish Board of Education*, U.S. District Court Judge Marcel Livaudais, Jr., rules that (1) it is unlawful to require teachers to read aloud disclaimers saying that the biblical version of creation is the only concept "from which students [are] not to be dissuaded," and (2) proposals for teaching "intelligent design" are equivalent to proposals for teaching "creation science." The court noted that "while encouraging students to maintain their belief in the Bible, or in God, may be a noble aim, it cannot be one in which the public schools participate, no matter how important this goal may be to its supporters." The Court awarded $49,444.50 to Herb Freiler's attorneys (see also Appendix D).

1997 Phillip Johnson tells *Christianity Today* that "[t]he important thing is not whether God created all at once [as scientific-creationism claims] or in stages [as progressive creationism or theistic evolution claims]. Anyone who thinks that the biological world is a product of a pre-existing intelligence . . . is a creationist in the most important sense of the word."

1997 *Nature* reports that the extinct *Pachyrhachis problematicus*—a primitive 3-foot-long snake—had short, well-developed legs.

1997 Pope John Paul II (Figure 67) issues a papal letter accepting modern evolutionary theory.

1997 Swedish molecular biologist Svante Pääbo (b. 1955) sequences mitochondrial DNA from Neanderthal fossils and concludes that Neanderthals are not direct ancestors of humans, but instead are a separate—and now extinct—species. Neanderthals co-occurred with *Homo sapiens*, but disappeared from the earth 28,000 years ago (see also Appendix C).

1998 A committee of 27 scientists and science educators, appointed by the state's commissioner on education, starts writing new science standards for Kansas. The rejection of these standards in favor of those favoring creationism later produced a controversy that brought ridicule to Kansas.

1998 In *State of Louisiana v. Richard J. Schmidt*, a Louisiana physician—accused of intentionally injecting his mistress with HIV from his patients—is convicted of attempted second-degree murder and sentenced to 50 years in prison. Central to the prosecution's case was showing the evolutionary relatedness of the HIV strains from the victim and patients of the physician, an analysis that represented the first use of phylogenetic data in a criminal case in the United States. The scientists who conducted the analysis noted the irony of prosecutors relying on evolutionary biology to convict a defendant in a state that had only 25 years earlier enacted antievolution legislation.

1998 NABT's revised "Statement on Teaching Evolution" supports the teaching of evolution while noting that "creation science, scientific creationism, intelligent-design theory, young-earth theory, [and other] creation beliefs have no place in the science classroom."

1998 Young-earth creationist and theologian Robert Reymond's *A New Systematic Theology of the Christian Faith* claims that "there is no reason to believe that the universe and the earth in particular are billions of years old . . . as many astronomers and geologists insist." Instead, Reymond revived Phillip Gosse's claim from 1857 that God created the earth "with an appearance of age." Critics again noted these claims mean that God is a tricky charlatan.

1998 A Gallup poll reports that 47% of Americans believe that "God created man pretty much in his present form at one time within the last 10,000 years."

1998 A survey of 500 high school biology teachers shows that 40% of respondents spend little or no time teaching evolution.

1998 In *Back to Genesis*, ICR's John Morris claims that "ICR very much appreciates the work of [ID advocates such as Phillip Johnson and Michael Behe] but we recognize that without biblical creationism they fall short of a God-pleasing mark. Any form of old-Earth thinking, theistic evolution, or progressive creation is so similar to secular evolution that their defense is ultimately a waste of time."

1998 Don Aguillard tells teachers attending the annual meeting of NABT that "the fight [for the teaching of evolution] has shifted from people like Susan Epperson (Figure 60) and myself to you. The task has shifted from constitutional challenges to grass-roots efforts."

1998 Historian and legal scholar Edward Larson is awarded the Pulitzer Prize in history for his 1997 book *Summer for the Gods: The Scopes Trial and America's Continuing Debate Over Science and Religion*. Larson's book examined the Scopes Trial from a legal perspective, an analysis Larson felt was lacking among the many publications devoted to this seminal event in the evolution-creationism controversy. Larson's subsequent books included *Evolution's Workshop: God and Science on the Galápagos Islands* (2001) and *Evolution: The Remarkable History of a Scientific Theory* (2004).

1998 In the popular *Darwin's Leap of Faith: Exposing the False Religion of Evolution*, television personalities John Ankerberg (b. 1945) and John Weldon argue that Darwin never considered his theory of evolution to be overly convincing and that he devised it because it was "convenient to his rejection of God."

1998 Money floods the coffers of antievolution organizations. For example, the budget of AiG for 1998 was $3,702,800 and that of ICR was $4,167,547. For comparison, the NCSE—an organization that advocates the teaching of evolution and the banishment of creationism from science classes—reported revenues of $258,957.

1998 

Rodney LeVake (b. 1954), a biology teacher at Faribault High School in Minnesota, is confronted by colleague Ken Hubert (b. 1955) about his teaching of evolution after other science teachers suspect that LeVake had purposely omitted the subject from his instruction. LeVake claimed that evolution is "impossible," that there was an "amazing lack of transitional forms in the fossil record," that evolution violated the Second Law of Thermodynamics, and that evolution was not science. In response, LeVake was reassigned to a ninth-grade physical science course. Claiming unfair treatment, LeVake—a self-described fundamentalist—was assisted by the ACLJ (an organization founded by televangelist

Pat Robertson to defend "the rights of believers"). On behalf of LeVake, the ACLJ sued the school and its administrators, claiming that LeVake was reassigned because his religious beliefs opposed evolution and asked the Court to give LeVake $50,000 (plus court costs) and declare "the district's policy, of excluding from biology teaching positions persons whose religious beliefs conflict with acceptance of evolution as an unquestionable fact, to be unconstitutional and illegal under the U.S. and Minnesota Constitutions." The District Court ruled against LeVake, noting that a teacher's right to free speech does not permit the teacher to circumvent the prescribed curriculum. The Minnesota Court of Appeals supported the original ruling, and the case ended in 2002 when the U.S. Supreme Court refused, without comment, to hear the case (*LeVake v. Independent School District #656*; see also Appendix D)

1998 NAS reaffirms its position that evolution is "the most important concept to modern biology" and that "there is no debate within the scientific community over whether evolution has occurred, and there is no evidence that evolution has not occurred."

1998 William Dembski's (b. 1960) *The Design Inference: Eliminating Chance Through Small Probabilities* introduces the use of an "explanatory filter" to identify design in nature. The irreducible complexity Michael Behe (Figure 69) claimed to have demonstrated was exactly the *specified complexity* Dembski required as evidence for design. Using a complicated formula that included the number of particles in the universe and a physical constant known as Planck time, Dembski proposed that patterns less likely than 0.5×10^{150} were evidence for design. Critics dismissed Dembski's claim as flawed and irrelevant. That same year, Dembski joined Baylor University as Associate Research Professor and soon thereafter was appointed by Baylor's president Robert Sloan to head the newly created Michael Polanyi Center (MPC) for Complexity, Information, and Design. The secretive nature of the creation of the MPC caused dissent among Baylor's faculty, especially when Dembski's agenda for the study of ID became apparent. In response, Sloan convened a committee of outside experts that recommended folding the MPC into an existing university entity. Dembski publicly declared the committee's findings a victory for ID over its opponents ("[they] have met their Waterloo"). When Sloan asked Dembski to retract the comments, Dembski refused, and Sloan removed Dembski as director.

Dembski, who had a multiyear contract with Baylor, remained at the school for five more years, during which time he worked off-campus writing books, later remarking, "[i]n a sense, Baylor did me a favor. I had a five-year sabbatical."

1999 A study published in *The Science Teacher* reports that only 57% of science teachers consider evolution to be a unifying theme in biology and that almost half of science teachers believe that there is much scientific evidence for creationism. Large percentages of biology teachers believed that creationism should be taught in public schools.

1999 In its "Resolution Opposing Creationism in Science Courses," the AIBS reaffirms its opposition to creationism and its support of evolution, which it describes as "the only scientifically defensible explanation for the origin of life and development of species." The resolution noted that "creationism is based on religious dogma stemming from faith rather than demonstrable facts . . . creationism should not be taught in any science classroom."

1999 Chinese paleontologist Xiao-Chun Wu and his colleagues discover *Sinornithocaurus*, a meat-eating dinosaur, in volcanic ash deposited 120 million years ago in China. *Sinornithocaurus* had a coat of "proto-feathers" conceivably used for insulation, but not for flight. This discovery supported the claim that birds evolved from dinosaurs. The same year, a fossil from China's Liaoning Province purportedly showing a feathered bird having a long dinosaur-like tail was unveiled. The fossil, bought at a gem show in Arizona, was described by the press as a "missing link" and named *Archaeoraptor* in an article published in *National Geographic*. Scientists later showed the fossil to be a fake; the head and upper body were from a primitive fossil bird (*Yanomis*) and the tail from *Microraptor*, a small, gliding theropod. Although scientists have found many examples of feathered dinosaurs, creationists continue to use the *Archaeoraptor* scandal to undermine evolutionary theory.

1999 After fatal shootings at Colorado's Columbine High School, Texas Congressman Tom DeLay (b. 1947) blames the tragedy on "school systems [that] teach the children that they are nothing but glorified apes who evolutionized out of some primordial soup of mud." Although this quote was attributed to DeLay, he read it in an article written by Addison Dawson for the *San Angelo Standard-Times*.

1999 William Dembski's *Intelligent Design: The Bridge Between Science & Theology* continues his advocacy of ID and specified complexity, casting the search for design within the framework of information theory. That same year, in an article titled "Signs of Intelligence: A Primer on the Discernment of Intelligent Design," Dembski tried to distinguish ID from creationism and claimed that "Naturalism is the disease: Intelligent design is the cure." Dembski also claimed that ID is biblical: "Indeed, Intelligent Design is just the Logos theology of John's gospel restated in the idiom of information theory."

1999 Encouraged by Governor Cecil Underwood (1922–2008), the Kanawha County School Board in West Virginia debates a proposal to begin teaching the biblical story of creation. The proposal was defeated, but a poll showed that almost 60% of the state's residents believed that the biblical story of creation is "the actual explanation for the origin of human life on earth."

1999 *George* magazine ranks the Scopes Trial as fourth in a list of the "one hundred greatest defining political moments" of the 20th century.

1999 In a letter to his followers, ICR's president John Morris asks "[w]ouldn't it be wonderful to remove the stumbling block of radioisotope dating?" Soon thereafter, ICR (Figure 64) announced a five-year plan to discredit radiometric dating.

1999 The Republican-dominated Kansas Board of Education endorses, by a vote of 6 to 4, a set of science education standards for the state's 305 public school districts. These standards—developed by the Creation Science Association for Mid-America—included no mention of biological macroevolution, the age of the earth, or the origin and early development of the universe. The board had earlier rejected standards developed by a 27-member panel of scientists and science educators. *Science* reported the story as "Kansas Dumps Darwin," and *The New York Times* described the vote with the headline "Board for Kansas Deletes Evolution from Curriculum—A Creationist Victory." AAAS described the decision as "a serious disservice to students and teachers." Soon after the Kansas decision, the Kentucky Department of Education deleted the word *evolution* from its educational guidelines, and conservative columnist and political commentator George Will (b. 1941) noted that "[e]very [political] party at any given time has a certain set of issues on its fringe that can make it look strange, and this is one that can make the Republicans look strange." An editorial in *Scientific American* suggested that college admissions boards question the qualifications of applicants from Kansas. However, antievolution crusader Phillip Johnson (who had given money and public support to a creationist school board candidate) called the board's decision "courageous," while biochemist and ID-advocate Michael Behe (Figure 69) called it "heartening."

1999 In the wake of the Kansas State Board of Education's decision to abandon the teaching of evolution, Phillip Johnson helps craft the "teach the controversy" strategy for promoting the teaching of ID. This tactic (i.e., "What educators in Kansas and elsewhere should be doing is to 'teach the controversy'") shifted the focus from specifically teaching ID to portraying evolution as a controversial theory in trouble, meaning that any discussion of evolution should include coverage of its (allegedly) serious problems and an examination of alternatives. "Teach the controversy" was frequently included in legislation to promote ID in science curricula, and the "Santorum Amendment" to the 2001 Elementary and Secondary Education Act Authorization Bill invoked standard "teach the controversy" rhetoric about science education while identifying evolution as a topic needing such special discussion.

1999 Presidents of AAAS and NSTA and the Chair of the National Research Council deny the Kansas State Board of Education copyright permission to reference or use text from their documents in Kansas's revised science standards.

1999 Adnan Oktar (b. 1956), an Islamic creationist who writes under the pseudonym "Harun Yahya," publishes *The Evolution Deceit*, a book that blasts the "dishonest philosophy" of "materialism" that "seeks to abolish the basic values on which the state and society rest." Soon thereafter, Oktar blamed war, poverty, pain, and massacres on the "selfish and pitiless world view" of Charles Darwin.

1999 American biologist Ken Miller's *Finding Darwin's God: A Scientist's Search for Common Ground between God and Evolution* presents evidence against creationism while describing Miller's religious views. Miller later testified in several evolution-related lawsuits, including *Kitzmiller et al. v. Dover Area School District*.

1999 *Scientific American* reports that less than 10% of the members of NAS believe in God. By contrast, 90% of Americans believe in God and claim that God played at least some role in creation.

1999 Scientists complete the sequencing of the first human chromosome (chromosome 21).

1999 Famed evolutionary biologist Stephen Jay Gould's (Figure 67) *Rocks of Ages* introduces his concept of "non-overlapping magesteria" (NOMA) to describe what Gould sees as the independence of science and religion as ways of knowing about the universe. Gould's development of NOMA was consistent with his contention that morality is independent of evolution, and he frequently bristled at the determinism he saw as inherent in sociobiological explanations of human behavior. NOMA was hailed by many as a reasonable assessment of the different realms of knowledge. Others, however, accused Gould of pandering, and ID-advocates called it a type of "apartheid" meant to separate God from the natural world (as Epicurus had done more than 2,000 years earlier). Richard Dawkins denigrated NOMA as demonstrating a "cowardly flabbiness of the intellect," and other critics claimed that Gould's suggestion excused religion from intellectual scrutiny and was faulty because it is impossible to separate creation myths from the value systems they support (e.g., many Christians defend their values through claims concerning how God created the universe).

1999 The "wedge document," developed by ID-advocates at the Center for the Renewal of Science and Culture (CRSC) of the Discovery Institute is surreptitiously posted online after being sent out for photocopying. The document outlined five-year and 20-year goals for the "overthrow of materialism and its cultural legacies." The strategy included three phases: scientific research, writing, and publication; publicity and opinion-making; and cultural confrontation and renewal. The CRSC initially denied authorship of the document, but later acknowledged that it had originated from the Discovery Institute as an "early fundraising proposal." The document was used as evidence equating ID with religion in *Kitzmiller et al. v. Dover Area School District* in 2005.

1999 The ACLU of Kansas and Western Missouri notifies Kansas school districts that it will consider legal action if the districts teach "creation science." Meanwhile, the Arizona House of Representatives debated "equal time" legislation. The legislation died in committee.

1999 The Idaho State Curricular Materials Selection Committee rejects the creationist text *Of Pandas and People.*

1999 The New Mexico State Board of Education makes its standards match national standards by mandating the teaching of evolution as part of its science curriculum.

1999 The Oklahoma State Textbook Committee votes to place a disclaimer in biology textbooks used in the state's public schools. The disclaimer instructed readers to consider evolution "as theory, not fact" because "no one was present when life first appeared on earth." Oklahoma Attorney General Drew Edmondson (b. 1946) later ruled that the disclaimer was unconstitutional because the Textbook Committee did not have the authority to make such a decision. Legislators responded by proposing bills that required science textbooks to acknowledge the existence of God, but these bills died before coming up for a vote.

1999 In *The Mythology of Modern Dating Methods*, John Woodmorappe denounces uniformitarianism because it hinders evangelism. The next year, antievolutionist William Dembski lamented that evolution "has blocked the growth of Christ . . . and people accepting the Scripture and Jesus Christ."

1999 The Tangipahoa Parish Board of Education (Louisiana) proposes an alternative disclaimer for teachers to read to students regarding the teaching of evolution. The plaintiffs (including Herb Freiler) challenged this new disclaimer, and the board asked for a hearing by all 15 members of the Fifth Circuit Court. This court ruled in favor of the plaintiffs in the *Freiler v. Tangipahoa Parish Board of Education* decision (1997), stating that a school board cannot require that a disclaimer be read immediately before the teaching of evolution in elementary and secondary classes. The U.S. Supreme Court (by a vote of 6 to 3) later affirmed the lower court's decision that ruled unconstitutional the Tangipahoa Parish School Board's policy of requiring teachers to read aloud a disclaimer whenever they teach evolution. Creationists on the Tangipahoa Board proclaimed "the war isn't over" (see also Appendix D). However, in 2000, the United States Supreme Court declined to review the case. Justices Rehnquist, Scalia, and Thomas dissented from the majority opinion.

1999 *Time* magazine publishes an article titled "Up from the Apes" that begins by stating "[d]espite the protests of creationists . . . science has long taught that human beings are just another kind of animal." Impassioned creationists denounced the article as propaganda.

1999 In an interview with the *Sunday Telegraph*, Richard Dawkins argues "evolution should be one of the first things you learn at school . . . and what do [children] get instead? Sacred hearts and incense. Shallow, empty religion."

2000 Evolutionist and ex-Christian Michael Ruse claims that evolutionists promote evolution as "a secular religion—a full-fledged alternative to Christianity, with meaning and morality."

2000 A bill introduced into the Kentucky House of Representatives that would ban the teaching of human evolution fails to become law.

2000 A national survey reports that only 57% of science teachers consider evolution to be a unifying theme in biology, that almost half believe that there is as much evidence for creationism as for evolution, that almost half fear raising controversy by teaching evolution, and that large percentages of biology teachers are creationists. The same survey showed that 79% of the population wants creationism taught in public schools and half believe that evolution is "far from being proven scientifically."

2000 Sue, the *T. rex* discovered a decade earlier by Sue Hendrickson (Figure 68), is unveiled at The Field Museum in Chicago and starts another wave of dinosaur-mania (Figure 70).

2000 Alabama judge and politician Roy Moore (b. 1947), already famous for having defied a federal court's order to remove the Ten Commandments from the wall of his courtroom and stop opening his court sessions with prayer, denounces evolution. Moore proclaimed that the solution to educational problems is the fear of God.

2000 At a ceremony announcing the completion of the first phase of the Human Genome Project, President Bill Clinton (b. 1946) notes that "today, we are learning the language in which God created life. . . . We are gaining ever more awe for the complexity, the beauty, the wonder of God's most divine and sacred gift." The project's director, Francis Collins, noted that "it is awe-inspiring to realize that we have caught the first look at our own instruction book, previously known only to God."

2000 District Court Judge Bernard Borene dismisses *Rodney LeVake v. Independent School District #656*. LeVake had argued for the right to

Figure 70 "Sue," the most famous real dinosaur in the world, resides at The Field Museum in Chicago. Sue (FMNH PR2081) is 42-feet long, 13-feet high at the hips, and weighed almost seven tons when alive. Sue is shown here with her discoverer, Sue Hendrickson, and Sue's dog Skywalker (see also Figure 68). The unveiling of Sue at The Field Museum produced worldwide publicity, soon after which the antievolution organization Answers in Genesis announced that it was "taking the dinosaurs back" for creationists. (© *2000 The Field Museum, GN89860_1c, Photographer George Papadakis*)

teach "evidence both for and against the theory" of evolution. Borene concluded that LeVake's right to free speech did not override the right of the school to set its curriculum. Rodney LeVake filed an appeal.

2000 In *Radioisotopes and the Age of the Earth*, Larry Vardiman of ICR claims "our unified premise is that observation and theory should always be subservient to a proper understanding of the Word of God."

2000 Astrophysicist Jan Hollis detects the simple sugar glycoaldehyde ($C_2H_4O_2$) in interstellar space (glycoaldehyde can react with other molecules to form ribose, the backbone of RNA). This discovery revived interest in the claim that the building blocks of life are abundant in our galaxy. By 2008, astronomers had detected more

than 150 different molecules in interstellar space, the largest of which is cyanodecapentayne ($HC_{11}N$), a molecule that does not occur naturally on the earth.

2000 The Catholic Church expresses its "profound sorrow" for executing Giordano Bruno 400 years earlier (Figure 5).

2000 Henry Morris claims that evolutionists promote promiscuity and lust, noting that the quest to produce offspring is a main goal of organisms under Darwinism.

2000 Hundreds of believers gather in northern Kentucky to dedicate land for the future site of the multi-million dollar Creation Museum and Family Discovery Center. According to Ken Ham, the CEO of AiG, this museum will undo "the damage done . . . when Clarence Darrow put William Jennings Bryan on the stand" at the Scopes Trial. John Whitcomb, Jr., promised that if the museum ever dishonored God, God would destroy it.

2000 In *Pfeifer v. City of West Allis*, Judge J. Adelman rules that Christopher Pfeifer's exclusion from a public library for a "creation science workshop" violates the First and Fourteenth Amendments of the U.S. Constitution.

2000 John McIntosh, a science teacher at Colton High School in California, is instructed to stop using his search for Noah's Ark as a way to allegedly teach the scientific method to his students.

2000 Kansas voters repudiate the State Board of Education's removal of evolution from the state's science standards by voting all but one of the antievolution candidates out of office in the state's Republican primary. The creationists blamed their defeats on elites from Washington, NAS, and other science groups. Creationist Phillip Johnson blamed the defeats on "very heavy-handed intimidation," and a Kansas student added a factual observation with an ironic twist: "No one was there that's still alive today that actually witnessed creation or evolution. It's just what a person believes. I mean, we have no right to say what exactly is true."

2000 National Heritage Academies, a chain of state-funded charter schools, experience rapid growth. Science classes at these schools were based on creationism.

2000 Oklahoma Attorney General William "Drew" Edmondson (b. 1946) rules that the state textbook committee cannot require an evolution disclaimer in all new biology books.

2000 *Orrorin tugenensis* is discovered by Martin Pickford and Brigitte Senut (b. 1954) in the Tugen Hills of Kenya. The six-million-year-old fossil—which had apelike teeth and forelimbs and humanlike thighs apparently adapted to bipedality—was named "Millenium Ancestor" and described as the earliest known member of the human family (see also Appendix C).

2000 Students at Thomas Jefferson High School in Lafayette, Indiana, petition the school board to have "special creation" taught in their biology classes. The board told students that creationism is not part of the state's science curriculum and therefore would not be taught.

2000 Supporters of ID, including Phillip Johnson and Michael Behe (Figure 69), present a three-hour briefing before Congress on the evidence for the origin and development of life and the universe as the work of an intelligent designer. Presenters also denounced Darwinian evolutionary theory and pointed out Darwinism's alleged damage to society. Senator Sam Brownback (b. 1956; R-Kansas) compared the current controversy about evolution to the one spawned by abolitionist John Brown. That same year, Johnson claimed in *The Wedge of Truth* that "God has influenced the creation on a regular basis."

2000 The Faculty Senate at Baylor University votes 26 to 2 to dissolve its Michael Polanyi Center, an ID think-tank directed by creationist William Dembski.

2000 Several major candidates for President of the United States declare their support for teaching creationism. Candidate George Walker Bush (b. 1948) claimed "on the issue of evolution, the verdict is still out on how God created the Earth." After becoming president in 2001, Bush continued to advocate the teaching of creationism in public schools.

2000 Randy Thornhill and Craig Palmer's controversial *A Natural History of Rape* invokes natural selection to claim that rape is an adaptation (i.e., to increase reproductive fitness) or a byproduct of other adaptive behaviors (e.g., dominance or aggression),

resulting from man's "evolved machinery for obtaining a higher number of mates."

2000 The Ottawa (Canada) *Citizen* reports that a new curriculum designed to avoid controversy will ensure that most students in Ontario go through elementary and high school without being taught evolution.

2000 The Showtime Network produces and broadcasts a new version of *Inherit the Wind.*

2000 NSF reports that almost half of the general public in the United States believes that humans and dinosaurs coexisted.

2000 In *The Vital Importance of Believing in Recent Creation*, Henry Morris claims that without special creation and a young earth, God cannot be the loving, personal, omniscient God described in the Bible.

2000 The Thomas B. Fordham Foundation publishes Lawrence Lerner's *Good Science, Bad Science: Teaching Evolution in the States*, which reports that 19 states do "a weak-to-reprehensible job of handling evolution in their science standards." Twelve states' standards did not include the word *evolution*. Lerner cautioned scientists and science teachers to remember "that most Americans believe that faith in God is the surest way to appreciate the wonder and grandeur of life itself."

2001 Arkansas state Representative Jim Holt asks young-earth creationist Kent "Dr. Dino" Hovind (b. 1953) to testify as an expert before the State Agencies and Governmental Affairs Committee to support Holt's bill requiring "that when public schools refer to evolution that it be identified as an unproven theory." In 1989, Hovind founded Creation Science Evangelism in Pensacola, Florida, and two years later opened an antievolution amusement park known as Dinosaur Adventure Land ("where dinosaurs and the Bible meet"). Hovind made many curious claims, including that evolution is impossible because "whenever a farmer crossbreeds a cow he expects to get a cow, not a kitten."

2001 The Public Broadcasting Service airs the seven-part, eight-hour series *Evolution*, which, according to *The Christian Science Monitor*, "does its best not only to explain Charles Darwin's theory of the origins of material life, but to take seriously conservative Christians' religious objections to it." The series received a wide viewership and acclaim from scientists and the popular press, but was not without its critics. For example, in response to pressure from State Senator Stan Hawkins, Idaho Public Television aired "The Young Age of the Earth" (produced by Earth Science Associates of Knoxville, Tennessee). The Discovery Institute released the 154-page "Getting the Facts Straight: A Viewer's Guide to PBS's *Evolution*," which claimed "[b]y systematically ignoring the bigger picture, *Evolution* distorts the issues and misleads its viewers." To counter classroom guides posted on PBS's Web site, Gary Luskin of the Intelligent Design and Evolution Awareness (IDEA) Center released the antagonistic "Ten Questions to Ask your Students about the PBS *Evolution* Series," and in an article titled "911 Rang Again," Ken Cumming of ICR compared the terrorist bombings of September 11, 2001, to the airing of the *Evolution* series, saying "while the public now understands from President Bush that 'We're at War' with religious fanatics around the world, they don't have a clue that America is being attacked from within through its public schools by a militant religious movement called Darwinists."

2001 Geoff Stevens, a lecturer with AiG, describes a lion killing and eating a water buffalo as "this is what sin looks like."

2001 In *Moeller v. Schrenko*, the Georgia Court of Appeals rules that using a biology textbook that states creationism is not science does not violate the Establishment Clause of the First Amendment to the Constitution.

2001 Physicist Nathan Aviezer's *Fossils and Faith* uses Piltdown Man and other fraudulent scientific claims to dismiss evolution and paleontology.

2001 Louisiana state legislator Sharon Broome (b. 1956) attracts national attention by introducing legislation linking evolution and Darwin with Hitler, racism, and eugenics. References to evolution and Darwin were later removed from Broome's legislation, leaving only a condemnation of racism.

2001 In *What Evolution Is*, Ernst Mayr claims that "evolution is the most important concept in biology. There is not a single Why? question that can be answered adequately without a consideration of evolution." In the foreword of Mayr's book, UCLA biologist Jared Diamond—who described humans as a third species of chimpanzee—was even more emphatic: "Evolution is the most profound and powerful idea to have been conceived in the last two centuries."

2001 Michael Behe (Figure 69), William Dembski, and Stephen Meyer edit *Science and Evidence for Design in the Universe*, a collection of essays addressing the role of design in nature.

2001 *Nature* announces the discovery of *Kenyanthropus platyops* with the headline "The Human Family Expands."

2001 A Gallup poll shows that 45% of Americas agree with the statement "God created human beings pretty much in their present form at one time within the last 10,000 years or so." Between 1982 and 2001, five Gallup polls produced similar results, despite major advances in paleontology and evolutionary biology during the same period.

2001 The Discovery Institute launches the Web site *A Scientific Dissent from Darwinism*, which asks those holding a Ph.D. in science, engineering, mathematics, or computer science to endorse the statement "We are skeptical of claims for the ability of random mutation and natural selection to account for the complexity of life." (As of late 2008, more than 750 signatures had been collected.) At the same time, the Discovery Institute published full-page advertisements in three well-known national periodicals with the same headline, as well as the names of about 100 scientists who endorsed the claim. The NCSE responded with its *Project Steve* parody, which asked Ph.D.-level scientists named "Steve" (or cognates such as Stephanie, Esteban, and so forth)—which should be about 1% of all scientists, and which was chosen in honor of the late Stephen Jay Gould (Figure 67)—to sign a letter supporting evolution. As of September 2009, *Project Steve* had gathered more than 1,100 names.

2001 The discovery of *Sahelanthropus tchadensis*, which lived 6–7 million years ago, produces what might be the earliest hominin (see also Appendix C). The fossil was discovered in central Chad, more than 2,400 kilometers west of the East African Rift, where almost all of the other earliest hominins were found. The brain of *Sahelanthropus* was chimp-sized, but its lower jaw, brow-ridges, and teeth all linked it with later hominins. Its discovery by Chadian student Ahounta Djimdoumalbaye, and announcement the following year by French paleontologist Michel Brunet (b. 1940) and his colleagues, prompted *The Christian Science Monitor* to claim that the fossil might be "the holy grail of anthropology: the missing link between humans and their ape forbearers." The discovery was soon named Toumaï, meaning "hope of life" in the local language. The following year, the skull of Toumaï appeared on the cover of *Nature*, under the headline "THE EARLIEST KNOWN HOMINID."

2001 The Kansas Board of Education votes 7 to 3 to restore evolution to the state's science curriculum, 18 months after excising all references to the origin of humans and the age of the earth at the urging of conservative Christians. Throughout the state, where more than 25% of the biology teachers are creationists, antievolutionists vowed to continue their efforts to include creationism in the state's science classes.

2001 The Michigan House of Representatives begins considering House Bill 4382, which would require science standards to refer to evolution as an unproven theory and to give equal time to "the theory that life is the result of the purposeful, intelligent design of the creator."

2001  U.S. Senator Rick Santorum (b. 1958) of Pennsylvania proposes a Sense of the Senate Amendment to the 2001 Elementary and Secondary Education Act Authorization Bill that specifically identifies evolution as a scientific concept that "generates so much continuing controversy." The amendment (written in large part by ID-advocate Phillip Johnson) was approved 91 to 8 by the Senate but was immediately attacked by scientists and educators. Although the amendment was removed from the final bill (eventually known as the No Child Left Behind Act), antievolutionists continued to claim that there was a federal mandate to teach "alternatives" to evolution.

2002 A BBC poll ranks Charles Darwin fourth among the "Ten Greatest Britons," right behind Diana, Princess of Wales, and ahead of Shakespeare and John Lennon.

2002 British Prime Minister Tony Blair (b. 1953) tells Parliament that he is pleased that creationism is taught with evolution in some British schools.

2002 William Dembski's *No Free Lunch: Why Specified Complexity Cannot Be Purchased Without Intelligence* claims to refute evolution by proposing that natural selection cannot perform better than random chance in generating specified complexity. Although scientists rejected and ridiculed Dembski's ideas, ID proponents viewed Dembski's work as a legitimate refutation of evolution.

2002 Francis Collins, director of the Human Genome Project, notes that science cannot "tell us much about the character of God."

2002 In *Icons of Evolution: Science or Myth? Why Much of What We Teach About Evolution is Wrong,* Jonathan Wells attacks what he calls the "icons" of evolution (e.g., the Miller-Urey abiogenesis experiments, the Galápagos finches, etc.). The final chapter of Wells's book mentions "dogmatic Darwinists" at least once on almost every page. Scientists dismissed Well's book as being misleading about the nature of science and the role of the "icons" in teaching biology.

2002 Utah's Wasatch Brewery introduces its "Evolution Amber Ale." The label of the popular brew, which showed a chimp morphing into a beer-drinking human, claimed that the ale was "Darwin approved" and "Created in 27 Days, Not 7."

2002 John E. Jones III (b. 1955) is appointed by President George W. Bush to the United States District Court for the Middle District of Pennsylvania; unanimous Senate confirmation followed in 2004. The following year, Jones famously ruled in *Kitzmiller et al. v. Dover Area School District* that ID is not science and that the Dover School Board included ID in the curriculum to promote religion.

2002 Judith Hooper's *Of Moths and Men* attacks the story of evolution in the peppered moth, where dark moths increased in frequency as tree trunks were blackened and lichens on trees were killed. The book suggested that H. B. D. Kettlewell committed several methodological and statistical errors, and concluded by questioning the importance of the study as an exemplar of evolution in action. Reviews in the scientific literature were negative, questioning both Hooper's methods and conclusions.

2002 The General Assembly of the Presbyterian Church reaffirms that "there is no contradiction between an evolutionary theory of human origins and the doctrine of God as Creator." That same year, the United Church Board for Homeland Ministries noted "the assumption that the Bible contains scientific data about origins misreads a literature which emerged in a pre-scientific age."

2002 The Cobb County, Georgia, school board votes to apply "warning" stickers to biology textbooks, a decision that results in *Selman v. Cobb County School District*. The warning read as follows: "This textbook contains material on evolution. Evolution is a theory, not a fact, regarding the origin of living things. This material should be approached with an open mind, studied carefully, and critically considered." In 2005, Federal Judge Clarence Cooper ordered that the stickers be removed (see also Appendix D).

2003 The Human Genome Project ends with all of its goals accomplished. Meanwhile, a study of 127 genes on chromosome 21 confirmed "chimpanzees are our closest relative to the exclusion of other primates."

2003 When the National Park Service begins selling young-earth creationist Tom Vail's *Grand Canyon: A Different View* at its bookstore at the rim of the canyon, hundreds of scientists—including the presidents of seven geologic societies—protest, claiming that the National Park Service is promoting "the anti-science movement known as young-Earth creationism."

2003 Roy Moore, the Chief Justice of the Alabama Supreme Court, is removed from office for refusing to remove a 5,280-pound granite monument to the Ten Commandments that he had secretly installed in the central rotunda of the state judicial building. The installation was filmed by Coral Ridge Ministries, which later sold videos of the installation to pay Moore's legal expenses. The monument was removed from the building in 2004.

2003 *Darwinism, Design, and Public Education* advances Stephen Meyer's (of the Discovery Institute) "teach the controversy" strategy for undermining the teaching of evolution.

2003 In Roseville, California, attorney Larry Caldwell complains that school textbooks include no "scientific alternatives" to evolution. After the school district refused to discuss the inclusion of Caldwell's antievolution materials, Caldwell claimed in a lawsuit that the school board unconstitutionally obstructed him from promoting and discussing his educational policies. All of Caldwell's

complaints were later dismissed by District Judge Frank Damrell, Jr., in 2005 and 2007.

2003 Student Michael Spradling complains that Texas Tech biology professor Michael Dini (b. 1954)—who bases recommendations for his former students partly on their acceptance of Darwinian evolution—is using religion as a basis for discriminating against students. After the Justice Department launched an investigation, Dini changed his criteria for recommendations, but still required that students provide a scientific explanation for the origin of humans. The Justice Department later dropped its investigation.

2004 Michael Behe (Figure 69) and David Snoke publish a paper in the peer-reviewed journal *Protein Science* noting "[a]lthough many scientists assume that Darwinian processes account for the evolution of complex biochemical systems, we are skeptical."

2004 Population geneticist Luigi Luca Cavalli-Sforza's (b. 1922) *Genes, Peoples, and Languages* claims that race is a flawed concept as applied to humans. Cavalli-Sforza—who guided the Human Genome Diversity Project—compared many of the thousands of identifiable human populations in the world and concluded that there is generally greater genetic variation among individuals within a race than between individuals across races and that "the idea of race in the human species serves no purpose."

2004 American biologist Michael Zimmerman launches the ecumenical Clergy Letter Project, which rejects creationism and affirms evolution as "a foundational scientific truth." By 2009, the project had been endorsed by almost 12,000 members of the Christian clergy, and a similar Rabbi Letter had been endorsed by 426 Jewish clergy.

2004 A 1-meter tall, 18,000-year-old hominid skeleton is discovered on the Indonesian island of Flores. It was named *Homo floresiensis*, but becomes best known as "hobbit" (after J. R. R. Tolkein's [1892–1973] book *The Hobbit* [1937]). Subsequent work suggested a direct ancestry from *Homo erectus* (see also Appendix C).

2004 Philosopher Barbara Forrest coauthors with biologist Paul Gross *Creationism's Trojan Horse: The Wedge of Intelligent Design*, a thorough critique of ID. Forrest later provided strong testimony in the *Kitzmiller v. Dover* lawsuit.

2004 Lisa Westberg Peters's (b. 1951) children's book, *Our Family Tree*, irritates some parents with its portrayal of human evolution. When schools learned that Peters's book discussed human evolution, several canceled her presentations or asked her to discuss other topics.

2004 *Pneumodesmus newmani*, the oldest known land animal yet discovered, is announced. *P. newmani* is a 428-million-year-old millipede that was 1 centimeter long.

2004 Stephen Meyer, director of the pro-ID Center for Science and Culture at the Discovery Institute in Seattle, publishes an article titled "The Origin of Biological Information and the Higher Taxonomic Categories" in the peer-reviewed *Proceedings of the Biological Society of Washington*. The article, which claimed evidence for the work of an intelligent designer by reviewing existing information (no new scientific data were presented) about the Cambrian Explosion, was approved for publication solely by managing editor and ID-supporter Richard Sternberg (i.e., there was no peer review). Publishers of the journal quickly issued a statement claiming that, given the opportunity, they "would have deemed the paper inappropriate for the pages of the *Proceedings*." ID advocates cited this article as evidence that ID is a legitimate area of scientific inquiry because of its appearance in a peer-reviewed science journal.

2004 Italian Education Minister Letizia Moratti (b. 1949) bans the teaching of evolution, claiming that it will encourage materialism. Two months later, Moratti's decision was rescinded.

2004 Serbian Education Minister Ljiljana Colić, an orthodox Christian, declares evolution to be "dogmatic" and bans its teaching unless creationism is also taught. The decision was soon reversed.

2004 Studies of *Orrorin tugenensis* push back the development of bipedalism in hominids to six million years ago (see also Appendix C).

2004 Geneticist Dean Hamer (b. 1951) publishes *The God Gene: How Faith is Hardwired into Our Genes,* based on his studies of the human gene *VMAT2,* which codes for a neurotransmitter. Hamer proposed that higher degrees of spirituality are associated with particular forms of the gene and that certain individuals may be spiritual because of genetic predispositions. Hamer's proposal was criticized by both theologians and scientists, the latter pointing out that "spirituality" is a vague concept and its association with specific forms of the *VMAT2* gene was weak.

2004 When Georgia State School Superintendent Kathy Cox removes references to evolution from the state's science curriculum, Jimmy Carter (b. 1924)—the 39th President of the United States—responds that "as a Christian, a trained engineer and scientist, and a professor at Emory University, I am embarrassed by Superintendent Kathy Cox's decision to censor and distort the education of Georgia's students. . . . There is no need to teach that stars can fall out of the sky and land on a flat Earth in order to defend our religious faith."

2005 In an article titled "OK, We Give Up" in its April Fool issue, *Scientific American* apologizes for promoting evolution by noting that "[a]s editors, we had no business being persuaded by mountains of evidence . . . [or] thinking that scientists understand their fields better than, say, U.S. senators or best-selling novelists do."

2005 Larry Brooher (b. 1958), a high school biology teacher in Bristol, Virginia, is ordered to stop teaching creationism in his biology class. Brooher had taught creationism and given students his homemade textbook *Creation Battles Evolution* without a recorded complaint for 15 years.

2005 In *Selman v. Cobb County School District*, U.S. District Judge Clarence Cooper (b. 1942) orders that the "warning stickers" used in Cobb County (Georgia) be removed because they violate the Establishment Clause of the U.S. Constitution. Judge Cooper wrote that the sticker "misleads students regarding the significance and value of evolution in the scientific community for the benefit of religious alternatives" and that "the sticker targets only evolution to be approached with an open mind, carefully studied and critically considered without explaining why it is the only theory being isolated as such" (see also Appendix D).

2005 Jeanne Caldwell, the wife of Larry Caldwell, alleges in a lawsuit that the government-funded *Understanding Evolution* Web site endorses religion and therefore violates the constitutional separation of church and state. The lawsuit, *Caldwell v. Caldwell et al.* (the first defendant was Roy Caldwell of the University of California Museum of Anthropology, the host of the Web site), was dismissed the following year. Caldwell's appeals ended in 2009 when the U.S. Supreme Court declined without comment to review the case.

2005 Ken Ham of AiG announces "[w]e're putting evolutionists on notice: we're taking the dinosaurs back. They're used to teach people that there's no God, and they're used to brainwash people."

2005 In response to the actions of the Kansas State Board of Education, physicist Bobby Henderson (b. 1980) facetiously promotes the Flying Spaghetti Monster as the designer of the universe (adherents to this new religion are "pastafarians"). In an open letter to the Kansas Board, Henderson wrote "I think we can all look forward to the time when these three theories are given equal time in our science classrooms across the country, and eventually the world; one third time for Intelligent Design, one third time for Flying Spaghetti Monsterism, and one third time for logical conjecture based on overwhelming observable evidence." The following year, Henderson's book *The Gospel of the Flying Spaghetti Monster* was a best-seller.

2005 AiG's income exceeds $13 million and that of ICR exceeds $7 million. NCSE brings in $1.2 million.

2005 In an article titled "Twilight for the Enlightenment?," the editor of *Science* likens the widespread rejection of evolution in the United States to pre-Enlightenment adherence to superstition and irrational beliefs about the world.

2005 Scientists confirm the close relationship of chimps and humans by documenting that the genomes of chimpanzees and humans differ by only 4%.

2005 Claiming it "has nothing to do with creationism," Australia's Campus Crusade for Christ sends free copies of the DVD "Unlocking the Mystery of Life: Intelligent Design" to all 3,000 schools in New South Wales. The same year, the Focus on the Family Christian group sent hundreds of similar DVDs to schools in New Zealand.

2005 American biologist David Sloan Wilson (b. 1949) coedits *The Literary Animal*, a collection of essays applying Darwinian concepts to literary analysis.

2005 A Pew Research Center poll reports that 64% of Americans support teaching creationism alongside evolution in classrooms. The following year, another Pew poll reported that 26% of Americans believe that life evolved solely through natural processes.

2005 AMNH (Figure 30) opens *Darwin*, an exhibit that describes the development of Darwin's ideas through use of live organisms, videos, and some of Darwin's manuscripts and possessions. Niles Eldredge (Figure 65) was the curator responsible for the exhibit's content.

2005 After several years of asking Tulsa Zoo officials to remove references to evolution (due to conflicts with a literal interpretation of the Bible) from its displays, Tulsa resident Dan Hicks proposes that a display illustrating the biblical story of creation be erected at the zoo, and offers $3,000 to support such an effort. Heated public debate followed, during which Hicks claimed that the "majority of Oklahomans are creationists" and that including Christianity in the zoo's exhibits was an "issue of fairness." The Park and Recreation Board, with the support of mayor Bill LaFortune, eventually voted 3 to 1 to accept a proposed 3-foot by 5-foot exhibit detailing the Genesis account of creation. AiG president Ken Ham announced that "we need more people like Dan Hicks who are willing to boldly lead the battle to tell people the truth concerning the creation of the universe." After the story made national news, donations to the zoo declined and the display was reconsidered by the board. Acknowledging significant public resistance (despite one poll showing that nearly 60% of those surveyed approved of the display), the board reversed itself; only LaFortune continued to support the creation display. Hicks grudgingly accepted defeat and remarked that it was time to "have a zoo that's just about animals."

2005 The separation of AiG from Creation Ministries begins a four-year legal battle that includes lawsuits in Australia and the United States, as well as accusations of financial mismanagement, "gutter tactics," and "ruthless" business decisions. When the dispute was finally settled in 2009, *The Australian* noted that since both organizations advocate a six-day creation, the four years that it took to conclude the case "must seem an eternity." In 2005, AiG—along with the creationist organization CRS—sponsored the Creation Mega Conference at Liberty University.

⚖ **2005** The Association of Christian Schools International, Calvary Chapel Christian School (Murrieta, CA), and six students at the school file a lawsuit charging that the University of California system violates the constitutional rights of applicants from Christian schools whose high school coursework is deemed inadequate preparation for college. The group specifically objected to the university system's policy of rejecting high school biology courses that use creationism-based textbooks (e.g., *Biology: God's Living Creation*) as "inconsistent with the viewpoints and knowledge generally accepted in the scientific community." In 2008, Federal Judge S. James Otero ruled in *Association of Christian Schools International et al. v. Roman Stearns et al.* that the university system's policies for evaluating applicants are constitutional. The plaintiffs immediately appealed the case.

2005 When the Kansas State Board of Education approves a draft of state science standards that allows equal time for evolution and ID, conservative columnist-psychiatrist Charles Krauthammer (b. 1950) notes "to justify the farce that intelligent design is science, Kansas has to corrupt the very definition of science . . . thus unmistakenly implying—by fiat of definition, no less—that the supernatural is an integral part of science."

2005 NCSE's Executive Director Eugenie Scott's (b. 1945) article in *California Wild* about creationism in California, which mentions Larry Caldwell's antievolution work in Roseville, prompts Caldwell to sue Scott and NCSE for libel. Caldwell later withdrew his lawsuit without comment.

2005 Cardinal Christoph Schönborn (b. 1945) writes in *The New York Times* that "evolution in the sense of common ancestry might be true, but evolution in the neo-Darwinian sense" is not. Schönborn later praised Charles Darwin for "producing one of that history's most influential works."

2005 President George W. Bush implies that there is a genuine scientific debate about ID and evolution when he proclaims that ID be taught with evolution in public schools "so people can understand what the debate is about." Officials at the Discovery Institute commended the president for his advocacy and began using his comments to raise money for people "under attack for questioning evolution."

2005 The Toumaï skull reappears on the cover of *Nature* with the headline "The Earliest Known Hominid." Whereas a 2002 cover of *Nature* had shown the skull hovering above the Djurab Desert, this cover showed a reconstructed skull overlooking swamplands in Botswana (see also Appendix C).

 **2005**

Kitzmiller et al. v. Dover Area School District is initiated by 11 parents (including Tammy Kitzmiller) of students in the Dover (Pennsylvania) school system when the school board requires science teachers to read, in the classroom, a statement proposing ID as an alternative to biological evolution (Figure 71; see also Appendix D). The 40-day trial included extended testimony from well-known individuals on both sides of the ID debate. On December 20, Judge John Jones (Figure 72) issued his 139-page opinion: "We find that the secular purposes claimed by the Board amount to a pretext for the Board's real purpose, which was to promote religion in the public school classroom, in violation of the Establishment Clause." Jones further proposed that application of both the *Lemon* and the endorsement tests allowed direct analysis of "the seminal question of whether ID is science. We have concluded that it is not. . . . [ID] is a religious view, a mere relabelling of creationism, and not a scientific theory" and that ID has "utterly no place in a science curriculum." Jones received death threats (he was protected by agents from the U.S. Marshal's Office for a while) and was denounced by conservatives such as commentator Ann Coulter and preacher Pat Robertson, who warned Dover's residents that "if there is a disaster in your area, don't turn to

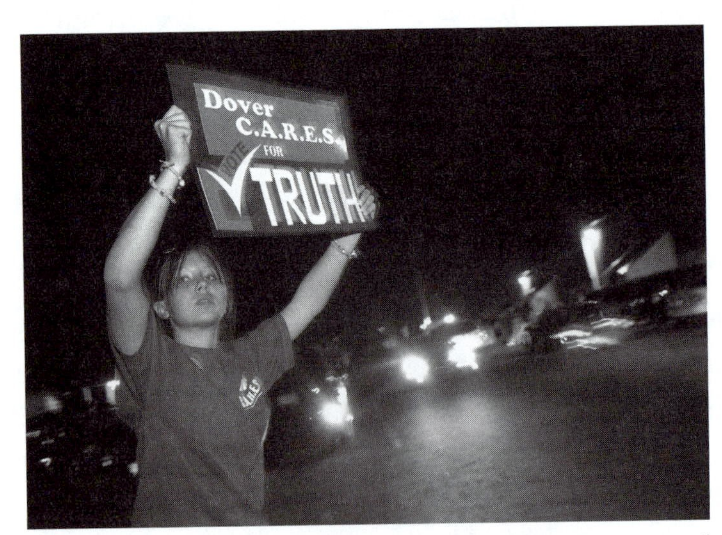

Figure 71 *Kitzmiller et al. v. Dover Area School District* generated much local and national interest. In this photo, Megan Kitzmiller—the daughter of Tammy Kitzmiller (the nominal plaintiff in *Kitzmiller*)—holds a sign supporting Dover Citizens Actively Reviewing Educational Strategies (Dover CARES). This photo was taken in Dover near a polling place on election day (November 8, 2005). (*Associated Press*)

Figure 72 Judge John Jones III (right) ruled in *Kitzmiller v. Dover* that ID is a religious view, not science. Judge Jones is shown here with Susan Epperson (middle; also see Figures 69 and 71) and her husband Jon (left). (*Susan and Jon Epperson*)

God. You just rejected him from your city, and don't wonder why he hasn't helped you when problems begin." Later, Jones—noting that the chief promoters of the Dover policy had been Republicans—lamented "I'm ashamed of some folks in the party I came from and the way they've demagogued. . . . It's pandering. It dumbs down the public."

2005 A BBC poll reports that 40% of people in the United Kingdom "think that religious alternatives to Darwin's theory of evolution should be taught as science in schools."

2005 Dutch Education Minister Maria van der Hoeven (b. 1949) suggests that dialogue between religious groups will be promoted by including ID in the curriculum. In response, a science writer in Amsterdam asked "[i]s Holland becoming the Kansas of Europe?"

2005 *Time* uses its cover to promote the "Evolution Wars" and the growing controversy about teaching ID. The article drew hundreds of 

impassioned responses. Meanwhile, in *Newsweek*, conservative columnist George Will noted "[t]he problem with intelligent-design theory is not that it is false but that it is not falsifiable. Not being susceptible to contradicting evidence, it is not a testable hypothesis. Hence it is not a scientific but a creedal tenet—a matter of faith, unsuited to a public school's science curriculum."

2006 Richard Dawkins's *The God Delusion* discusses his view that religion is a violent and destructive force. The book received widespread attention and was followed by books by other authors (including some biologists) that similarly treated religion as a pernicious influence (e.g., Sam Harris's *The End of Faith* and Christopher Hitchens's *God Is Not Great*). Simultaneously, evolutionary biologists started viewing religion as an evolved trait whose existence could be understood as conferring a benefit on individuals or groups.

2006 The anti-ID film *Flock of Dodos* receives its first public screening in Kansas, a state again in the midst of a controversy over the teaching of evolution in its public schools. Randy Olson, a marine biologist, made *Flock of Dodos* to examine the claims of the ID movement as well as the relatively weak response the scientific community had historically mounted to challenges to its authority.

2006 Young-earth creationist D. James Kennedy (1929–2007), a staunch antievolutionist and founder of Coral Ridge Ministries and Knox Theological Seminary, publishes *Darwin's Deadly Legacy*, which links evolution to Hitler (claiming Darwin's theory "fueled Hitler's ovens") and denounces evolution as "the big lie" that "resulted in the deaths of more people than have been killed in all of the wars in the history of mankind." During his multi-decade crusade against evolution, Kennedy claimed that there is a conspiracy among scientists to "repress" antievolution facts and suggested that life has no meaning to evolutionists. Kennedy produced many antievolution publications, including *The Crumbling of Evolution* (1983), *The Collapse of Evolution* (1981), and *Evolution's Bloopers and Blunders* (1986).

2006 In his memoir *Brother Astronomer*, Vatican astronomer Guy Consolmagno (b. 1952) denounces creationism as "a kind of paganism."

2006 A poll in the United Kingdom shows that 48% of respondents accept evolution, 39% accept creationism, and 13% are unsure about life's history.

2006 Ernie Fletcher (b. 1952), Republican governor of Kentucky, advocates improving science education by teaching ID. Fletcher—a Primitive Baptist who claimed he had a "thorough understanding of the science and theory of evolution"—grouped ID with "2 + 2 = 4" as "a self-evident truth" and denounced the Kentucky Academy of Science, stating that it "disappoints and astounds me that the so-called intellectual elite are so concerned about accepting self-evident truths that nearly 90% of the population understands." (Kentucky already had a law, Kentucky Revised Statute 158.177, that allowed the teaching of creationism in Kentucky schools; in 2009 that law was still on the books.) In 2006, Fletcher—the state's first Republican governor in 30 years—was indicted on misdemeanor charges that were later dismissed after a judge ruled Fletcher could not be tried while in office. The grand jury's findings claimed that Fletcher had approved a "widespread and coordinated plan" to trade state jobs for support. On November 5, 2007, Fletcher ordered the Ten Commandments be displayed alongside other historical documents in the state Capital. On the following day, Fletcher was soundly defeated in his bid for reelection.

2006 Famed biologist E. O. Wilson, writing in *USA Today*, claims that disagreements between science and fundamentalist Christianity are "unsoluble. . . . The two world views—science-based explanations and faith-based religions—cannot be reconciled."

2006 Japanese fishermen catch a bottlenose dolphin having an extra pair of hind limbs. These so-called *atavisms*—that is, the reappearance of a trait last seen in a remote ancestor but not observed since—provide evidence of evolution from a terrestrial precursor. As Charles Darwin had noted, "these characters, like those written . . . with invisible ink, lie ready to be evolved whenever the organism is disturbed."

2006 An episode of *The Simpsons* titled "The Monkey Suit" caricatures creationism as an intellectual joke. Meanwhile, when Peter Griffin—a character on *Family Guy*—was questioned about his intelligence, he learned that he is ranked above "creationist" and below "retarded."

2006 In his final book, *Some Call It Science*, Henry Morris—a giant of American fundamentalism and the most influential creationist of the 20th century—laments that "the gospel of new life in Christ

has been replaced by the Darwinian 'gospel of death,' the belief that millions of years of struggle and death has changed pond scum into people." That same year, following a series of strokes, the 87-year-old Morris died in California. To the end of his life, Morris denounced evolution while maintaining "there is no real scientific evidence that evolution is occurring at present or ever occurred in the past." Morris profoundly influenced religion in 20th-century America.

2006 Georgia becomes the first state to approve Bible classes as electives in public schools. In the first year of the program (the 2007–2008 school year), only 37 of the state's 440 high schools offered Bible study as an elective.

2006 In a report in *Science*, Jon Miller and colleagues demonstrate that public acceptance of evolution is lower in the United States than it is in all other European countries and Japan. Only 40% of adults in the United States agreed with the statement "human beings, as we know them, developed from earlier species of animals." This presents a stark contrast to the 75%–80% in agreement from Iceland, Denmark, Sweden, France, the United Kingdom, and Japan. The authors highlighted fundamentalist religious beliefs as the strongest corollary to antievolution convictions. Specifically, they cited fundamentalist Christianity as pitting the Bible against modern science.

2006 The cover of *Nature* reports the discovery of *Selam*, an infant *Australopithecus afarensis*, by a team of scientists led by Ethiopian anthropologist Zeresenay Alemseged (b. 1969). *Selam*, the most complete hominin infant thus far discovered, was found in the Dikika region of the Afar region only a few miles from where Lucy was discovered. *Selam* is often called *Dikika Baby*, *Lucy's Child*, and *Little Lucy*.

2006 Francis Collins (b. 1940), director of the National Human Genome Research Institute of the U.S. National Institutes of Health, publishes *The Language of God: A Scientist Presents Evidence for Belief* that melds science with his religious beliefs (e.g., Collins claims that DNA sequences are "God's instruction book"). Collins described the sequencing of the human genome as an "occasion of worship" and believed that God and science should not be viewed as threats to each other.

2006 Geologist Kenneth Hurst, the father of two students at California's Frazier Mountain High School, sues the local school district, claiming that a course taught by soccer coach Sharon Lemburg promoting ID and young-earth creationism as an alternative to evolution is unconstitutional. *Hurst v. Newman* (Steve Newman was a member of the El Tejon Unified School District Board of Trustees) ended two weeks later when the parties agreed that courses titled "Philosophy of Design" or "Philosophy of Intelligent Design" would be terminated within 10 days and never offered again. Evangelist Pat Robertson proclaimed the decision to be "terribly wrong" because "there are inexplicable gaps in the so-called evolution theory."

2006 Russian schoolgirl Maria Shraiber and her father sue to circumvent Russia's compulsory teaching of evolution, claiming "Darwin only presented a hypothesis that has not been proved by him or anyone else." The following year, a St. Petersburg judge dismissed the case.

2006 The conservative League of Polish Families, a partner in the government of Jaroslaw Kaczynski (b. 1949), launches an antievolution campaign, which is opposed by the Polish Academy of Sciences. In the next election, the League lost all of its government seats.

2006 In *Godless: The Church of Liberalism* (2006), conservative commentator Ann Coulter denounces evolution as irrational, godless religion "less scientifically provable than the story of Noah's ark." Coulter claimed that evolution is not science (e.g., it is no better supported than Coulter's mocking "Flatulent Raccoon Theory" for the origin of life) and represents part of the "religion" of liberalism. William Dembski was in "constant correspondence" with Coulter as she wrote the book.

2006 Michael Behe (Figure 69) claims, "a lot of scientists . . . will stand up and say, 'Sure we don't know how the complexity of the cell evolved.' But if you want to get somebody [in science] to stand up and say, 'I think the complexity of the cell points to design,' then the numbers go down to me and a couple of others." Behe believed that it is "implausible that the designer is a natural entity."

2006 The Discovery Institute announces that it has "put over $4 million towards scientific and academic research into evolution and intelligent design."

2006 While working at Ellesmere Island in the Canadian Arctic, University of Chicago paleontologist Neil Shubin discovers fossils of a Paleozoic (375-million-year-old) fish, *Tiktaalik roseae* (in the language of the local Inuit, *tiktaalik* means "large fish in a stream"). The 1.3-meter-long *Tiktaalik*, called a "fishapod," had gills, scales, and a snout (like fish), but also—like land animals—had a flexible neck, eyes atop a broad alligator-like skull, wrists, and limb-like fins having flexible elbow joints and five fingers. *Tiktaalik* is the sort of transitional fossil predicted by Darwin's theory. The discovery was not entirely accidental, for Shubin understood that to find a fossil means to find an appropriately aged rock likely to contain that fossil. Ellesmere Island has large amounts of exposed fossil-bearing rocks of red siltstone that date back to 375 million years ago, just about the time when the transition in animal life from water to land was suspected to have occurred. Moreover, 375 million years ago Ellesmere Island was a subtropical river delta (continental drift later moved it to its present location).

2006 When the creationist group Truth in Science sends antievolution packets to secondary schools in the United Kingdom, the British Centre for Science Education is founded "to oppose the tide of creationism in the United Kingdom."

2006 Alfred Russel Wallace's admirers erect a monument outside the church where Wallace was baptized and near where he was born.

2006 Young-earth creationist Kent "Dr. Dino" Hovind is convicted of 58 federal charges, including tax evasion. Hovind's wife, Jo, was convicted of 44 of the charges. Hovind had ridiculed the government's ability to sentence him to prison, claiming he could "make their lives miserable." However, at his sentencing Hovind cried and begged for leniency, proposing that his followers would pay his fines. After reminding attendees that Hovind's case "is not and has never been about religion," Judge Rodgers fined Hovind $611,954 ($604,876 in restitution to the Internal Revenue Service and $7,078 to cover the prosecution costs for Hovind's trial) and sentenced him to 10 years in federal prison. Hovind's appeal to the U.S. Court of Appeals for the 11th Circuit was denied.

2007 Islamic creationist Adnan Oktar (Figure 73) publishes the lavishly produced and visually stunning *The Atlas of Creation*, a compendium of alleged evidence for divine creation. Oktar sent copies to teachers, scientists, and others in France, the Netherlands, and

Figure 73 Adnan Oktar is an influential Islamic creationist who published the lavishly illustrated *The Atlas of Creation*. (*Harun Yahya International*)

the United States. Oktar headed the Foundation for Scientific Research (Bilim Arastirma Vakfi), an organization that originally endorsed young-earth creationism, but later discarded that perspective because it is not endorsed by the Qur'an. Oktar also denounced ID as insufficiently Islamic. In 2008, Oktar was sentenced to three years in a Turkish prison for "creating an illegal organization for personal gain."

2007 The Farnan family sues the Capistrano Unified School District (Orange County, California) and its history teacher, James Corbett, alleging that some of his comments—especially those about divine intervention, prayer, "the spaghetti monster," and creationism—are hostile to religion and an endorsement of irreligion in a public school classroom, thereby violating the First Amendment rights of their son Chad. In 2009, District Judge James Selna ruled in *C.F. v. Capistrano Unified School District* that Corbett's description of creationism as "superstitious nonsense" is constitutionally impermissible. Corbett had been listed as a defendant in John Peloza's earlier lawsuit, which claimed that Corbett had a "class-based animus against practicing Christians" and tried to "force" Peloza, through "harassment and intimidation," to teach evolution.

2007 Michael Behe's (Figure 69) *The Edge of Evolution: The Search for the Limits of Darwinism*, a follow-up to *Darwin's Black Box*, propounds

Behe's views on the limited creative power of mutation and natural selection. Like *Darwin's Black Box*, *The Edge of Evolution* was harshly criticized and dismissed by scientists.

2007

Near Petersburg, Kentucky, AiG opens a 60,000 square foot, $27 million Creation Museum filled with life-sized dinosaur models, live exhibits, and other collections that "proclaim the authority of the Bible from its very first verse" (Figure 74). The museum claimed that believing in young-earth creationism is the only true path to salvation and that all else undermines the word of God. In the museum, Methuselah—Noah's grandfather who died just before the Flood at age 969—warns visitors of God's upcoming judgment. Exhibits compare "human reason" with "God's word" and show children cavorting with dinosaurs, which were aboard Noah's Ark after being created on the sixth day of the biblical creation week. The Creation Museum, which tells visitors "that there is a Creator, and that this Creator is Jesus Christ, who is our Savior," also endorses James Ussher's claim that creation occurred in 4004 BC (Figure 8), and portrays war, death, and the pains of childbirth as the wages of primal sin. Museum visitors learn that most fossils were created by the Flood (God's worldwide

Figure 74 In 2007, Ken Ham's Answers in Genesis—an evangelical organization that promotes young-earth creationism—opened the $27 million Creation Museum in Kentucky. The museum, which shows humans cavorting with 40-foot-long dinosaurs, promotes young-earth creationism as a solution to societal and personal problems. This modest sign marks the entrance to the expansive museum and associated grounds. (*Randy Moore*)

judgment), that all animals were vegetarian before Adam's sin, and that young-earth creationism is the only view of creationism that does not destroy churches. Critics claimed that the museum equates religious faith with ignorance. The museum hosted more than 360,000 visitors during its first year of operation.

2007 ICR (Figure 64) launches *International Journal for Creation Research*, a new research journal.

2007 Pentecostal Bishop Boniface Adoyo of Kenya—the leader of the 10-million-member Evangelical Association of Kenya—leads a protest of the first public showing of Turkana Boy (a fossil of *Homo ergaster*). When Adoyo claimed that he "did not evolve from Turkana Boy or anything like it," famed fossil-hunter Richard Leakey (b. 1944), whose team had discovered Turkana Boy in 1984, responded that Adoyo is a distant relative of Turkana Boy, "whether the bishop likes it or not" (see also Appendix C).

2007 Christina Comer, the director of the science curriculum for the Texas Education Agency (TEA), forwards an e-mail announcing a talk by Barbara Forrest about the history and scientific shortcomings of ID. Less than two hours later, Comer was instructed to send a disclaimer because TEA was to "remain neutral" about evolution and creationism. Comer was later forced to resign her job.

2007 In the first of several Republican presidential-candidate debates, three of the candidates—Sam Brownback, Mike Huckabee, and Tom Tancredo—raised their hands when asked "[i]s there anyone on the stage who does not . . . believe in evolution?" Although Arizona Senator John McCain, who was eventually his party's nominee, did not raise his hand, he was quick to say that when he visits the Grand Canyon (Figure 35), "the hand of God is there also."

2007 In another example of evolution in action, the Centers for Disease Control and Prevention announces that gonorrhea (caused by the bacterium *Neisseria gonorrhaeae*) has become so resistant to fluoriquinolones (e.g., Cipro) that the drugs should no longer be used to treat it. This left only one class of antibiotics—cephalosporins—to treat the nearly 350,000 cases of the disease reported annually in the United States. Penicillin, the previous treatment for gonorrhea, was abandoned because of the evolution of antibiotic resistance in the 1980s. Similarly, methicillin-resistant *Staphylococcus aureus* (MRSA),

which kills more than 19,000 Americans per year (more than AIDS), now accounts for more than 65% of *Staphylococcus* infections, up from just 2% in 1974. Today, 5%–10% of patients admitted to hospitals acquire a bacterial infection during their stay; the probability that these infections will resist at least one antibiotic has risen steadily during the past few decades.

2007 The influential D. James Kennedy of Coral Ridge Ministries claims "Darwin's ideas, which provoked laughter and lampoons in virtually every newspaper of his own day, and is a theory for which to this day there is virtually no reliable scientific evidence, have become the cornerstone of modern humanism."

2007 In February, a newly elected Kansas State Board of Education rejects the antievolution guidelines accepted in draft form in 2005. This was the fourth major change in the state's treatment of evolution in the previous eight years.

2007 Meave Leakey (b. 1942) and her colleagues describe the remains of *Homo habilis* and *Homo erectus* a short distance apart in the same rock layer. Their discovery suggested that *H. erectus* probably did not descend from *H. habilis* (as previously thought) and that the two species coexisted in the same area for up to 500,000 years (see also Appendix C).

2007 The First Conference on Creation Geology attracts hundreds of the fundamentalist avant-garde to Cedarville, Ohio.

2007 Author and commentator William F. Buckley (1925–2008)—the patriarch of American conservatism—endorses ID.

2007 ICR (Figure 64) concludes its eight-year *Radioisotopes and the Age of the Earth* (RATE) study by claiming that accelerated radioactive decay "such as during the Genesis flood, the Fall of Adam, or early Creation week" explains the apparent conflict between the Bible-based claim that the earth is young and the science-based conclusion that the earth is ancient. ICR later added "details regarding the Flood are beginning to unfold," "scientific discoveries continue to erode Darwinism," and "there is now abundant evidence that man and dinosaurs lived at the same time."

2008 A poll reports that 58% of Canadians accept evolution, while 22% believe that God created humans in their present form within the last 10,000 years; another 20% are unsure. A similar Gallup poll reported that 50% of respondents in the United States preferred pro-evolution responses, and 44% preferred "God created human beings pretty much in their present form at one time within the last 10,000 years."

2008 Democratic presidential candidate Barack Obama states in an interview with the *York Daily Record* (Pennsylvania), "I'm a Christian. . . . I also believe our schools are there to teach worldly knowledge and science. I believe in evolution, and I believe there's a difference between science and faith."

2008 In an article titled "Mitochondria, the Missing Link Between Body and Soul: Proteomic Prospective Evidence" published in *Proteomics*, Mohamad Warda and Jin Han state the goal of their research is "to disprove the endosymbiotic hypothesis of mitochondrial evolution," which is to be "replaced in this work by a more realistic alternative." Warda and Han then purport to show that "proteomics overlapping between different forms of life are more likely to be interpreted as a reflection of a single common fingerprint initiated by a mighty creator." The paper was quickly attacked as an example of non-scientific thinking and a failure of the peer-review system. The publisher later retracted the article due to "a substantial overlap of the content of this article with previously published articles in other journals."

2008 Young-earth creationist Jerry Bergman, who was denied tenure at Bowling Green State University, publishes *Slaughter of the Dissidents: The Shocking Truth about Killing the Careers of Darwin Doubters.* Bergman claimed that biologists who embrace creationism face difficulties in their careers because of their religious views, not their competence as biologists.

2008 ICR sells its 4,000-square-foot Museum of Creation and Earth History in Santee, California, to Tom Cantor, the founder and owner of Scantibodies Laboratory, Inc. The museum, which was renamed the Creation and Earth History Museum, is now operated by the non-profit Life and Light Foundation, and continues to promote young-earth creationism.

2008 A survey published in *The Orthodox Christian Church Today* of the two largest Orthodox denominations in the United States reports that 35% of respondents want public schools to teach evolution but not creationism, and 33% want public schools to teach creationism but not evolution (another 32% are undecided). Almost equal percentages either agreed (41%) or disagreed (38%) with the statement "[e]volutionary theory is compatible with the idea of God as Creator."

2008 AiG urges people not to be swayed by scientific evidence and claims that "God's word" is undermined by pornography, abortion, homosexuality, lawlessness, "Man's opinion," and evolution. AiG later affirmed its claim that school violence is caused by the teaching of evolution.

2008 After being forced to resign for forwarding an e-mail announcing a talk about evolution and creationism, Christina Comer sues in federal district court for reinstatement and payment of her legal fees. In 2009, Comer's suit was dismissed, and she appealed the decision

2008 *Expelled: No Intelligence Allowed*, a documentary promoted by the Discovery Institute, opens nationwide and promises to "expose the frightening agenda of the 'Darwinian Machine.'" The movie claimed that conspiring scientists suppress criticisms of evolution and that evolution is responsible for societal ills such as the Holocaust. The film was a critical failure that received poor reviews; for example, *The New York Times* described it as "a conspiracy-theory rant masquerading as investigative inquiry . . . an unprincipled propaganda piece that insults believers and nonbelievers alike." *Expelled* was also a box-office flop; the movie grossed only $7.9 million and was out of theatres in only eight weeks. *Expelled*'s star, comedian Ben Stein, dismissed critics as "the self-appointed atheist elite."

2008 Officials at the Centers for Disease Control and Prevention estimate that 70% of hospital-acquired bacterial infections in the United States—which kill 90,000 Americans per year—are resistant (as a result of evolution) to at least one drug. The World Health Organization predicted that some diseases, including pneumonia, tuberculosis, and malaria, could have "no effective therapies within the next 10 years."

2008 Over the protests of professional scientific organizations such as AAAS, Louisiana Governor Bobby Jindal (b. 1971) signs into law Senate Bill 733 (Louisiana Science Education Act), which invites teachers—in the name of "academic freedom"—to hold "an open and objective discussion of scientific theories being studied, including but not limited to evolution, the origins of life, global warming, and human cloning." Critics claimed that the legislation does little but protect teachers who want to teach creationism. Several scientific organizations moved their conventions to other states.

2008 Protein from a 68-million-year-old *Tyrannosaurus rex* is shown to resemble that of chickens and ostriches, further supporting the claim that birds are the living descendents of dinosaurs.

2008 Science teacher John Freshwater of Mount Vernon, Ohio, is fired for teaching creationism, posting Bible verses in his classroom, and using an electrical device to brand crosses into students' arms. Freshwater had been teaching creationism for more than a decade. Many students protested Freshwater's firing by bringing Bibles to school.

2008 ICR's application to grant graduate degrees in Texas is unanimously rejected by the Texas Higher Education Coordinating Board. ICR president John Morris responded to the rejection by claiming that ICR has "a higher mandate (Mark 12:30)" and asked its supporters to pray for their success as they considered appealing the decision. In 2009, ICR sued to reverse the decision (*Institute for Creation Research Graduate School v. Paredes et al.*). Later in 2008, ICR launched its National Creation Science Foundation that provided money "to advance the study of origins science." All research funded by the foundation "must be conducted from a young-Earth, global Flood perspective, and investigators must abide by the biblical and creation science tenets" of ICR. The foundation's first grant was given to young-earth creationist Steve Austin.

2008 Jeff Bada and Adam Johnson report in *Science* their analyses of residue in vials from Stanley Miller's 1953 experiments on abiogenesis. Miller reported producing five amino acids in an *in vitro* simulation of the early earth; Bada and Johnson detected 22.

2008 Famed young-earth creationist John Whitcomb, Jr.—who launched the modern creationism movement with coauthor Henry Morris with *The Genesis Flood*—claims that evolution is a "flagrant example of mankind's willful suppression of the truth." According to Whitcomb, evolution is not a science "because it contradicts biblical revelation."

2008 NCSE releases the 3rd edition of *Voices for Evolution*, a compilation of statements from 176 educational, religious, scientific, and civil liberties organizations supporting the teaching of evolution.

2008 Following a flood of public complaints, the Cincinnati Zoo cancels a promotion with AiG's Creation Museum.

2008 In an interview about his decision in *Kitzmiller v. Dover*, Judge John Jones III (Figure 72; see also Appendix D) notes that Michael Behe (Figure 69)—the lead witness for the defendants—"did not distinguish himself," school board members lied under oath, and ID is "dressed-up creationism."

2008 When biologist and Anglican priest Michael Reiss (b. 1960)—the education director of the Royal Society—urges scientists to explain to students why creationism is not science and why evolution is, Britain's *The Times* reports the story as "Leading Scientist Urges Teaching of Creationism in Schools." The headline was repeated by creationists everywhere. Reiss resigned because the story "led to damage to the Society's reputation."

2008 The Creation Science Association for Mid-America, which had advised some members of the Kansas Board of Education about its science education standards in 1999, describes evolutionary biologists as "totally deluded or ignorant," "totally incompetent," "dangerous," "liars," and "unproductive leaches" who should not be allowed to vote or roam free. The Association also claimed that evolutionary biologists—who will be killed by Muslims—"have caused more misery, and killed and tortured more people, in the last 90 years than all the wars of the last 2,000 years."

2008 As the 400th anniversary of Galileo's telescope approaches, Pope Benedict XVI recasts Galileo as a man of faith who helped the faithful "contemplate with gratitude the Lord's work."

2008 Nicolai Copernicus's skull is identified by comparing DNA from his putative skull to that of strands of hair in one of Copernicus's books.

2008 In a paper titled "Biomolecule Formation by Oceanic Impacts on Early Earth" in *Nature Geoscience*, Yoshihiro Furukawa claims that "organic molecules necessary for life's origins were (created) by oceanic impacts of extraterrestrial objects."

2008 Anticipating the bicentennial of Charles Darwin's death, Rev. Malcolm Brown (the Church of England's director of mission and public affairs) urges the Church to apologize to Darwin. Although Andrew Darwin (Charles's great-great-grandson) claimed that such apologies were usually made "to make the person or organization making the apology feel better," Darwin's great-grandson Horace Barlow noted that "they buried him in Westminster Abbey, which I suppose was an apology of sorts."

2009 On the eve of Charles Darwin's 200th birthday, a Gallup poll reports that only 39% of Americans say they "believe in the theory of evolution," while a quarter say they do not believe in the theory, and another 36% do not have an opinion either way.

2009 Antievolution legislation is considered in the legislatures of nine states during the first half of 2009. Since 2001, an average of 10 antievolution/pro-creationism bills were offered each year in state legislatures, with a high of 17 appearing in 2006. Southern states were active in these efforts (both Alabama and Louisiana introduced eight bills since 2001), but the leader over this period was Michigan (nine bills). The content of these bills ranged from "teach the controversy"—aimed at promoting discussion of the supposed weaknesses of evolutionary explanations and consideration of "alternatives"—to bills urging students to "think critically" about a list of subjects (which always includes evolution). Most proposed bills were never brought to a vote.

2009 The Creation Science Association of Mid-America announces that old-earth creationism is "a position clearly not taken by Jesus or His Father."

2009 The Texas Freedom Network Education Fund reports that 94% of Texas scientists claim that so-called "weaknesses of evolution" are not valid scientific objections to evolution. Nevertheless, the Texas Board of Education adopted standards requiring students to examine "all sides of scientific evidence." The board's chairperson, dentist and young-earth creationist Don McLeroy (b. 1946), implored fellow creationists "to stand up to [the] experts" while branding scientists as "atheists," parents who want their kids to learn about evolution as "monsters," and clergy who see no conflict between religion and science as "morons." The Texas Senate later replaced McLeroy as chair of the Texas Board of Education with conservative newspaper-editor Gail Lowe.

2009 The bicentennial of Charles Darwin's birth produces a flood of books, exhibits, symposia, merchandise, collectibles (including a £2 coin), and celebrations honoring Darwin's impact on science and society. February editions of scientific journals (e.g., *Nature, Science*) and popular magazines (e.g., *National Geographic, Scientific American*) chose cover art to commemorate the bicentennial of Charles Darwin's birth (Figure 75). Famed biologist E. O. Wilson remarked, "Charles Darwin's *On the Origin of Species* can fairly be ranked as the most important book ever written."

2009 In *Darwin's Sacred Cause,* Adrian Desmond and James Moore argue that Charles Darwin's interest in evolution can be traced to his and his family's hatred of slavery. Darwin believed that evolution disproved the claim by slave owners that human races are fundamentally different.

2009 Historian Ronald L. Numbers edits the collection *Galileo Goes to Jail, and Other Myths about Science and Religion.* Several of the chapters (myths) in the book—e.g., the Theory of Organic Evolution Is Based on Circular Reasoning, Darwin and Haeckel Were Complicit in Nazi Biology, the Scopes Trial Ended in Defeat for Antievolutionism—are tied to the evolution-creationism controversy.

2009 The Concern Group for Hong Kong Science Education, which describes itself as "the most resourceful bilingual portal for the promotion of education of evolution and real sciences in Hong Kong," establishes an online petition to challenge the governmental Education Bureau's curricular statement—"in addition to Darwin's theory, students are encouraged to explore other explanations for evolution and the origins of life, to help illustrate the dynamic nature of scientific knowledge." The statement was removed.

2009 An analysis of the genome of 53 populations of people indicates that there are only three generic groups of *Homo sapiens*: African, East Asian, and European.

2009 Eighth-grade science teacher John Freshwater, facing dismissal for allegedly preaching in class, teaching creationism, and branding students with a Tesla coil, sues Mount Vernon City School District, claiming it violated his constitutional and civil rights (*Freshwater v. Mount Vernon Board of Education et al.*). Freshwater demanded reinstatement and $1,000,000 for damages.

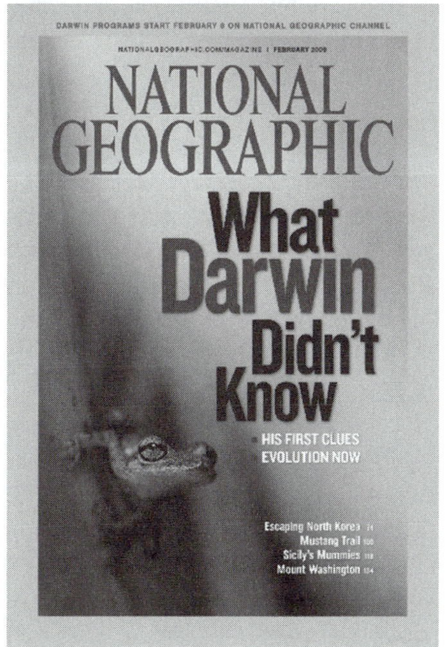

Figure 75 Among the journals paying tribute to the bicentennial of Charles Darwin's birth were (a) *The American Biology Teacher* (From *The American Biology Teacher*, Volume 71, Number 2 (2009). © Finney Creative, Inc.), (b) *The Smithsonian* (*February 2009 cover. Reprinted by permission from Smithsonian Magazine*), (c) *National Geographic* (*February 2009 cover, © 2009 National Geographic*), and (d) *Nature* (*Reprinted by permission from Macmillan Publishers Ltd:* Nature *456, 7220 [20 November 2008] Cover. Copyright 2008*). *Nature* dedicated four journal covers to Darwin between the months of October 2008 and February 2009.

2009 Svante Pääbo announces a draft genome of Neanderthals, concluding that our extinct cousins made "very little, if any" contribution to genes of modern humans (see also Appendix C).

2009 To help satisfy financial judgments associated with Kent Hovind's felony convictions in 2006, U.S. District Judge Casey Rodgers clears the way for the government's seizure of Hovind's Dinosaur Adventure Land, "a theme park and science museum that gives God the glory for His creation."

2009 Scientists claim that *Ida*, a 47-million-year-old cat-size primate found in 1983 in the Messel Shale Pit (a World Heritage Site) near Frankfurt, Germany, is a common ancestor of all later monkeys, apes, and humans. The controversial fossil, named *Darwinus masillae*, is 95% complete and has four legs, a long tail, fingertips, and nails.

2009 A survey by the Pew Research Center reports that although 97% of scientists accept that humans and other organisms have evolved over time, only 61% of the public agrees. The same poll reported that 87% of scientists and 32% of the public agree that humans have evolved.

2009 A survey conducted by the British Council's Darwin Now program reports that in 10 countries that were surveyed, there is a broad acceptance that science and faith do not have to conflict. In the United States, 53% of respondents agreed that "it is possible to believe in a God and still hold the view that life on earth, including human life, evolved over time as a result of natural selection."

2009 Researchers at NCSE repeat Lawrence Lerner's 2000 study and report, in the journal *Evolution Education and Outreach*, that although state science standards tend to cover evolution more extensively than they did a decade ago, "certain types of creationist language are becoming more common in state standards." Only 11 states had standards with an unsatisfactory treatment of evolution.

2009 Anthropologists announce the discovery of Ardi, a 4-foot-tall female *Ardipithecus ramidus* who lived in Ethiopia 4.4 million years ago. Ardi replaces Lucy as the oldest known ancestor in the human lineage, and demonstrates that hominins were bipedal a million years before Lucy.

Appendix A

Estimates of Earth's Age

Claimed Age of Earth (in Years)	Date of Claim	Advocate	Basis of Claim
1,972,949,091	120–150 BCE	Hindu priests	Religion
5,698	169	Theophilus	Biblical chronology
6,321	5th century	St. Augustine	Biblical chronology
5,918	1644	John Lightfoot	Biblical chronology
5,994	1650	James Ussher	Biblical chronology
5,983	ca. 1620	Johannes Kepler	Movement of solar apogee
>2 billion	1748	Benoît de Maillet	Decline of sea level
75,000	1774	Comte de Buffon	Cooling of Earth
38–96 million	1860	John Phillips	Sediment accumulation
20–400 million	1862	William Thomson	Cooling of Earth
10–500 million	1862	William Thomson	Cooling of Sun
100 million	1869	Thomas Huxley	Sediment accumulation
>56 million	1879	George Darwin	Orbital period of the moon
28 million	1892	Alfred Wallace	Sediment accumulation
80–90 million	1899	John Joly	Sodium accumulation in ocean
>1.3 billion	1917	Arthur Holmes	Cooling of Earth
1.6–3.9 billion	1927	Arthur Holmes	Radiometric dating
3.4 billion	1929	Ernest Rutherford	Radiometric dating
4.55 billion	1956	Claire Patterson	Radiometric dating

Based on Dalrymple, G.B. 1991. *The Age of the Earth.* Stanford, CA: Stanford University.

Appendix B

The Geological Timescale

Most of the major geological time spans were named in the 1800s by geologists who were either indifferent or hostile to evolution. Some of these names commemorate areas where rocks of a particular age—and containing distinctive fossils—were first collected and studied. For example, the Cambrian Period was named in 1835 by Adam Sedgwick while he was studying rocks from Northern Wales (*Cambria* is a Roman-Latin term for *Welsh*). Similarly, the Permian was named after the town of Perm on the western edge of the Ural Mountains, and in 1799 Alexander von Humboldt applied the name Jurassic to the predominantly limestone rocks exposed in the Jura Mountains of France and Switzerland.

The names of some other geological time spans commemorate the main types of substances deposited at a particular time. For example, the Carboniferous was named in 1822 for the coal beds and associated strata of north-central England, and the Cretaceous was named for the chalk deposited as the Cretaceous progressed (*Kreta* is Latin for *chalk*). As first noted by William Smith in 1796, different geological ages can be distinguished and recognized by their distinctive fossils.

Finally, other periods are named in a numerical scheme. For example, north of the Jura Mountains, the strata below the Jurassic consist of limestone sandwiched between sandstone and clay. In 1834, Friedrich von Alberti named this three-layered zone the Triassic.

The International Stratigraphic Chart

Era	Period and Symbol* (Year Named)	Epoch (Year Named)	Millions of Years (Ma) from Start to Present**	Major Events
Cenozoic (Cz) "Recent Life"	Quaternary (Q)*** (1829)	Holocene (1833)	0.0117	Human activities reduce biological diversity to the lowest levels since the Mesozoic Flowering plants dominate the land Rise of agriculture and civilization
		Pleistocene (1839)	1.806	Modern humans appear Continents in modern positions Repeated glaciations and lowering of sea level Apelike ancestor of humans appear in Africa
	Neogene (N) (1856)	Pliocene (1833)	5.332	Increasingly dry, cool climate First hominins appear
		Miocene (1833)	23.03 ± 0.5	
	Paleogene (Pg) (1856)	Oligocene (1854)	33.9 ± 0.1	Radiation of mammals, birds, snakes, angiosperms, teleost fishes, pollinating insects Continents near modern positions; climate increasingly cool
	Eocene		55.8 ± 0.2	
	Paleocene		66.5 ± 0.3	

The Cretaceous-Tertiary (K-T) mass extinction eliminates more than 60% of all species. This is the most famous mass extinction, not because of its magnitude (the Permian mass extinction wiped out many more species), but because of its most famous victims: dinosaurs. Crocodiles, lizards, birds, and mammals are relatively unaffected.

The International Stratigraphic Chart *(continued)*

Mesozoic (Mz) "Middle Life"	Cretaceous (K) (1822)	145.5±4.0	Increasing diversity of mammals, birds, social insects, and angiosperms Continued radiation of dinosaurs Primitive birds replace pterosaurs Breakup of Gondwana
	Jurassic (J) (1799)	199.6 ± 0.6	First birds, lizards, and angiosperms Gymnosperms dominate land Diversification of dinosaurs Mammals common but small Breakup of Pangaea into Gondwana and Laurasia
	The Triassic mass extinction eradicates about 20% of all species, thereby opening niches that allow the diversification of dinosaurs.		
	Triassic (Tr) (1834)	251.0 ± 0.4	Increasing diversity of marine life Continents begin to separate Gymnosperms rise to dominance Diversification of reptiles First dinosaurs First mammals and crocodiles
Paleozoic (Pz) "Ancient Life"	Permian (P) (1841)	299.0 ± 0.8	Glaciations and low sea level Diversification of insects, including beetles and flies Decline of amphibians Continents aggregate into Pangaea, creating the Appalachians Appearance of gymnosperms
	The Permian mass extinction kills more than 90% of all species. The Permian mass extinction is the most devastating mass extinction.		

(continued)

The International Stratigraphic Chart *(continued)*

Era	Period and Symbol* (Year Named)	Epoch (Year Named)	Millions of Years (Ma) from Start to Present**	Major Events
	Carboniferous (C) (1822)		359.2 ± 2.5	Extensive forests of early vascular plants dominate land Early orders of winged insects Appearance of reptiles Diversification of amphibians Gondwana and small northern continents form
	Devonian (D) (1839)		416.0 ± 2.8	First amphibians, seed plants, insects, ferns, tress, and ammonoids Diversification of bony fishes Early tetrapods First wingless insects
	The Devonian mass extinction eliminates as many as two-thirds of all species. Most of these extinctions are marine species; there are far fewer losses on land.			
	Silurian (S) (1835)		443.7 ± 1.5	Origin of jawed fishes Earliest vascular plants First millipedes and arthropleurids on land
	Ordovician (O) (1879)		488.3 ± 1.7	Diversification of invertebrates; arthropods and mollusks dominate the sea Early fish, corals, and echinoderms First green plants and fungi on land Ice Age at end of period

The International Stratigraphic Chart *(continued)*

	The Ordovician-Cambrian mass extinction eliminates most marine species; many groups lose more than half of their species. Biologists know relatively little about the cause and impact of this mass extinction because most animals were soft-bodied and therefore were unlikely to become fossils.		
	Cambrian (Є) (1835)	542.0 ± 1.0	First appearance of many animal phyla in the Cambrian Explosion Diversification of marine animals and algae Possible "Snowball Earth period" Abundant marine invertebrates First shelled organisms First invertebrates Trace fossils of animals Multicellular animals
Proterozoic "First Life"		2500	Earliest eukaryotes Evolution of aerobic respiration Atmosphere becomes oxygenic Diversification of bacteria Photosynthetic bacteria produce O_2
Archaean "Old"		4000	First fossils Origin of life
Hadaean (informal) "Hell"		Lower limit not defined	Oldest known rocks and minerals Formation of the Earth (ca. 4550 Ma)

* Paleontologists often refer to *faunal stages* rather than *geological periods*.
** These dates are accurate ± 1%, but the boundary dates continue to change as geologists learn more about rocks and refine their methods of radiometric dating. The dates listed here are from the *International Stratigraphic Chart of the International Commission on Stratigraphy 2004.*
*** The definitions of the Quaternary and the Pleistocene are under revision. The historic "Tertiary" includes the Paleogene and Neogene, but now has no official rank.

Appendix C

Major Species of Known Hominins

Much evidence documents the evolution of modern humans (*Homo sapiens*). Although some dates and relationships between species remain unresolved (e.g., whether Turkana Boy is *Homo erectus* or *H. ergaster*), most species are well documented; for example, there are more than 120 individuals of *A. africanus* (of which "Taung Child" is but one), 160 individuals of *A. afarensis* (of which "Lucy" is but one), 150 individuals of *Homo erectus*, 90 individuals of *Australopithecus robustus*, and 500 individuals of *H. neanderthalensis*.

Age of Hominin (Ma)*	Hominin (Nickname)	Date of Discovery	Location of Discovery	Type Specimen**
7.0–6.0	*Sahelanthropus tchadensis* (Toumaï)	2001	Toros-Menalla, Chad	Adult cranium
6.0	*Orrorin tugenensis* (Millennium Ancestor)	2001	Tugen Hills, Kenya	Adult mandible
5.8–5.2	*Ardipithecus kadabba*	1996	Middle Awash, Ethiopia	Jawbone
4.5–4.3	*Ardipithecus ramidus* (Root Ape, Ardi)	1992	Aramis, Ethiopia	Adult teeth
3.5–3.3	*Kenyanthropus platyops* (Flat-faced Man)	1998	Lake Turkana, Kenya	Adult cranium
4.2–3.9	*Australopithecus anamensis*	1965	Kanapoi, Kenya	Adult cranium

(continued)

Age of Hominin (Ma)*	Hominin (Nickname)	Date of Discovery	Location of Discovery	Type Specimen**
2.5–?	*Australopithecus garhi*	1997	Middle Awash, Ethiopia	Adult cranium
3.9–3.0	*Australopithecus afarensis* (Lucy, A.L. 288-1)	1974	Hadar, Ethiopia	Adult mandible
3.0–2.4	*Australopithecus africanus* (Taung Child, Mrs. Ples)	1924	Taung, South Africa	Immature skull
2.7–2.3	*Paranthropus aethiopicus* (Black Skull)	1985	West Turkana, Kenya	Adult mandible
2.9–1.4	*Paranthropus boisei* (*Australopithecus boisei*) (Nutcracker Man, Zinj, Dear Boy)	1959	Olduvai, Tanzania	Adult Cranium
1.9–1.4	*Paranthropus robustus* (*Australopithecus robustus*)	1938	South Africa	Adult cranium
2.2–1.6	*Homo habilis* (Handyman)	1960	Olduvai, Tanzania	Adult mandible
2.3–1.7	*Homo rudolfensis*	1972	Lake Turkana, Kenya	Adult cranium
1.8	*Homo georgicus*	2000	Dmanisi, Republic of Georgia	Adult mandible
1.9–1.5	*Homo ergaster* (Turkana Boy)	1971	Lake Turkana, South Africa	Adult mandible
1.7–0.04	*Homo erectus* (Java Man, Peking Man)	1891	Trinil, Java	Adult skullcap
0.7–0.2	*Homo heidelbergensis* (Mauer Jaw)	1907	Mauer, Germany	Adult mandible
0.3–0.03	*Homo neanderthalensis* (Neanderthal Man)***	1856	Neander Valley, Germany	Partial adult skeleton
0.07–0.01	*Homo floresiensis* (Hobbit)	2003	Flores, Indonesia	Partial adult skeleton
0.2–0.0	*Homo sapiens*			Carl Linnaeus

*Geological time is typically measured in millions of years, or mega-annums (Ma), where the word "ago" is taken as read (e.g., 65 Ma can mean 65 million years ago).

**The type specimen of a species is a particular specimen to which the species name was first properly applied. The rules governing the naming of a species are described by the *International Code of Zoological Nomenclature*.

***Around 1900, German orthography changed and the silent *h* in some words was eliminated. Thus, "Neanderthal Man" and "Neandertal Man" are both appropriate, but this change did not apply to the spelling of the specific epithet, *H. neanderthalensis*.

Appendix D

Major Legal Decisions Involving the Teaching of Evolution and Creationism

1925 *State of Tennessee v. John Thomas Scopes*

In the original "Trial of the Century," coach and substitute science-teacher John Scopes was convicted of the misdemeanor of teaching human evolution in a public school in Tennessee. Scopes's trial, which William Jennings Bryan described as "a duel to the death" between evolution and Christianity, remains the most famous event in the history of the evolution-creationism controversy. The "Scopes Monkey Trial" also provided a framework for the movie and play *Inherit the Wind.*

1927 *John Thomas Scopes v. State of Tennessee*

The Tennessee Supreme Court upheld the constitutionality of a Tennessee law forbidding the teaching of human evolution, but urged that Scopes's conviction be set aside. This decision ended the legal issues associated with the Scopes Trial, and the ban on teaching human evolution in Tennessee, Mississippi, and Arkansas remained unchallenged for more than 40 years.

289 S.W. 363 (Tenn. 1927)

1968 *Epperson v. Arkansas*

The U.S. Supreme Court struck down an Arkansas law making it illegal to teach human evolution. As a result of this decision, all

laws banning the teaching of human evolution in public schools were overturned by 1970.
393 U.S. 97 (1968)

1972

Willoughby v. Stever
The D.C. Circuit Court of Appeals ruled that government agencies such as the National Science Foundation can use tax money to disseminate scientific findings, including evolution.
Civil Action No. 1574-72 (D.D.C. August 25, 1972), aff'd mem., 504 F. 2d 271 (D.C. Cir. 1974), cert. denied, 420 U.S. 927 (1975)

1973

Wright v. Houston Independent School District
The 5th Circuit Court of Appeals ruled that (1) the teaching of evolution does not establish religion, (2) there is no legitimate state interest in protecting particular religions from scientific information "distasteful to them," and (3) the free exercise of religion is not accompanied by a right to be shielded from scientific findings incompatible with one's beliefs.
366 F. Supp. 1208 (S.D. Tex. 1972), aff'd, 486 F.2d 137 (Fifth Cir. 1973), cert. denied sub. nom. Brown v. Houston Independent School District, 417 U.S. 969 (1974)

1975

Daniel v. Waters
The 6th Circuit Court of Appeals overturned the Tennessee law requiring equal emphasis on evolution and the Genesis version of creation.
515 F.2d 485 (6th Cir. 1975)

1977

Hendren v. Campbell
The County Court in Marion, Indiana, ruled that it is unconstitutional for a public school to adopt creationism-based biology books because these books advance a specific religious point of view.
Superior Court No. 5, Marion County, Indiana, April 14, 1977

1980

Crowley v. Smithsonian Institution
The D.C. Circuit Court of Appeals ruled that the federal government can fund public exhibits that promote evolution. The government is not required to provide money to promote creationism.
636 F.2d 738 (D.C. Cir. 1980)

1981

Segraves v. State of California
The Sacramento Superior Court ruled that the California State Board of Education's Science Framework, as qualified by its

antidogmatism policy, sufficiently accommodates the views of Segraves, contrary to his claim that discussions of evolution prohibit his and his children's free exercise of religion. The state's antidogmatism policy specified that science discussions focus on "how" and that speculation about origins not be presented dogmatically.
Sacramento Superior Court #278978

1982 *McLean v. Arkansas Board of Education*
An Arkansas federal district court ruled that creation science has no scientific merit or educational value as science. Laws requiring equal time for "creation science" are unconstitutional.
529 F. Supp. 1255, (E.D. Ark. 1982)

1987 *Edwards v. Aguillard*
The U.S. Supreme Court overturned the Louisiana law requiring public schools that teach evolution to also teach "creation science," noting that such a law advances religious doctrine and therefore violates the First Amendment's establishment of religion clause.
482 U.S. 578 (1987)

1990 *Webster v. New Lenox School District #122*
The 7th Circuit Court of Appeals ruled that a teacher does not have a First Amendment right to teach creationism in a public school. A school district can ban a teacher from teaching creationism.
No. 122, 917 F.2d 1004 (Seventh Cir. 1990)

1994 *Peloza v. Capistrano Unified School District*
The 9th Circuit Court of Appeals ruled that evolution is not a religion and that a school can require a biology teacher to teach evolution.
37 F.3d 517 (Ninth Cir. 1994)

1996 *Hellend v. South Bend Community School Corporation*
The 7th Circuit Court of Appeals ruled that a school must direct a teacher to refrain from expressions of religious viewpoints (including creationism) in the classroom.
93 F.3d 327 (Seventh Cir. 1996), cert. denied, 519 U.S. 1092 (1997)

1999 *Freiler v. Tangipahoa Parish Board of Education*
The 5th Circuit Court of Appeals ruled that it is unlawful to require teachers to read aloud a disclaimer stating that the biblical view of creationism is the only concept from which students are

not to be dissuaded. Such disclaimers are "intended to protect and maintain a particular religious viewpoint."
185 F.3d 337 (Fifth Cir. 1999), cert. denied, 530 U.S. 1251 (2000)

2000 *LeVake v. Independent School District #656*
A Minnesota state court ruled that a public school teacher's right to free speech as a citizen does not permit the teacher to teach a class in a manner that circumvents the prescribed course curriculum established by the school board. Refusing to allow a teacher to teach the alleged "evidence against evolution" does not violate the free speech rights of a teacher.
625 N.W.2d 502 (MN Ct. of Appeal 2000), cert. denied, 534 U.S. 1081 (2002)

2001 *Moeller v. Schrenko*
The Georgia Court of Appeals ruled that using a biology textbook that states creationism is not science does not violate the Establishment or the Free Exercise Clauses of the Constitution.
554 S.E.2d 198 (GA Ct. of Appeal 2001)

2005 *Selman et al.v. Cobb County School District*
The U.S. District Court for the Northern District of Georgia ruled that it is unconstitutional to paste stickers claiming that, among other things, "evolution is a theory, not a fact," into science textbooks. Such stickers convey "a message of endorsement of religion" and "aid the belief of Christian fundamentalists and creationists."
No. 1:02CV2325 (N. D. Ga. Filed August 21, 2002, decided January 13, 2005)

2005 *Kitzmiller et al. v. Dover Area School District*
The U.S. District Court for the Middle District of Pennsylvania ruled that (1) "the overwhelming evidence . . . established that intelligent design (ID) is a religious view, a mere re-labeling of creationism, and not a scientific theory," and, instead, is nothing more than creationism in disguise, (2) the advocates of ID wanted to "change the ground rules of science to make room for religion," and (3) "ID is not supported by any peer-reviewed research, data, or publications." The judge also noted the "breathtaking inanity" of the school board's policy and the board's "striking ignorance" of ID and made the following point: "It is ironic that several of

[the members of the School Board], who so staunchly and proudly touted their religious convictions in public, would time and again lie to cover their tracks and disguise the real purpose behind the ID Policy."

No. 4:04CV02688 (M. D. Pa.), December 20, 2005

Glossary

adaptation An inherited trait that confers a fitness benefit by improving an organism's chances of survival and reproduction.

agnosticism The doctrine or belief that neither accepts nor rejects a god. An agnostic defers belief or disbelief in a god until there is more evidence. Some agnostics simply claim that they do not know whether god(s) exists, whereas others do not believe that the question of whether god exists can be answered.

allele A variant of a gene. Evolution is measured as changes in the allelic frequencies in a population over time.

altruism Self-sacrificing behavior that benefits others at a cost to the altruist. In biology, altruistic behavior decreases the actual or potential reproductive success of the altruist while increasing the actual or potential reproductive success of others.

Alvarez event See *K-T event*.

analogous trait A trait shared by two or more species but not present in their common ancestor. An example of an analogous trait is the streamlined shape of fish and whales; this trait is shared not from common ancestry, but instead from similar selective pressures. In *On the Origin of Species,* Charles Darwin described analogous traits as "analogical or adaptive resemblances." Also called *convergently evolved trait.* Compare with *homologous trait.*

antediluvial The period before the biblical flood described in Genesis. Compare with *diluvial.*

anthropic principle A cosmological principle stating that the universe is constrained by necessity to allow human existence. According to the *weak form* of the principle, we would not exist if conditions were not right for our existence. According to the *strong form* of the principle, conditions are right for human existence so that we will exist.

antibiotic resistance The process by which bacteria evolve, via natural selection, immunity to antibiotics.

appearance of age The claim that God created the world in six literal days relatively recently, and that this world was "mature" from its birth (i.e., it did not have to grow or develop from a simple beginning). That is, Adam and Eve were never infants, trees were created with tree rings, and geologic sediments were created with fossils already in them. Advocates of this idea have included Phillip Gosse, Henry Morris, and John Whitcomb, Jr. Also known as *ideal-time creationism.*

Archaea One of the two prokaryotic domains of life, the other being Bacteria.

Archaeopteryx The oldest and most primitive fossil bird yet discovered. *Archaeopteryx*, one of the most famous fossils in the world, lived 150 million years ago; it was about 6 inches high, 12–18 inches long, and weighed 3–7 pounds. The first *Archaeopteryx* fossil was discovered in 1861 in Upper Jurassic limestone in Germany. *Archaeopteryx* ("ancient feather"), which was named in 1863 by Richard Owen, had feathers, wings, and hollow bones like a bird, and teeth, legs, and a bony tail like a small coelurosaur (a subgroup of theropods that includes *Tyrannosaurus, Velociraptor*, and *Microraptor*).

archetype An abstract concept of a primitive form or body plan from which a group of organisms presumably developed. Anatomist Richard Owen described an archetype as a "divine idea" and "primal pattern."

argument from design The argument that order in nature is evidence that nature was created by a divine power. This argument was stated most famously by William Paley in his book *Natural Theology* (1802) and developed in the eight *Bridgewater Treatises* (1833–1834). Whereas the authors of these works stated explicitly that they sought scientific evidence from nature to demonstrate the character and goodness of God, the modern Intelligent Design (ID) movement lacks such candor. Indeed, although ID is a version of the argument from design, the modern ID movement is primarily a political and marketing tool to promote particular religious views in school curricula.

artificial selection The deliberate selection of organisms by humans, performed to emphasize desirable or useful traits.

atavism The reappearance in an organism of a trait last seen in a remote ancestor but not seen more recently. Atavisms are similar to *vestigial traits*, but occur only in rare individuals instead of entire species. Examples of atavisms include extra limbs on dolphins, teeth in birds, and tails on humans.

atheism The doctrine or belief that there is no god.

australopithecines A subfamily (Australopithecinae) consisting of a single genus (*Australopithecus*) of early African bipedal hominins who lived 2.0–4.4 million years ago. Australopithecines walked upright, had relatively small brains (less than 500 cm^3), and did not use stone tools. The most famous australopithecine is *Lucy*. Also see Appendix C.

Bacteria One of two prokarytic domains of life, the other being Archaea.

baraminology The system of "discontinuity systematics" for classifying life into groups called *baramins* having no common ancestry. Baraminology, which was devised by young-earth creationist Frank Marsh in 1941, is based on a literal reading of *kinds* in Genesis. Compare with *phylogeny*.

Batesian mimicry A type of mimicry discovered by Henry Bates' studies of butterflies in which the model has a trait (e.g., it is poisonous or distasteful) that makes it unattractive to a predator, but the mimic lacks the unattractive trait.

Beagle The British gunship on which Charles Darwin spent the years 1831–1836. During this time, Darwin sailed more than 40,000 miles, most of which were focussed on South America. Without his experiences aboard the *Beagle*, it is unlikely that Darwin would have discovered evolution by natural selection.

biblical deluge See *Noachian flood*.

biblical literalism Adherence to a literal and explicit interpretation of the Bible, especially regarding creation, Noah's flood, lifespans of biblical patriarchs, and Jesus' miracles. Also called *Biblicism*.

big bang An expansion of dense matter that marked the origin of the universe.

binomial A two-part, Latinized name of a species. For example, modern humans are members of *Homo sapiens* ("wise man"). In this binomial, *Homo* is the genus, and *sapiens* is the specific epithet. Although humans are the only extant species of *Homo*, some genera are more diverse: *Canis* includes the gray wolf (*Canis lupus*) and the coyote (*Canis latrans*). Today's classification schemes, which continue to use binomials, try to reflect evolutionary relatedness and are said to be phylogenetic. The foundation of binomial nomenclature was established in the 1753 edition of Linnaeus' *Species Plantarum*.

biogenic law The name given in 1866 by Ernst Haeckel to describe recapitulation. See *ontogeny recapitulates phylogeny*.

biogeography The study of the geographic distribution of organisms and the changes in those distributions over time. Biogeography comprises a major source of evidence for evolution.

biological species concept The definition of a species as a group of organisms that can interbreed. See *species*.

biostratigraphy A branch of stratigraphy that correlates and assigns relative ages to rock strata based on fossils present in the strata. Also see *faunal succession*.

bipedal Walking on hind limbs, especially in an upright, human manner.

Butler law Tennessee's 1925 law forbidding public-school teachers to deny the literal biblical account of man's origin and to teach in its place the evolution of man from lower orders of animals. This law (Tennessee HB 185), which was repealed in 1967, was the basis for John Scopes' misdemeanor conviction in the famous 1925 "Monkey Trial" in Dayton, Tennessee. Also see Appendix D.

Cambrian Explosion An event at the beginning of the Cambrian period lasting 40–50 million years in which there was an abrupt (on a geological scale) increase in the diversity of animal species. Many of the major extant animal phyla appear in the fossil record during the Cambrian Explosion, which is often called the "big bang of animal evolution." Also see Appendix B.

catastrophism The geological theory espoused in the 18th and 19th centuries by Georges Cuvier, Louis Agassiz, and others that the earth's geologic features have resulted primarily from sudden, widespread (worldwide), violent, or unusual catastrophes (presumably caused by capricious natural forces) that are beyond our current experiences (e.g., a worldwide flood). These catastrophes were followed by the creation of new forms of life. Compare with *uniformitarianism*.

chromosome A threadlike structure made of DNA and proteins that carries genetic information.

coevolution Reciprocal adaptation occurring between species. Coevolution has occurred between flowers and their pollinators, predators and their prey, and parasites and their hosts.

common ancestor The most recent ancestral form or species from which two species evolved.

concordism See *day-age creationism.*

continental drift The movement of continents in geologic time over the earth's surface due to *plate tectonics.*

convergently evolved trait See *analogous trait.*

co-option See *exaptation.*

creation Creationists' term for the beginning when God created life and the universe. For young-earth creationists, creation occurred 6,000–10,000 years ago as is literally described in Genesis; for old-earth creationists, creation occurred billions of years ago, and Genesis is interpreted allegorically.

creationism The religious belief in a supernatural deity or force that intervenes, or has intervened, directly in nature. Some forms of creationism (e.g., young-earth creationism) stipulate that each species (or perhaps higher taxon) was created separately and in essentially its present form rather than by natural processes such as evolution. Western civilization is largely dominated by Christian stories of creation and there are many interpretations of the Genesis story of creation (e.g., *theistic evolution, young-earth creationism, intelligent-design creationism,* and *gap creationism*).

creation science A type of creationism which attempts to use scientific methods to prove the Genesis account of creation while disproving modern theories of evolution and cosmology. Creation science was originated by George McCready Price and popularized by *The Genesis Flood: The Biblical Record and Its Scientific Implications* (1961) by John Whitcomb, Jr., and Henry Morris. In *McLean v. Arkansas Board of Education,* Judge William Overton ruled that creation science has no educational value as science.

Cretaceous system The group of stratified rocks normally below the oldest Tertiary deposits and above the Jurassic system. The Cretaceous system closed the Mesozoic era. The system, which is famous for its abundant chalk, is usually represented by a K (for *Kreide,* which is German for *chalk*). The Cretaceous ended with the *K-T event.* Also see Appendix B.

Darwinism The concept proposed by Charles Darwin that biological evolution has produced many highly adapted species through natural selection acting on hereditary variation in populations. See *natural selection.*

Darwin's finches A group of 14 species in four genera (*Geospiza, Camarhynchus, Certhidea,* and *Pinaroloxias*) of Passerine birds found on the Galápagos Islands, representatives of which were collected by Charles Darwin and others during the voyage of the *Beagle.* Darwin did not mention finches in *On the Origin of Species,* and the term "Darwin's finches" was not popular until the publication of David Lack's *Darwin's Finches* (1947). Interestingly, Galápagos mockingbirds of the genus *Nesomimus* exhibit even greater island-specific endemism than do the finches.

Darwinist Someone who espouses Darwinian evolution as the mechanism that has produced life's diversity.

day-age creationism The belief that the six days of creation described in Genesis were not ordinary 24-hour days determined by the rotation of the earth with respect to the Sun, but instead were long periods of time—even thousands or millions of years. Day-age creationism originated with the harmonizing effects of Buffon, Cuvier, and others and was endorsed by William Jennings Bryan during his questioning by Clarence Darrow at the Scopes Trial. Day-age creationism is also known as *concordism*.

deep time The extreme antiquity of the earth, as reflected by the geologic record, but in contrast to the young age of the earth based on biblical literalism. See Appendices A and B.

deism The belief that God works through fixed laws of nature. This term is usually used to describe an intellectual movement in the 17th and 18th centuries that accepted the evidence of a creator but rejected the belief in a supernatural deity who interacts with humans (i.e., that God created the universe but had nothing more to do with it).

descent with modification The phrase used by Charles Darwin to refer to the process by which natural selection favors some variations, resulting in their becoming more common in the next generation. *Descent with modification* was Darwin's term for *evolution*.

diluvial Of or relating to a flood or floods, especially the biblical flood described in Genesis.

dinosaurs A subgroup of reptiles that dominated the Mesozoic and evolved into birds. Nonavian dinosaurs were victims of the mass extinction that closed the Cretaceous.

dispensationalism The belief in historical periods having different "dispensations" (i.e., covenantal relationships) with God. Dispensationalism, a type of *premillennialism*, was first popularized by *The Scofield Reference Bible* (1909) and later by Hal Lindsey's *The Late, Great Planet Earth* (1970) and the *Left Behind* series of books coauthored by antievolutionist Tim LaHaye. Dispensationalism emphasizes prophecy and end-times theology.

DNA Deoxyribonucleic acid. DNA stores genetic information in all cellular organisms. DNA and the three-letter genetic code are examples of a *universal homology* and evidence of a common ancestry for all living organisms.

Down House The large house and 18-acre estate on Luxted Road southwest of London in which Charles Darwin and his family lived for more than 40 years. Darwin wrote all of his major books and papers at Down House. Today, Down House is a public museum.

drift See *genetic drift*.

endosymbiotic hypothesis The proposal that eukaryotic cells, including organelles such as mitochondria and chloroplasts, arose from symbiotic associations of once free-living prokaryotic cells.

The Enlightenment The historic period in Western civilization, beginning in the 18th century, that promoted reason (including science) as the primary way of understanding nature.

Establishment Clause The clause in the First Amendment to the U.S. Constitution stating: "Congress shall make no law respecting an establishment of religion."

Eukarya The domain of life that includes all eukaryotic organisms.

eugenics A social philosophy and practice which advocates the improvement of human hereditary traits through social intervention (e.g., the forced sterilization of "unfit" people is a type of "negative eugenics"). Eugenics was especially popular in the United States and Europe before World War II.

evolution Change in the genetic makeup of a population over time. Evolution occurs over successive generations, not in the lifetimes of individual organisms. Charles Darwin referred to evolution as *descent with modification.*

evolutionary developmental biology (evo-devo) A discipline of biology that compares the development of different organisms to determine ancestral relationships and the developmental mechanisms that produce evolutionary change.

exaptation A term coined in 1992 by Stephen Gould and Elisabeth Vrba to refer to a trait evolved for one function and later co-opted (and perhaps further evolved) for another. A trait that may be beneficial, but was favored by natural selection for a different function. Also called *co-option.*

extinction The permanent disappearance of a species or a group of species.

faunal succession A scientific principle originated in 1816 by William Smith based on the observation that fossils in sedimentary rock strata are arranged in a specific, reliable order that can be identified over wide areas. For example, fossilized humans are never found in the same strata as dinosaurs because the two species lived millions of years apart. The principle of faunal succession, which allows geologists to identify and date strata by the fossils within the strata, is explained by the theory of evolution. Faunal succession became the basis for *biostratigraphy.*

fitness The relative success of an organism, in an evolutionary context, measured by its ability to contribute genes to future generations.

fixity of species The perspective that species do not change through time (at least not outside the bounds of their "kind") and do not give rise to new species. A belief in the unchanging nature of species has historically been associated with the belief that each species was specially created by God.

flood geology The claim originated by George McCready Price that a biblical, worldwide flood (as described in Genesis 6–9) accounts for geological formations and the fossil record (e.g., that helpless invertebrates were buried first, while larger animals fled or floated to what became higher strata in the geologic record). Price's idea was developed by John Whitcomb, Jr., and Henry Morris in *The Genesis Flood* (1961), the book that began the modern creationism movement. Flood geology, which is also referred to as *diluvial geology* and *creation geology*, is based on biblical literalism and is a hallmark of *young-earth creationism.*

fossil A preserved remnant, impression, or trace of an organism that lived in the past. Today, there are more than 250,000 known species of fossil animals, most of which are shelled creatures that lived in shallow seas. Although Charles Darwin's ideas for *On the Origin of Species* were based almost entirely on living organisms, today fossils are a primary source of evidence for Darwin's theory. Few organisms become fossils.

framework creationism The belief first advocated in 1892 by John Davis that the seven days of creation are a poetic or literary device (i.e., a topical or symbolic framework) rather than a chronologic description for presenting creation.

Free Exercise Clause The clause accompanying the Establishment Clause of the First Amendment to the U.S. Constitution. Together, they read, "Congress shall make no law respecting an establishment of religion, or prohibiting the free exercise thereof."

fundamentalism The strict maintenance of ancient or fundamental doctrines (including those involving creation stories and other events and rules) described in documents or traditions of a religion. In the United States, fundamentalism usually refers to a form of Christianity based on a strict and literal interpretation of the Bible.

fundamentalist A person who subscribes to fundamentalism. In the United States, the term *fundamentalist* was coined by Baptist preacher Curtis Laws in 1920 to describe someone willing to "cling to the great fundamentals" and "do battle royal" for the faith. Laws' original definition was broad and required neither biblical inerrancy nor dispensationalism.

Galápagos archipelago The volcanic islands in the Pacific Ocean 620 miles west of Ecuador (the country to which they belong) visited by Charles Darwin in September 1835. Finches and other organisms collected at these islands by Darwin and others later helped Darwin formulate his theory of evolution by natural selection.

gap creationism The religious belief that there was a large time-period between the creation stories described in Genesis 1:1 and Genesis 1:2. Gap creationism posits that the creation described in Genesis 1:1 was destroyed before Genesis 1:2, when God created Adam, Eve and the rest of the world in six days. Advocates of gap creationism have included William Buckland and Thomas Chalmers. Gap creationism, a type of *old-earth creationism*, is also known as *ruin-restoration creationism* and *restitution creationism*.

gene A unit of heredity. Most genes are specific sequences of DNA (or RNA in some viruses) that contain information needed to make a protein.

gene pool All of the genes in a population.

genetic code The collection of 64 triplet sequences (codons) in messenger RNA molecules, where each triplet codes for a specific amino acid. Because only 20 different amino acids exist in proteins, the genetic code is redundant (multiple codons specify the same amino acid) but unambiguous (a specific codon only codes for a particular amino acid).

genetic drift A change in the allelic frequencies of a small population due to chance rather than to natural selection. Especially in small populations, drift can cause populations to evolve in nonadaptive ways.

genetics The science of heredity; the study of genes and their relationship to the traits of organisms.

geocentrism The claim that the earth, and not the Sun, is the center of the solar system. Compare with *heliocentrism*.

geologic time The 4.6-billion-year timescale used by scientists to describe events in the earth's history. See Appendix B.

God of the gaps The view that God's existence can be deduced from unexplained phenomena, or gaps, in scientific knowledge.

gradualism The proposition that large phenotypic differences have evolved through many slightly different intermediate steps. Also called *phyletic gradualism*. Compare with *punctuated equilibrium*.

Great Chain of Being A hierarchical system of classification in which divinely created organisms have unchangeable positions that reflect their degrees of perfection. This popular idea, which originated with the ancient Greeks, was used in the Middle Ages and Renaissance to place nature in a religious context, as well as to justify social inequities and the divine rights of kings and nobility. The influence of the Great Chain of Being persists with claims that some organisms are "higher" or "lower" than other organisms. Today, we know that evolution is not about links between steps in a hierarchal ladder or chain, and that no species is "higher" or "lower" than others. That is, cockroaches, jellyfish, and ferns are not "lower" or "older" than humans, nor are they more "primitive." *Anything* alive today has been subject to evolution by natural selection just as long as *everything* alive today has been subject to evolution by natural selection. The Great Chain of Being is also called *scala naturae.*

group selection The idea that the frequency of alleles can increase in a population because of the benefits those alleles bestow on groups, regardless of the alleles' effect on the fitness of individuals within that group.

heliocentrism The claim that the Sun, and not earth, is the center of the solar system. Compare with *geocentrism.*

hominid A member of the taxonomic family Hominidae that includes humans, gorillas, chimpanzees, and orangutans. Compare with *hominin.* Also see Appendix C.

hominin A member of the taxonomic tribe Hominini that includes the human lineage; any human relative whose last common ancestor post-dates the divergence of humans and chimpanzee lineages. Aside from modern humans, all hominins (e.g., *Sahelanthropus, Australopithecus*) are now extinct. Compare with *hominid.* Also see Appendix C.

homologous trait A similarity observed in related species that results from common ancestry. In *On the Origin of Species,* Charles Darwin referred to homologous traits as "real affinities." Compare with *analogous trait.*

humanism A philosophical perspective that embraces rationalism, rejects religious explanations, and claims that values and morality arise from rational human thought rather than from divine or supernatural matters.

ideal-time creationism See *appearance of age.*

inheritance of acquired traits The claim that traits acquired during a parent's lifetime can be passed to offspring. This was the basis for the first (albeit inaccurate) recorded scientific theory of evolution. The inheritance of acquired traits, which is associated with Jean-Baptiste Lamarck, was popular until the establishment of modern genetics. Also known as the *use-disuse theory*, *soft inheritance*, and *Lamarckism.*

intelligent-design creationism (ID) The nonscientific belief that nature is too complex to be explained only by natural causes and that the order, complexity, and design in nature are proof of an "intelligence," which is assumed by most advocates of ID to be God. Although ID—a successor to the "creation science" movement—is a version of the *argument from design*, the modern ID movement is considered by many to be primarily a political and marketing tool that recasts creationism to promote particular religious views in school curricula. In *Kitzmiller et al. v. Dover Area School District*, federal judge John E. Jones III ruled in 2005 that ID is "a mere relabeling of creationism, and not a scientific theory" (see also Appendix D).

irreducible complexity The claim that some biological systems are too complex to have evolved from simpler systems.

just-so stories Untestable explanations for adaptations. This term is derived from Rudyard Kipling's *Just-So Stories* (1902), a children's book that includes fanciful stories about how animals got their traits.

K-T event An ancient cataclysm involving at least one large meteorite that hit the earth near the present-day Yucatán Peninsula approximately 65 million years ago at the end of the Cretaceous (the third and final period of the Mesozoic era). This event triggered a mass extinction, the most famous victims of which were the nonavian dinosaurs. Also called the *Alvarez event* for its early proponents, Luis and Walter Alvarez. See Appendix B.

kin selection A type of selection that acts on indirect fitness (i.e., fitness gains from the increased reproduction of relatives).

kind In creation science, organisms whose existence began at creation and who have no common ancestors.

Lamarckism A discredited but historically influential proposal of Jean-Baptiste Lamarck claiming that changes in traits during an organism's life can be passed to the organism's offspring. Also see *inheritance of acquired traits.*

Lemon test A legal principle stating that a law must have a secular purpose, must not advance or inhibit religion, and must not lead to "excessive government entanglements" with religion. All three of these "prongs" of the Lemon test must be satisfied if a law is to be constitutional under the Establishment Clause.

LUCA Last universal common ancestor. If the evolution of all life is shown in a tree of life, the organism represented by the ancestral node (i.e, the putative forebearer of all life) is LUCA. Just as humans and chimps share a common ancestry, all modern forms of life share a common history back to the divergence that produced the three domains of life (i.e., as far back as LUCA, estimated to be 3.6–4.1 billion years ago). LUCA was not necessarily the first living organism or the only common ancestor of life, but rather the most recent one (or group of organisms).

Lucy Nickname for a partial (47 of 206 bones) female skeleton of a fossil hominin discovered by David Johanson in 1974 in Ethiopia. Lucy was 4 feet tall, lived approximately 3.2 million years ago, and was a member of *Australopithecus afarensis*. Lucy, who is formally known as "A. L. 288-1," is regarded by some paleontologists as an ancestor of all subsequent *Australopithecus* and *Homo* species. Also see Appendix C.

macroevolution A vague term used to describe major evolutionary change involving groups above the level of species.

mass extinction The collective extinction of large numbers of species in a relatively short geologic period of time. There have been several mass extinctions during the earth's history, the largest of which occurred 250 million years ago, at the end of the Permian; this mass extinction ("The Great Dying") eliminated more than 90% of the species on the earth. However, the most famous mass extinction occurred 65 million years ago at the end of the Cretaceous. That extinction, which killed two-thirds of existing species, is best known for its most famous victims: the nonavian dinosaurs. Also see Appendix B.

materialism The philosophical position that the universe consists only of measurable physical objects.

microevolution Changes in gene frequencies within species or populations.

missing link A popular term used during Charles Darwin's era to denote hypothetical organisms that linked different groups of organisms, and especially humans with anthropoid apes. Most biologists no longer use the term *missing link* because it implies that organisms are linked by a hierarchal chain or ladder, when, in fact, organisms share common ancestors. The use of *missing link* also inaccurately implies that if a certain fossil has not yet been found, then evolution cannot be valid. See *transitional form*.

Modern Synthesis The integration of Charles Darwin's theory of evolution by natural selection with other scientific disciplines—especially those involving paleontology and contemporary genetics—to produce a comprehensive theory of evolution. The Modern Synthesis was devised largely between 1937 and 1947.

modernism A movement that modified traditional beliefs in response to modern ideas, especially in the Protestant church in the late 19th and early 20th centuries. Modernism, which included the acceptance of evolution, was denounced by many religious leaders.

molecular clock The clocklike regularity of the change of a molecule (e.g., a gene) or a whole genotype over geologic time.

mutation A random change in genetic information. Mutations, which can be caused by mistakes during replication or by damage from external agents such as chemicals or radiation, produce the genetic changes that provide new genetic information. Mutations create new alleles that may—or may not—be manifest in an organism's phenotype. Heritable mutations provide the genetic variation within populations upon which natural selection acts.

natural selection An evolutionary mechanism that produces differences in survival and reproduction among organisms with different heritable traits. Natural selection is the mechanism for adaptive evolutionary change proposed by Charles Darwin in *On the Origin of Species* (1859).

natural theology Theology or knowledge of God based on observations of nature rather than on divine revelation.

naturalism The philosophical position that the universe can be explained wholly by natural processes without reference to supernatural forces.

Neanderthal A hominin that was similar to, but distinct from, modern humans. Neanderthals lived in Western Asia and Europe from 150,000 to 30,000 years ago. Neanderthal fossils were discovered in 1856 in Germany's Neander Valley (*Neander Thal* in Old German). In 1901, when German spelling was made more consistent with pronunciation, *Thal* was changed to *Tal*; this is why Neanderthals are sometimes referred to as Neandertals. Also see Appendix C.

neo-creationism The repackaging of creation science after *Edwards v. Aguillard* (1987) so as to include "alternative scientific explanations" to evolution. This repackaging was aimed at avoiding the legal problems associated with creation science.

neo-Darwinism The modern claim that natural selection, acting on randomly generated genetic variation, is the major cause of evolution. Charles Darwin was less emphatic than Alfred Russel Wallace about the preeminence of natural selection among other mechanisms of evolutionary change. Also see *Modern Synthesis*.

neo-Lamarckism The claim that responses to environmental stimuli can be inherited and transmitted by natural selection to future generations.

Noachian Flood Genesis 7 reports that "all the fountains of the great deep broken up, and the windows of heaven were opened" and "the rain was upon the earth forty days and forty nights," which ultimately meant that "every living substance was destroyed which was upon the face of the ground." Before the Flood, Noah and his family herded "two and two of all flesh" of "every beast after his kind" into the Ark; when the rain ended, organisms aboard the Ark were released and repopulated the earth. The reality of the Noachian Flood is a critical component of most young-earth creationists' perspective of the earth's history, and so-called "Flood geologists" interpret geologic formations (e.g., the Grand Canyon) as proof of the action of the biblical Flood. Also known as the *biblical deluge*.

old-earth creationism The claim that the earth is billions of years old, during which time the changes in the earth and the earth's organisms have been directed by God. Examples of old-earth creationism include *progressive creationism, gap creationism*, and *day-age creationism*. Unlike *young-earth creationists*, old-earth creationists accept the scientific evidence documenting an ancient earth and universe.

ontogeny The development of an organism over its lifetime, from zygote to death. Compare with *phylogeny*.

ontogeny recapitulates phylogeny Development repeats evolution. The idea that one can see the assemblage of evolutionary stages in embryonic development (i.e., that embryological development is a flashback to past ancestral events). Also known as the *biogenic law*.

orthogenesis The refuted hypothesis that variations in evolution follow a unilinear, innate direction and are not merely sporadic and fortuitous. Advocates of classic orthogenesis (e.g., Lamarck) argue that life has an inherent tendency to evolve in a unilinear way because of some "driving force." Also known as *progressive evolution*.

paleontology The scientific study of fossils to reconstruct the history of life.

Pangaea A supercontinent that existed 250 million years ago consisting of most or all of today's continents. The breakup of Pangaea during the late Jurassic caused the isolation (and therefore the separate evolution) of different groups of organisms from each other.

pangenesis A hereditary proposal of Charles Darwin and others in which small "pangenes" or "gemmules" are produced by all of an organism's tissues and are sent to reproductive structures, where they are incorporated into gametes. A change in the amount of a specific pangene resulting from the use or disuse of a particular structure was proposed to explain the inheritance of acquired traits.

panspermia The claim that life came to the earth from somewhere else in the universe.

phyletic gradualism See *gradualism*.

phylogenetics The study of evolutionary relationships between groups of organisms. Compare with *baraminology*.

phylogeny The evolutionary history of a species or higher taxonomic group of organisms. Compare with *ontogeny*.

Piltdown Man A fraudulent fossil consisting of a human skull and an ape's jaw, allegedly discovered in England and presented in 1912 as a genuine hominin from the early Pleistocene. Although Piltdown Man was shown in 1953 by scientists to be a hoax, many creationists continue to cite Piltdown Man as evidence that evolutionary biology is fraudulent.

plate tectonics The theory that the earth's crust consists of movable plates that can join or separate over geologic time. The movements of plates explains continental drift, earthquakes, volcanoes, mountain building, and some aspects of biogeography.

population A geographically localized group of interbreeding organisms that share a gene pool.

postmillennialism The belief, based on Revelation 20, that Christ will return to the earth after an age of Christian prosperity called the Millennium. Many postmillennialists believe that most of the Bible's end-times prophecies have already been fulfilled.

premillennialism The belief that Christ will return to the earth to inaugurate and reign over a millennial kingdom. This belief is termed *premillennialism* because it assumes that humans are now awaiting Christ. Premillennialism is based largely on a literal interpretation of Revelation 20:1–6. Most fundamentalists and conservative Christians endorse dispensational premillennialism (end-times theology). Also see *dispensationalism* and *postmillennialism.*

principle of superposition The claim by Nicolas Steno in 1669 that in undisturbed sedimentary rock, older strata are lower in the geologic column than younger strata.

progressive creationism The religious belief that the earth is billions of years old, that God created *kinds* of animals sequentially, and that the fossil record accurately represents life's history because different animals and plants appeared at different times rather than having been created all at once. Earlier forms of life are not genetically related to later ones, for the *kinds* of organisms represent separate creations by God. Progressive creationists believe that evolution occurs, but only within *kinds* of organisms.

progressive evolution See *orthogenesis.*

prophetic-day creationism The belief popularized by P. J. Wiseman's *Creation Revealed in Six Days* (1948) that the six days of creation were ordinary days during which God described the successive creation events to Moses or some other seer. Also called *revelatory creationism.*

punctuated equilibrium A concept developed by Niles Eldredge in 1971 (and popularized by Eldredge and Stephen Jay Gould in 1972) suggesting that the tempo of evolution is more sporadic than gradual. Populations evolve rapidly into new species, after which there are long periods of equilibrium with little evolutionary change. Compare with *gradualism.*

radiometric dating A common way of estimating the age of a fossil or rock by analyzing the elemental isotopes and the products of their decay within the accompanying rock.

restitution creationism See *gap creationism.*

RNA world The name given by Nobel laureate Walter Gilbert (b. 1932) to the concept that pieces of RNA having catalytic and self-replicating abilities predated protein synthesis before life appeared on the earth.

saltation Large-scale evolutionary change between successive generations generally leading to speciation. Prior to an adequate understanding of the underlying genetics of evolution, it was widely held that evolutionary novelties (including species) could arise only through major mutational changes occurring in a single generation. Integration of an understanding of inheritance and the nature of the underlying genetic variation of traits has allowed an understanding of how speciation can occur through gradual genetic change across many generations.

scala naturae See *Great Chain of Being.*

Scopes Trial The misdemeanor trial (*State of Tennessee v. John Thomas Scopes*) in July 1925, in Dayton, Tennessee, of coach and teacher John Scopes for allegedly violating the Butler Law, which banned the teaching of human evolution in Tennessee's public schools. Scopes was convicted, but when his conviction was set aside two years later by the Tennessee Supreme Court, he was not retried. The Scopes Trial is the most famous event in the history of the evolution-creationism controversy. Also see Appendix D.

secular Activities, attitudes, or other things that are not based on religion or spiritual issues.

secular humanism An outlook or philosophy that advocates human rather than religious values. The belief that humanity can be moral and achieve fulfillment without belief in any god or other supernatural idea. After U.S. Supreme Court Justice Hugo Black in *Torcaso v. Watkins* (1961) described secular humanism as a religion that does not "teach what would generally be considered a belief in the existence of god," creationists began claiming that evolution promotes the religion of secular humanism.

selection pressure Environmental forces (e.g., the scarcity of food) that result in the differential survival and reproduction of some organisms having traits that provide resistance.

sexual selection Differences in fitness as a result of differences in the ability to obtain mates.

Social Darwinism A trend in social theory which holds that Darwin's theory of evolution by natural selection can substantiate a political ideology (e.g., that ruthless egoism is the most successful policy) and critique human social institutions. Although biological evolution in its pure form is descriptive (i.e., it tells us, without judgment, what has happened to life on the earth), Social Darwinism is *prescriptive.*

sociobiology The biological study of social behavior (including that of humans), with particular reference to the adaptive features of those behaviors.

special creation The belief that all forms of life were created by God as separate, distinct species. This belief implies that species do not change through time and that there are no evolutionary relationships between different species.

speciation The evolution of new species. Speciation usually proceeds by the splitting of one lineage from a parental stock, and not by a slow transformation of large parental stocks.

species The fundamental unit of biological classification commonly defined as a group of organisms that can interbreed to produce fertile, viable offspring (i.e., the *biological species concept*). Species are considered by most biologists to be real biological entities—unlike higher levels of taxonomic organization (e.g., genera, orders, families) that are human devices to catalog biological diversity—that have arisen due to specific evolutionary histories. Although the biological species concept is used most widely, it has limited applicability to many types of organisms, especially those that are extinct or that do not reproduce sexually, and there have been dozens of species concepts devised. This has led, for example, to recent proposals that the identification of unbranching lineages—that is, groups of organisms that have a unique, shared evolutionary history—should be used to define species, although this information may be difficult to collect for many organisms.

spontaneous generation The supposed production of life from nonliving matter and without biological parentage.

Sputnik A series of Soviet artificial satellites, the first of which (*Sputnik I*, launched October 4, 1957) was the first artificial satellite to orbit the earth. *Sputnik I* ignited the "space race" within the Cold War, stimulated the United States to improve science education, and ultimately helped return evolution to the high school curriculum and biology textbooks.

strata Layers of rock.

survival of the fittest A phrase coined by philosopher Herbert Spencer in 1852 to describe the competition for survival and preeminence in a population. Although Charles Darwin later used *survival of the fittest* as a synonym for *natural selection*, it is a metaphor seldom used by modern biologists.

systema naturae The system of nature. A classification system proposed by Carolus Linnaeus in which all organisms are organized into succeedingly less inclusive groups.

taxonomy The theory and practice of naming and classifying groups of organisms, according to their similarities.

teleology The belief that things are designed for, or directed toward, some final result. According to this view, animals will have ears because they *want* to hear.

theism The belief in the existence of a god or gods, especially the belief that a god created the universe, intervenes in it, and sustains a personal relationship with its creatures.

theistic evolution The belief that God uses evolution to bring about the universe according to his plan. The hand of God was needed for the creation of the human soul. Theistic evolution, which is the position of the Catholic Church, is the view of creation taught at most mainstream Protestant seminaries.

theory In science, as opposed to common usage, a theory is a well-substantiated explanation of some aspect of the natural world that usually incorporates many confirmed observational and experimental facts. A scientific theory makes predictions consistent with what we see. A scientific theory is not a guess; on the contrary, a scientific theory is widely accepted within the scientific community (e.g., the germ theory claims that certain infectious diseases are caused by microorganisms). Scientific theories do not become facts; scientific theories *explain* facts.

trait A characteristic, condition, or property of an organism or population.

transitional form An organism having anatomical features intermediate between those of two major groups of organisms in an evolutionary sequence. Transitional forms show evolutionary sequences between lineages by having characteristics of ancestral and newer lineages. Since all populations are in evolutionary transition, a transitional form represents a particular evolutionary stage that is recognized in hindsight. *Archaeopteryx* is a transitional form between birds and dinosaurs, and *Tiktaalik* is a transitional form between land-dwelling tetrapods evolved from fish ancestors. Compare with *missing link*.

transmutation The concept of evolution—especially when leading to the production of new species—was generally referred to as "transmutation" for hundreds of years.

For example, Jean-Baptiste Lamarck referred to his proposal of evolution via inheritance of acquired characteristics as his "transmutation hypothesis," and Charles Darwin (who did not use the word *evolution* in *On the Origin of Species*) developed many of his ideas about evolution in his "transmutation notebooks."

type specimen The original specimen upon which the description of a new species is based. Examples of type specimens include Taung Child (*Australopithecus africanus*), Nutcracker Man (Zinj; *Australopithecus boisei*, OH 5), and Java Man (*Homo erectus*, Trinil 2).

unconformity The surface between two successive geologic strata representing a missing interval in the geologic record. Scotland's Siccar Point, which helped James Hutton appreciate the earth's vast age, is an unconformity.

uniformitarianism A theory suggested by James Hutton and developed by Charles Lyell summarized by the phrase "the present is the key to the past" —that is, that the earth's geologic features have developed over long periods through a variety of slow geologic processes involving common events such as rain, volcanic activity, and wind. Uniformitarianism does not imply that change occurred at a uniform rate or deny the occurrence of localized catastrophes. Indeed, much evidence attests to occasional catastrophic events in the earth's history, most famously the impact of a meteor or comet off the Yucatán Peninsula approximately 65 million years ago that led to the extinction of much Mesozoic life (including nonavian dinosaurs). Such events that depart from gradual geologic change do not necessarily support young-earth creationism or biblical catastrophes such as a worldwide flood. Compare with *catastrophism*.

universal homology A similar trait found in all organisms. Gene replication, transcription, and translation—the basis of molecular biology—are examples of universal homologies.

vestigial trait Traits of organisms that are identifiable and often characteristic of that species but which either have no apparent function or have a function different from that for which they evolved. Vestigial traits reflect the evolutionary history of a lineage where recent selective forces favored the change or loss of once-useful characteristics (adaptations). Examples of vestigial traits in humans include the coccyx and muscles that move our ears.

vitalism The claim that the origin and characteristics of life depend on a force or principle distinct from known physical and chemical forces.

wedge strategy A political and social action plan of the Discovery Institute to disredit evolution while promoting Christianity. The wedge metaphor, attributable to Phillip Johnson, involves a metal wedge (i.e., opposition to evolution) splitting a log (i.e., scientific materialism) and represents a public relations program for inserting supernaturalism in the public's understanding of science.

young-earth creationism The claim that earth is 6,000–10,000 years old, that the six days of creation described in Genesis each lasted 24 hours, and that catastrophic events such as Noah's Flood produced the Grand Canyon and other geological features. Young-earth creationists, who reject many aspects of modern biology, geology, physics, and other sciences, are biblical literalists who believe that modern organisms were created by God, and that life did not evolve. Young-earth creationism, which involves divine intervention that suspends the laws of nature, also contradicts theistic evolution, which claims that God works through natural laws.

Bibliography

Adams, Mark B., ed. 1994. *The Evolution of Theodosius Dobzhansky*. Princeton, NJ: Princeton University Press.

Allen, Garland E. 1978. *Thomas Hunt Morgan: The Man and His Science.* Princeton, NJ: Princeton University Press.

Allen, Mea. 1967. *The Hookers of Kew, 1785–1911.* London: Joseph.

Alters, Brian and Sandra Alters. 2001. *Defending Evolution in the Classroom: A Guide to the Creation/Evolution Controversy*. Sudbury, MA: Jones & Bartlett.

Alvarez, Walter. 1997. *T. Rex and the Crater of Doom.* Princeton, NJ: Princeton University Press.

Applegate, Debby. 2006. *The Most Famous Man in America: The Biography of Henry Ward Beecher.* New York: Doubleday.

Ayala, Francisco J. 2007. *Darwin's Gift to Science and Religion.* Washington, DC: Joseph Henry Press.

Ayala, Francisco J. 1997. *Genetics and the Origin of Species: From Darwin to Molecular Biology, 60 Years After Dobzhansky*. Washington, DC: National Academy of Sciences.

Baker, Catherine and James Miller, eds. 2006. *The Evolution Dialogues: Science, Christianity, and the Quest for Understanding.* Washington, DC: American Association for the Advancement of Science.

Baker, J. R. 1978. *Julian Huxley, Scientist and World Citizen, 1887–1975: A Biographical Memoir.* Paris: UNESCO.

Bates, Henry Walter. 1863. *The Naturalist on the River Amazons: The Search for Evolution.* London: John Murray.

Bannister, Robert. 1979. *Social Darwinism: Science and Myth in Anglo-American Thought.* Philadelphia: Temple University Press.

Baugh, Carl. 1999. *Why Do Men Believe Evolution Against All Odds?* Bethany, OK: Bible Belt Publishing.

Baxter, Stephen. 2004. *Ages in Chaos.* New York: Forge.
Behe, Michael. 1996. *Darwin's Black Box: The Biochemical Challenge to Evolution.* New York: Free Press.
Behe, Michael. 2007. *The Edge of Evolution.* New York: Free Press.
Berra, Tim. 1990. *Evolution and the Myth of Creationism.* Stanford, CA: Stanford University Press.
Bird, Roland T. 1985. *Bones for Barnum Brown: Adventures of a Dinosaur Hunter.* Ft. Worth, TX: Texas Christian University Press.
Bird, Wendell. 1991. *The Origin of Species Revisited.* Nashville: Thomas Nelson.
Birx, H. James. 1993. *Interpreting Evolution: Darwin and Teilhard de Chardin.* Amherst, NY: Prometheus Books.
Bjornerud, Marcia. 2005. *Reading the Rocks: The Autobiography of the Earth.* Cambridge, MA: Westview Press.
Black, Edwin. 2003. *War Against the Weak: Eugenics and America's Campaign to Create a Master Race.* New York: Four Walls Eight Windows.
Blunt, Wilfred. 2004. *Linnaeus: The Compleat Naturalist.* London: Francis Lincoln.
Bodry-Sanders, Penelope. 1998. *African Obsession: The Life and Legacy of Carl Akeley.* Jacksonville, FL: Batax Museum.
Boller, Paul, Jr. 1969. *American Thought in Transition: The Impact of Evolutionary Naturalism, 1865–1900.* Chicago: Rand McNally.
Boulter, Michael. 2009. *Darwin's Garden: Down House and the Origin of Species.* London: Constable and Robinson.
Bowden, Jean K. 1989. *John Lightfoot: His Work and Travels.* London: Bentham-Moxon Trust at the Royal Botanic Gardens, Kew.
Bowler, Peter J. 1983. *The Eclipse of Darwinism.* Baltimore: Johns Hopkins University Press.
Bowler, Peter J. 1988. *The Non-Darwinian Revolution. Reinterpreting a Historical Myth.* Baltimore: Johns Hopkins University Press.
Bowler, Peter J. 1990. *Charles Darwin: The Man and His Influence.* Cambridge, UK: Blackwell Scientific.
Bowler, Peter J. 2007. *Monkey Trials and Gorilla Sermons.* Cambridge, MA: Harvard University Press.
Bowler, Peter J. 2003. *Evolution: The History of an Idea.* 3rd ed. Berkeley: University of California Press.
Box, Joan Fisher R. 1978. *R. A. Fisher: The Life of a Scientist.* New York: Wiley.
Bramwell, Valerie and Robert M. Peck. 2008. *All in the Bones: A Biography of Benjamin Waterhouse Hawkins.* Philadelphia: The Academy of Natural Sciences of Philadelphia.
Brown, Don. 2003. *Rare Treasure: Mary Anning and Her Remarkable Discoveries.* New York: Houghton Mifflin.
Browne, Janet. 1995, 2002. *Charles Darwin: A Biography.* 2 vols. New York: Knopf.
Burkhardt, Richard W. 1995. *The Spirit of the System: Lamarck and Evolutionary Biology.* Cambridge, MA: Harvard University Press.
Campbell, John and Stephen Meyer. 2004. *Darwinism, Design and Public Education.* East Lansing, MI: Michigan State University Press.
Carlson, Elof Axel. 2004. *Mendel's Legacy: The Origin of Classical Genetics.* Cold Spring Harbor, NY: Cold Spring Harbor Laboratory Press.
Carpenter, Joel. 1997. *Revive Us Again: The Reawakening of American Fundamentalism.* New York: Oxford University.

Carroll, Sean. 2006. *Endless Forms Most Beautiful.* New York: Norton.

Carroll, Sean. 2006. *The Making of the Fittest: DNA and the Ultimate Forensic Record of Evolution*. New York: W.W. Norton.

Carroll, Sean. 2009. *Into the Jungle: Great Adventures in the Search for Evolution*. San Francisco: Pearson.

Caudill, Edward. 1989. *Darwinism in the Press: The Evolution of an Idea*. Hillsdale, NJ: L. Erlbaum.

Chambers, Paul. 2002. *Bones of Contention: The Fossil That Shook Science*. London: John Murray.

Chase, Allan. 1977. *The Legacy of Malthus.* New York: Knopf.

Clark, Harold W. 1966. *Crusader for Creation: The Life and Writings of George McCready Price*. Omaha, NE: Pacific Press.

Clark, Ronald. 1969. *J.B.S.: The Life and Work of J. B. S. Haldane*. London: Hodder and Stoughton.

Collier, Bruce, and James MacLachlan. 1999. *Charles Babbage and the Engines of Perfection*. New York: Oxford University Press.

Conkin, Paul K. 1998. *When All the Gods Trembled: Darwinism, Scopes, and American Intellectuals*. Lanham, MD: Rowman and Littlefield.

Corey, Michael. 2001. *The God Hypothesis: Discovering Design in Our "Just Right" Goldilocks Universe*. Lanham, MD: Rowman & Littlefield.

Cotkin, George. 1992. *Reluctant Modernism: American Thought and Culture, 1880–1900*. New York: Twayne.

Coyne, Jerry. 2009. *Why Evolution is True*. New York: Viking.

Dalrymple, G. Brent. 1991. *The Age of the Earth.* Stanford, CA: Stanford University Press.

Darrow, Clarence. 1932. *The Story of My Life*. New York: Charles Scribner's Sons.

Dart, Raymond. 1959. *Adventures with the Missing Link.* New York: Viking.

Darwin, Charles. 1859. *On the Origin of Species by Means of Natural Selection; Or the Preservation of Favoured Races in the Struggle for Life*. London: John Murray.

Darwin, Charles. 1871. *The Descent of Man and Selection in Relation to Sex*. London: John Murray.

Darwin, Charles. 1993. *Autobiography of Charles Darwin 1809–1882.* Edited by N. Barlow. London: Collins.

Darwin, Charles. 2002 [1835]. *The Voyage of the Beagle*. New York: Barnes & Noble edition.

Davis, Percival and Dean H. Kenyon. 1993. *Of Pandas and People: The Central Question of Biological Origins*, 2nd Edition. Dallas: Haughton Publishing.

Dawkins, Richard. 1976. *The Selfish Gene*. Oxford, UK: Oxford University Press.

Dawkins, Richard. 1982. *The Extended Phenotype: The Gene as the Unit of Selection*. San Francisco: Oxford.

Dawkins, Richard. 1986. *The Blind Watchmaker: Why the Evidence of Evolution Reveals a Universe without Design.* New York: W.W. Norton.

Dawkins, Richard. 1995. *River Out of Eden: A Darwinian View of Life*. New York: Basic Books.

De Camp, L. Sprague. 1968. *The Great Monkey Trial*. Garden City, NY: Doubleday.

Dean, Dennis R. 1999. *Gideon Mantell and the Discovery of Dinosaurs*. Cambridge: Cambridge University Press.

Degler, Carl N. 1991. *In Search of Human Nature: The Decline and Revival of Darwinism in American Social Thought*. New York: Oxford University Press.

Dembski, William. 1998. *The Design Inference*. Cambridge: Cambridge University Press.

Dembski, William. 1999. *Intelligent Design: The Bridge Between Science and Theology*. Downers Grove, IL: InterVarsity.

Dembski, William. 2002. *No Free Lunch: Why Specified Complexity Cannot be Purchased without Intelligence*. Lanham, MD: Rowman Littlefield.

Dennett, Daniel C. 1995. *Darwin's Dangerous Idea: Evolution and the Meaning of Life*. New York: Simon & Schuster.

Desmond, Adrian J. 1997. *Huxley: Evolution's High Priest*. New York: Viking Penguin.

Dobzhansky, Theodosius. 1937. *Genetics and the Origin of Species.* New York: Columbia University Press.

Dupree, A. Hunter. 1988. *Asa Gray: American Botanist, Friend of Darwin.* Baltimore: Johns Hopkins.

Eisley, Loren. 1961. *Darwin's Century: Evolution and the Men Who Discovered It.* Garden City, NY: Doubleday.

Eldredge, Niles. 1983. *The Monkey Business: A Scientist Looks at Creationism*. New York: Washington Square.

Eldredge, Niles. 2000. *The Triumph of Evolution: and the Failure of Creationism*. New York: Freeman.

Endersby, Jim. 2008. *Imperial Nature: Joseph Hooker and the Practices of Victorian Science*. Chicago: University of Chicago.

Engelman, Laura. 2001. *The BSCS Story: A History of the Biological Sciences Curriculum Study*. Colorado Springs, CO: Biological Sciences Curriculum Study.

Erwin, Douglas H. 2006. *Extinction: How Life on Earth Nearly Ended.* Princeton: Princeton University Press.

Fairbanks, Daniel J. 2007. *Relics of Eden: The Powerful Evidence of Evolution in Human DNA.* Amherst, NY: Prometheus Books.

Falwell, Jerry. 1981. *The Fundamentalist Phenomenon.* Garden City, NY: Doubleday.

Fastovsky, David E. 1996. *The Evolution and Extinction of the Dinosaurs*. Cambridge; New York: Cambridge University Press.

Fiffer, Steve. 2000. *Tyrannosaurus Sue: The Extraordinary Saga of the Largest, Most Fought Over T. rex Ever Found.* New York: W.H. Freeman.

Firstenberger, William A. 2005. *In Rare Form: A Pictorial History of Baseball Evangelist Billy Sunday*. Iowa City: University of Iowa Press.

Fisch, Menachem and Simon Schaffer, eds. 1991. *William Whewell: A Composite Portrait*. Oxford: Clarendon Press.

Fischer, Robert B. 1997. *God Did It, But How?* 2nd ed. Ipswich, MA: American Scientific Affiliation.

Forrest, Barbara & Paul R. Gross. 2004. *Creationism's Trojan Horse: The Wedge of Intelligent Design*. New York: Oxford University Press.

Fortey, Richard. 2008. *Dry Storeroom No. 1: The Secret Life of the Natural History Museum*. New York: Alfred A. Knopf.

Fortey, Richard. 2009. *Fossils: The History of Life*. New York: Sterling.

Freeman, Michael. 2004. *Victorians and the Prehistoric World.* New Haven, CT: Yale University Press.

Freeman, Scott and Jon C. Herron. 2007. *Evolutionary Analysis*, 3rd ed. Upper Saddle River, NJ: Pearson, Prentice Hall.

Furniss, Norman F. 1954. *The Fundamentalist Controversy, 1918–1931*. New Haven: Yale University Press.

Futuyma, Douglas J. 1995. *Science on Trial: The Case for Evolution*. Revised edition. New York: Pantheon Books.

Futuyma, Douglas J. 2005. *Evolution*. Sunderland, MA: Sinauer.

Gatewood Willard B., Jr. 1969. *Controversy in the Twenties: Fundamentalism, Modernism, and Evolution*. Nashville: Vanderbilt University Press.

Giberson, Karl W. 2008. *Saving Darwin: How to be a Christian and Believe in Evolution*. New York: HarperCollins.

Gillham, Nicholas W. 2001. *A Life of Sir Francis Galton*. New York: Oxford University Press.

Gillispie Charles C. 1951. *Genesis and Geology: A Study in the Relations of Scientific Thought, Natural Theology, and Social Opinion in Great Britain, 1790–1850*. New York: Harper & Row.

Ginger, Ray. 1958. *Six Days or Forever? Tennessee vs. John Thomas Scopes*. New York: Oxford University Press.

Gish, Duane T. 1972. *Evolution: The Fossils Say No!* San Diego, CA: Creation-Life.

Gish, Duane T. 1995. *Evolution: The Fossils Still Say No!* El Cajon, CA: Institute for Creation Research.

Godfrey, Laurie and Andrew Petto. 2008. *Scientists Confront Creationism*. New York: Norton.

Gonzalez, Guillermo and Jay Richards. 2004. *The Privileged Planet: Hour Our Place in the Cosmos is Designed for Discovery*. Washington, DC: Regnery.

Gould, Stephen J. 1977. *Ever Since Darwin: Reflections in Natural History*. New York: Penguin.

Gould, Stephen J. 1980. *The Panda's Thumb: More Reflections in Natural History*. New York: Norton.

Gould, Stephen J. 1981. *The Mismeasure of Man*. New York: Norton.

Gould, Stephen J. 1989. *Wonderful Life: The Burgess Shale and the Nature of History*. New York: Norton.

Gould, Stephen J. 1999. *Rocks of Ages: Science and Religion in the Fullness of Life*. New York: Ballantine.

Grafen, Alan and Mark Ridley. 2007. *Richard Dawkins*. Oxford: Oxford University Press.

Graham, Roderick. 2004. *The Great Infidel: A Life of David Hume*. Edinburgh: John Donald.

Grant, Peter R. 1999. *Ecology and Evolution of Darwin's Finches*. Princeton: Princeton University Press.

Greene, John C. 1959. *The Death of Adam: Evolution and Its Impact on Western Thought*. Ames: Iowa State University Press.

Ham, Kenneth. 1987. *The Lie: Evolution*. Green Forest, AR: Master Books.

Ham, Kenneth. 2006. *The New Answers Book*. Green Forest, AR: Master Books.

Hankins, Barry. 1996. *God's Rascal: J. Frank Norris and the Beginnings of Southern Fundamentalism*. Lexington, KY: University of Kentucky.

Hankins, Barry. 2008. *Evangelicalism and Fundamentalism: A Documentary Reader*. New York: New York University Press.

Haught, John F. 2001. *Responses to 101 Questions on God and Evolution*. New York: Paulist Press.

Hayward, James L. 1998. *The Creation/Evolution Controversy: An Annotated Bibliography*. Metuchen, NJ: Scarecrow.

Hazen, Robert M. 2005. *Genesis: The Scientific Quest for Life's Origins*. Washington, DC: Joseph Henry Press.

Healey, Edna. 2002. *Emma Darwin: The Wife of an Inspirational Genius*. London: Headline Book Publishers.

Henig, R. M. 2002. *The Monk in the Garden: The Lost and Found Genius of Gregor Mendel, the Father of Genetics*. New York: Houghton Mifflin.

Herbert, Sandra. 2005. *Charles Darwin, Geologist*. Ithaca, NY: Cornell University Press.

Himmelfarb, Gertrude. 1959. *Darwin and the Darwinian Revolution*. New York: Norton.

Hodge, Charles. 1994 [1874]. *What is Darwinism?* Grand Rapids, MI: Baker.

Hoeveler, J. David. 2007. *The Evolutionists: American Thinkers Confront Charles Darwin, 1860–1920*. New York: Rowman & Littlefield Publishers.

Hull, David L. 1973. *Darwin and His Critics: The Reception of Darwin's Theory of Evolution by the Scientific Community*. Cambridge, MA: Harvard University Press.

Humes, Edward. 2007. *Monkey Girl: Evolution, Education, Religion, and the Battle for America's Soul*. New York: Ecco.

Hunter, Cornelius G. 2003. *Darwin's Proof: The Triumph of Religion over Science*. Grand Rapids, MI: Brazos Press.

Huxley, Julian. 1963. *Evolution: The Modern Synthesis*. London: Allen and Unwin.

Irvine, William. 1955 [1963]. *Apes, Angels, and Victorians: The Story of Darwin, Huxley, and Evolution*. New York: Time Reading Program.

Isaak, Mark. 2005. *The Counter-Creationism Handbook*. Westport, CT: Greenwood Press.

Jackson, Patrick W. 2006. *The Chronologers' Quest: The Search for the Age of the Earth*. Cambridge, UK: Cambridge University Press.

Jardine, L. 2004. *The Curious Life of Robert Hooke*. New York: HarperCollins.

Johanson, Donald, Blake Edgar, and David Brill. 2006. *From Lucy to Language: Revised, Updated, and Expanded*. New York: Simon and Schuster.

Johnson, Phillip E. 1991. *Darwin on Trial*. Downers Grove, IL: InterVarsity.

Johnson, Phillip E. 1997. *Defeating Darwinism by Opening Minds*. Downers Grove, IL: InterVarsity.

Johnson, Phillip E. 2000. *The Wedge of Truth: Splitting the Foundations of Naturalism*. Downers Grove, IL: InterVarsity.

Jordonova, Ludmilla J. 1984. *Lamarck*. Oxford: Oxford University Press.

Kazin, Michael. 2007. *A Godly Hero: The Life of William Jennings Bryan*. New York: Anchor Books.

Kennedy, James Gettier. 1978. *Herbert Spencer*. Boston: Twayne Publishers.

Keynes, Randall. 2001. *Annie's Box: Charles Darwin, His Daughter, and Human Evolution*. London: Fourth Estate.

Keynes, Randall. 2003. *Fossils, Finches, and Fuegians*. Oxford: Oxford University Press.

Kitcher, Philip. 1982. *Abusing Science: The Case Against Creationism*. Cambridge, MA: MIT Press.

Koestler, Arthur. 1971. *The Case of the Midwife Toad*. London: Hutchinson.

Kofahl, Robert E. 1977. *The Handy Dandy Evolution Refuter*. San Diego, CA: Beta.

Lack, David. 1947. *Darwin's Finches*. Cambridge: Cambridge University Press.

LaFollette, Marcel. 2008. *Reframing Scopes: Journalists, Scientists, and Lost Photographs from the Trial of the Century*. Lawrence: University Press of Kansas.

Lamarck, Jean Baptiste. 1809. *Philosophie Zoologique*. Paris: Dentu.

Laporte, Leo. 2000. *George Gaylord Simpson: Paleontologist and Evolutionist*. New York: Columbia University Press.

Larson, Edward J. 1997. *Summer for the Gods: The Scopes Trial and America's Continuing Debate over Science and Religion*. New York: Basic Books.

Larson, Edward J. 2001. *Evolution's Workshop: God and Science on the Galápagos Islands*. New York: Basic Books.

Larson, Edward J. 2004. *Evolution: The Remarkable History of a Scientific Theory*. New York: Modern Library.

Lawrence, Jerome & Robert E. Lee. 1955. *Inherit the Wind*. New York: Bantam.

Le Guyader, Herve and Marjorie Grene. 2004. *Étienne Geoffroy Saint-Hilaire: A Visionary Naturalist*. Chicago: University of Chicago Press.

Lebo, Lauri. 2008. *The Devil in Dover: An Insider's Story of Dogma v. Darwin in Small-Town America*. New York: New Press.

Leeming, David and Margeret Leeming. 1994. *A Dictionary of Creation Myths*. New York: Oxford University Press.

LeMahieu, D.L. 1976. *The Mind of William Paley*. Lincoln: University of Nebraska Press.

Lienesch, Michael. 2007. *In the Beginning: Fundamentalism, The Scopes Trial and the Making of the Antievolution Movement*. Chapel Hill: University of North Carolina Press.

Lightman, Bernard. 2007. *Victorian Popularizers of Science*. Chicago: University of Chicago Press.

Lindley, David. 2004. *Degrees Kelvin: A Tale of Genius, Invention, and Tragedy*. New York: Joseph Henry.

Livingston, David N. 1987. *Darwin's Forgotten Defenders: The Encounter Between Evangelical Theology and Evolutionary Thought*. Grand Rapids, MI: Eerdmans.

Lovejoy, Arthur O. 1936. *The Great Chain of Being: A Study in the History of an Idea*. New York: Harper & Brothers.

Lubenow, Marvin. 1992. *Bones of Contention*. Grand Rapids, MI: Baker.

Lurie, Edward. 1988. *Louis Agassiz: A Life in Science*. Baltimore: Johns Hopkins University Press.

Lurquin, Paul F. and Linda Stone. 2007. *Evolution and Religious Creation Myths: How Scientists Respond*. New York: Oxford University Press.

Lyell, Charles. 1830–1833. *Principles of Geology*. John Murray: London.

Malthus, Thomas R. 1798, rev. ed. 1803. *An Essay on the Principle of Population, As It Affects the Future Improvement of Society, with Remarks on the Speculations of Mr. Godwin, M. Condorcet, and Other Writers*. London: J. Johnson.

Marsden, George M. 1980. *Fundamentalism and American Culture*. New York: Oxford University Press.

Marsh, Frank L. 1976. *Variation and Fixity in Nature*. Mountain View, CA: Pacific Press.

Mawer, Simon. 2006. *Gregor Mendel: Planting the Seeds of Genetics*. New York: Harry N. Abrams.

Mayr, Ernst. 1991. *One Long Argument: Charles Darwin and the Genesis of Modern Evolutionary Thought*. Cambridge, MA: Harvard University.

Mayr, Ernst. 2001. *What Evolution Is*. New York: Basic Books.

Mayr, Ernst and William B Provine. 1980. *The Evolutionary Synthesis: Perspectives on the Unification of Biology*. Cambridge, MA: Harvard University Press.

McCalla, Arthur. 2006. *The Creationist Debate: The Encounter Between the Bible and the Historical Mind*. New York: Continuum International.

McCoy, Roger M. 2006. *Ending in Ice: The Revolutionary Idea and Tragic Expedition of Alfred Wegener*. New York: Oxford University Press.

McIver, Tom. 1988. *Anti-evolution: An Annotated Bibliography*. Jefferson, NC: McFarland.

McGowan, Christopher. 2001. *The Dragon Seekers*. Cambridge, MA: Perseus.

Meacham, Standish. 1970. *Lord Bishop: The Life of Samuel Wilberforce*. Cambridge, MA: Harvard University Press.

Miller, Kenneth. 1999. *Finding Darwin's God: A Scientist's Search for Common Ground between God and Evolution*. New York: HarperCollins.

Mindell, David P. 2006. *The Evolving World: Evolution in Everyday Life*. Cambridge, MA: Harvard.

Moore, John A. 2002. *From Genesis to Genetics: The Case of Evolution and Creationism*. Berkeley: University of California Press.

Moore, Randy. 2002. *Evolution in the Courtroom*. Santa Barbara, CA: ABC-CLIO.

Moore, Randy and Mark D. Decker. 2008. *More than Darwin: An Encyclopedia of the People and Places of the Evolution-Creationism Controversy*. Westport, CT: Greenwood.

Moore, Randy and Janice Moore. 2006. *Evolution 101*. Westport, CT: Greenwood.

Morell, Virginia. 1995. *Ancestral Passions: The Leakey Family and the Quest for Humankind's Beginnings*. New York: Simon & Schuster.

Morris, Henry M. 1974. *Scientific Creationism*. El Cajon, CA: Master Books.

Morris, Henry M. 1975. *The Troubled Waters of Evolution*. San Diego, CA: Creation Life.

Morris, Henry M. 1982. *Evolution in Turmoil*. San Diego, CA: Creation-Life.

Morris, Henry M. 1985. *Creation and the Modern Christian*. El Cajon, CA: Master Books.

Morris, Henry M. 1989. *The Long War Against God: The History and Impact of the Creation/Evolution Conflict*. Grand Rapids, MI: Baker.

Morris, Henry M. 1993. *History of Modern Creationism*. Santee, CA: Institute for Creation Research.

Morris, Henry M. & John D. Morris. 1996. *The Modern Creation Trilogy: Scripture and Creation, Science and Creation, Society and Creation*. Green Forest, AR: Master Books.

National Academy of Sciences. 1984. *Science and Creationism: A View from the National Academy of Sciences*. Washington, DC: National Academy Press.

National Academy of Sciences. 1998. *Teaching About Evolution and the Nature of Science*. Washington, DC: National Academy Press.

National Academy of Sciences. 2008. *Science, Evolution, and Creationism*. Washington, DC: National Academy Press.

Nelkin, Dorothy. 1982. *The Creation Controversy: Science or Scripture in the Schools?* New York: Norton.

Newell, Norman. 1982. *Creation and Evolution: Myth or Reality*. New York: Columbia University Press.

Newman, Horatio Hackett. 1938. *Evolution, Genetics, and Eugenics*, 3rd edition. Chicago: University of Chicago Press.

Nichols, Peter. 2003. *Evolution's Captain: The Dark Fate of the Man Who Sailed Charles Darwin Around the World*. New York: HarperCollins.

Numbers, Ronald L. 1998. *Darwinism Comes to America*. Cambridge, MA: Harvard University Press.

Numbers, Ronald L. 2006. *The Creationists: From Scientific Creationism to Intelligent Design*. Cambridge, MA: Harvard University Press.

Numbers, Ronald L. 2007. *Science and Christianity in Pulpit and Pew*. New York: Oxford University Press.

O'Connor, Ralph. 2007. *The Earth on Show: Fossils and the Poetics of Popular Science, 1802–1856*. Chicago: University of Chicago Press.

O'Donnell, James. 2005. *Augustine: A New Biography*. New York: HarperCollins.

Osborn, H. F. 1926. *Evolution and Religion in Education: Polemics of the Fundamentalist Controversy of 1922 to 1926*. New York: Charles Scribner's Sons.

Owen, Richard. 1894. *The Life of Richard Owen*. London: John Murray.

Paley, William. 1802. *Natural Theology: Or, Evidences of the Existence and Attributes of the Deity, Collected from the Appearances of Nature*. London: R. Fauldner.

Palumbi, Stephen R. 2001. *The Evolution Explosion: How Humans Cause Rapid Evolutionary Change*. New York: Norton.

Pennock, Robert T. 1999. *Tower of Babel: The Evidence Against the New Creationism*. Boston: MIT Press.

Pennock, Robert T., ed. 2001. *Intelligent Design Creationism and its Critics: Philosophical, Theological, and Scientific Perspectives*. Cambridge, MA: MIT Press.

Perakh, M. 2003. *Unintelligent Design*. Amherst, NY: Prometheus.

Pigliucci, Massimo. 2002. *Denying Evolution: Creationism, Scientism, and the Nature of Science*. Sunderland, MA: Sinauer.

Plate, Robert. 1964. *The Dinosaur Hunters: Othniel C. Marsh and Edward D. Cope*. New York: McKay.

Pleins, J. David. 2003. *When the Great Abyss Opened: Classic and Contemporary Readings of Noah's Flood*. New York: Oxford University Press.

Price, George M. 1924. *The Phantom of Organic Evolution*. New York: Fleming H. Revell.

Price, George M. 1926. *Evolutionary Geology and the New Catastrophism*. Mountain View, CA: Pacific Press Pub. Association

Pringle, Peter. 2008. *The Murder of Nikolai Vavilov: The Story of Stalin's Persecution of One of the Great Scientists of the Twentieth Century*. New York: Simon and Schuster.

Prothero, Donald R. 2007. *Evolution: What the Fossils Say and Why It Matters*. New York: Columbia University Press.

Provine, William B. 1986. *Sewall Wright and Evolutionary Biology*. Chicago: University of Chicago Press.

Quammen, David. 2006. *The Reluctant Mr. Darwin: An Intimate Portrait of Charles Darwin and the Making of his Theory of Evolution*. New York: W.W. Norton.

Raby, Peter. 2001. *Alfred Russel Wallace: A Life*. Princeton: Princeton University Press.

Raven, Charles. 1942. *John Ray, Naturalist, His Life and Works*. London: Cambridge University Press.

Rea, Tom. 2001. *Bone Wars: The Excavation and Celebrity of Andrew Carnegie's Dinosaur*. Pittsburgh: University of Pittsburgh Press.

Reed, Roy. 1997. *Faubus: The Life and Times of an American Prodigal*. Fayetteville: University of Arkansas.

Regal, Brian. 2002. *Henry Fairfield Osborn: Race, and the Search for the Origins of Man*. Burlington, VT: Ashgate.

Ridley, Matt. 1996. *Evolution*. Cambridge, MA: Blackwell Science.

Ridley, Matt. 2006. *Francis Crick: Discoverer of the Genetic Code*. New York: Atlas Books.

Roberts, Jon H. 1988. *Darwin and the Divine in America: Protestant Intellectuals and Organic Evolution, 1859–1900*. Madison: University of Wisconsin Press.

Roger, Jacques & Sarah Lucille Bonnefoi. 1997. *Buffon: A Life in Natural History*. Ithaca, NY: Cornell University Press.

Rose, Michael. 2000. *Darwin's Spectre: Evolutionary Biology in the Modern World*. Princeton: Princeton University Press.

Rosenberg, Ellen M. 1989. *The Southern Baptists: A Subculture in Transition*. Knoxville: University of Tennessee Press.

Ross, Hugh. 1994. *Creation and Time: A Biblical and Scientific Perspective on the Creation-Date Controversy*. Colorado Springs, CO: NavPress.

Ross, Hugh. 1998. *The Genesis Question: Scientific Advances and the Accuracy of Genesis*. Colorado Springs, CO: NavPress.

Roughgarden, Joan. 2006. *Evolution and Christian Faith: Reflections of an Evolutionary Biologist*. Washington, DC: Island Press.

Rowland, Ingrid D. 2008. *Giordano Bruno: Philosopher, Heretic*. New York: Farrar, Straus & Giroux.

Rudwick, Martin J. S. 1997. *Georges Cuvier, Fossil Bones, and Geological Catastrophes*. Chicago: University of Chicago Press.

Rudwick, Martin J.S. 2005. *Bursting the Limits of Time: The Reconstruction of Geohistory in the Age of Revolution*. Chicago: University of Chicago Press.

Rudwick, Martin J.S. 2008. *Worlds Before Adam: The Reconstruction of Geohistory in the Age of Reform*. Chicago: University of Chicago.

Ruse, Michael. 1979. *The Darwinian Revolution: Science Red in Tooth and Claw*. Cambridge, MA: Harvard University Press.

Ruse, Michael. 1996. *But Is It Science?: The Philosophical Question in the Creation/Evolution Controversy*. Amherst, NY: Prometheus Books.

Ruse, Michael. 2003. *Darwin and Design: Does Evolution Have a Purpose?* Cambridge, MA: Harvard University Press.

Ruse, Michael. 2005. *The Evolution-Creation Struggle*. Cambridge, MA: Harvard University Press.

Ruse, M., and J. Travis, eds. 2009. *Evolution: The First Four Billion Years*. Cambridge, MA: Harvard University Press.

Russell, C. Allyn. 1976. *Voices of American Fundamentalism*. Philadelphia, PA: Westminster Press.

Russett, Cynthia Eagle. 1976. *Darwin in America, the Intellectual Response: 1865–1912*. San Francisco: Freeman.

Sager, Carrie, ed. 2008. *Voices for Evolution*. Berkeley, CA: National Center for Science Education.

Sapp, Jan. 2003. *Genesis: The Evolution of Biology*. New York: Oxford University Press.

Sarkar, Sahotra. 2007. *Doubting Darwin? Creationist Designs on Evolution*. Oxford: Blackwell Publishing.

Sawyer, G. J., and V. Deak. 2007. *The Last Human: A Guide to 22 Species of Extinct Humans*. New Haven: Yale University Press.

Scopes, John T., and J. Presley. 1967. *Center of the Storm: Memoirs of John T. Scopes*. New York: Holt, Rinehart & Winston.

Scott, Eugenie C. 2009. *Evolution vs. Creationism: An Introduction*, 2nd ed. Westport, CT: Greenwood Press 2009.

Scott, Eugenie and Glenn Branch, eds. 2006. *Not in Our Classrooms: Why Intelligent Design Is Wrong for Our Schools*. Boston, MA: Beacon Press.

Secord, James A. 2000. *Victorian Sensation: The Extraordinary Publication, Reception, and Secret Authorship* of Vestiges of the Natural History of Creation. Chicago: University of Chicago Press.

Semonin, Paul. 2000. *American Monster: How the Nation's First Prehistoric Creature Became a Symbol of National Identity*. New York: New York University Press.

Shanks, Niall. 2007. *God, the Devil, and Darwin: A Critique of Intelligent Design Theory*. New York: Oxford University Press.

Shermer, Michael. 2002. *In Darwin's Shadow: The Life and Science of Alfred Russel Wallace*. Oxford: Oxford University Press.

Shipley, Maynard. 1927. *The War on Modern Science: A Short History of the Fundamentalist Attacks on Evolution and Modernism*. New York: Alfred Knopf.

Shipman, Pat. 1998. *Taking Wing: Archaeopteryx and the Evolution of Bird Flight*. New York: Simon & Schuster.

Shipman, Pat. 2002. *The Man Who Found the Missing Link: Eugene Dubois and His Lifelong Quest to Prove Darwin Right*. Cambridge, MA: Harvard University Press.

Shorto, Russell. 2008. *Descartes' Bones: A Skeletal History of the Conflict Between Faith and Reason*. New York: Doubleday.

Shubin, Neil. 2008. *Your Inner Fish: A Journey Into the 3.5-Billion-Year History of the Human Body*. New York: Pantheon Books.

Simpson, George Gaylord. 1944. *Tempo and Mode in Evolution*. New York: Columbia University Press.

Slack, Gordy. 2007. *The Battle Over the Meaning of Everything: Evolution, Intelligent Design, and a School Board in Dover, PA*. San Francisco: Jossey-Bass.

Smith, Wilbur M. 1951. *A Watchman on the Wall: The Life Story of Will H. Houghton*. Grand Rapids, MI: Eerdmans.

Soyfer, Valery N. 1994. *Lysenko and the Tragedy of Soviet Science*. New Brunswick, NJ: Rutgers University Press.

Spencer, Frank. 1990. *Piltdown: A Scientific Forgery*. London: British Museum (Natural History).

Stamos, David. 2009. *Evolution and the Big Questions*. Malden, MA: Blackwell Publishing.

Stone, Irving. 1941. *Clarence Darrow for the Defense*. New York: Doubleday, Doran & Company.

Stoner, Don. 1997. *A New Look at an Old Earth: Resolving the Conflict Between the Bible and Science*. Eugene, OR: Harvest House.

Strahler, Arthur. 1999. *Science and Earth History: The Evolution/Creation Controversy*. Buffalo, NY: Prometheus Press.

Sutton, Matthew. 2007. *Aimee Semple McPherson and the Resurrection of Christian America*. Cambridge, MA: Harvard University Press.

Tattersall, Ian. 2008. *The World from Beginnings to 4000 BCE*. New York: Oxford University Press.

Tattersall, Ian. 2009. *The Fossil Trail: How We Know What We Think We Know about Human Evolution*. Second ed. New York: Oxford University Press.

Teachout, Terry. 2003. *The Skeptic: A Life of H. L. Mencken*. New York: Perennial.

Thackray, John, and Bob Press. 2001. *The Natural History Museum: Nature's Treasurehouse*. London: Natural History Museum.

Thomson, Keith. 2005. *Before Darwin*. New Haven, CT: Yale University Press.

Thorndike, Johnathan L. 1999. *Epperson v. Arkansas: The Evolution-Creationism Debate*. Springfield, NJ: Enslow Publishers.

Tompkins, Jerry R. 1965. *D-Days at Dayton: Reflections on the Scopes Trial*. Baton Rouge, LA: Louisiana State University Press.

Toumey, Christopher. 1994. *God's Own Scientists*. New Brunswick, NJ: Rutgers University Press.

Trollinger, William Vance, Jr. 1990. *God's Empire: William Bell Riley and Midwestern Fundamentalism*. Madison, WI: University of Wisconsin.

Trumball, Charles G. 1920. *The Life Story of C. I. Scofield*. New York: Oxford University Press.

Ussher, James. 1650 [2003]. *Annals of the World: James Ussher's Classic Survey of World History*. (Modern English republication, ed. Larry & Marion Pierce). Green Forest, AR: Master Books.

Vail, Tom. 2003. *Grand Canyon: A Different View*. Green Forest, AR: Master Books.

Van Oosterzee, Penny. 1997. *Where Worlds Collide: The Wallace Line*. Ithaca, NY: Cornell University Press.

Walker, Samuel. 1999. *In Defense of American Liberties: A History of the ACLU*. Carbondale: Southern Illinois University Press.

Wallace, Alfred Russel. 1880. *Island Life, or The Phenomena and Causes of Insular faunas and Floras: Including a Revision and Attempted Solution of the Problem of Geological Climates*. London: Macmillan.

Wallace, Alfred Russel. 1905. *My Life. A Record of Events and Opinions*. London: George Bell.

Walsh, John Evangelist. 1996. *Unraveling Piltdown: The Science Fraud of the Century and Its Solution*. New York: Random House.

Walters, S.M. and E.A. Stow. 2001. *Darwin's Mentor: John Stevens Henslow*. New York: Cambridge University Press.

Webb, George F. 1994. *The Evolution Controversy in America*. Lexington: University Press of Kentucky.

Weiner, Jonathan. 1994. *The Beak of the Finch: A Story of Evolution in Our Time*. New York: Alfred Knopf.

Wells, Jonathan. 2002. *Icons of Evolution*. Washington, DC: Regnery.

Whitcomb, John C., Jr. and Henry Morris. 1961. *The Genesis Flood*. Philadelphia, PA: Presbyterian and Reformed.

Whittington, Henry. 1985. *The Burgess Shale*. New Haven, CT: Yale University Press.

Wilson, David B. and Warren D. Dolphin. 1996. *Did The Devil Make Darwin Do It?: Modern Perspectives On the Creation-Evolution Controversy*. Ames: Iowa State University Press.

Wilson, David S. 2002. *Darwin's Cathedral: Evolution, Religion, and the Nature of Society*. Chicago: University of Chicago Press.

Wilson, David S. 2007. *Evolution for Everyone*. New York: Bantam Dell.

Wilson, Edward O. 2000. *Sociobiology: The New Synthesis—25th Anniversary Edition*. Cambridge, MA: Harvard University Press.

Winchester, Simon. 2001. *The Map that Changed the World: William Smith and the Birth of Modern Geology*. New York: HarperCollins.

Witham, Larry A. 2002. *Where Darwin Meets the Bible: Creationists and Evolutionists in America*. New York: Oxford University Press.

Woodmorappe, John. 1996. *Noah's Ark*. Santee, CA: Institute for Creation Research.

Woodward, Thomas. 2003. *Doubts about Darwin: A History of Intelligent Design*. Grand Rapids, MI: Baker Books.

Wright, Sewell. 1968–1978. *Evolution and the Genetics of Populations*. Chicago: University of Chicago Press.

Young, David A. and Ralph F. Stearley. 2008. *The Bible, Rocks and Time: Geological Evidence for the Age of the Earth.* Downers Grove: InterVarsity Press.
Young, Matt and Paul K. Strode. 2009. *Why Evolution Works (and Creationism Fails).* Piscataway: Rutgers University Press.
Young, Matt and Taner Edis. 2004. *Why Intelligent Design Fails: A Scientific Critique of the New Creationism.* New Brunswick, NJ: Rutgers University Press.
Zimmer, Carl. 2001. *Evolution: The Triumph of an Idea.* New York: HarperCollins.
Zimmerman, Virginia. 2008. *Excavating Victorians.* Albany, NY: State University of New York.

Index

Note: Some of the index entries are quotations. These are alphabetized under the first word of the quotation. Books are indexed by title only and presented in italic font; author names are not included in the index entries.

About the Authors

RANDY MOORE (b. 1954) (right) earned a Ph.D. in biology from UCLA, after which he worked as a biology professor at several large universities. He edited *The American Biology Teacher* for 20 years, teaches courses about evolution and creationism, and has written several books about the evolution-creationism controversy (e.g., *More Than Darwin: The People and Places of the Evolution-Creationism Controversy*, with Mark Decker). Today, Randy is H.T. Morse-Alumni Distinguished Teaching Professor of Biology at the University of Minnesota. He enjoys music, running, and visiting sites associated with the evolution-creationism controversy.

MARK DECKER (b. 1960) (left) has a Ph.D. in conservation biology from the University of Minnesota, where he is now Associate Director for Scholarship and Teaching in the Biology Program. Mark is interested in all aspects of science teaching, particularly science literacy among non-science college majors. Away from the office, he plays guitar, skydives, and enjoys outdoor activities with his family.

SEHOYA COTNER (b. 1969) (middle) earned a Ph.D. in conservation biology from the University of Minnesota. Currently a member of the Biology Program at the University of Minnesota, Sehoya's research is focused on science education, specifically, effective strategies for teaching about evolution and the nature of science. When she's not teaching or learning about evolution, she likes biking, hiking, and playing *Chutes & Ladders* with her family.

Behind the authors is the front page of the July 22, 1925 issue of *The Des Moines Register* announcing the verdict in the Scopes Trial and William Jennings Bryan's biblical challenges to Clarence Darrow. For more about the Scopes Trial, see page 194.